CRATER LAKE: AN ECOSYSTEM STUDY

CRATER LAKE, OREGON
(Courtesy News and Communication Services, Oregon State University)

SIXTY-NINTH ANNUAL MEETING
of the
PACIFIC DIVISION/AMERICAN ASSOCIATION FOR THE ADVANCEMENT OF SCIENCE
held at
OREGON STATE UNIVERSITY, CORVALLIS, OREGON
June 18-22, 1988

CRATER LAKE

AN ECOSYSTEM STUDY

Edited by

Ellen T. Drake[1]
Gary L. Larson[2]
Jack Dymond[1]
Robert Collier[1]

[1]*College of Oceanography*
[2]*College of Forestry, Cooperative Park Studies Unit*
Oregon State University, Corvallis, Oregon 97331

San Francisco, California
1990

Library of Congress Cataloging in Publication Data:
Catalog card number: 89-63688

ISBN 0-934394-07-5

This volume was computer typeset in Times type using
Xerox Ventura Publisher®, Professional Extension® from
Xerox Corporation running on an AST 386C® microcomputer.
Camera-ready copy was printed on an Apple Laserwriter Plus®.

Manufactured in the United States of America by the Allen Press, Lawrence, Kansas 66044

TABLE OF CONTENTS

Introduction: *Ellen T. Drake* 5

Status of the ten-year limnological study of Crater Lake, Crater Lake National Park: *Gary L. Larson* 7

The geologic setting of Crater Lake: *Charles R. Bacon and Marvin A. Lanphere* 19

Sedimentary history of Crater Lake Caldera, Oregon: *John H. Barber, Jr. and C. Hans Nelson* 29

The chemistry of Crater Lake sediments: Definition of sources and implications for hydrothermal activity: *Jack Dymond and Robert Collier* 41

An alternative hypothesis to explain temperature anomaly and other data observed at Crater Lake, Oregon: *Joseph G. La Fleur* 61

Chemical and physical properties of the water column at Crater Lake, OR: *Robert Collier, Jack Dymond, James McManus, and John Lupton* 69

The ecology and chemistry of caldera springs of Crater Lake National Park: *Stanley V. Gregory, Randall C. Wildman, Linda R. Ashkenas, and Gary A. Lamberti* 81

Chemical and isotopic compositions of waters from Crater Lake, Oregon, and nearby vicinity: *J. Michael Thompson, Manuel Nathenson, and L. Douglass White* 91

Chemical balance for major elements in water in Crater Lake, Oregon: *Manuel Nathenson* 103

Chemistry of Crater Lake, Oregon, and nearby springs in relation to weathering: *Manuel Nathenson and J. Michael Thompson* 115

Crater Lake climate and lake level variability: *Kelly T. Redmond* 127

CRATER LAKE ECOSYSTEM

Secchi disk, photometry, and phytoplankton data from Crater Lake: Long-term trends and relationships: *Clifford N. Dahm, Douglas W. Larson, N. Stan Geiger, and Lois K. Herrera* 143

Phytoplankton species distribution in Crater Lake, Oregon, 1978-1980: *N. Stan Geiger and Douglas W. Larson* 153

Spatial and temporal patterns in the phytoplankton of Crater Lake (1985-1987): *Mary K. Debacon and C. David McIntire* 167

Sampling strategy and a preliminary description of the pelagic zooplankton community in Crater Lake: *Elena Karnaugh* 177

Ecology of kokanee salmon and rainbow trout in Crater Lake: *Mark W. Buktenica and Gary L. Larson* 185

Limnological response of Crater Lake to possible long-term sewage influx: *Douglas W. Larson, Clifford N. Dahm, and N. Stan Geiger* 197

Summary of Crater Lake studies and comparison with the early stages of eutrophication of Lake Tahoe: *Charles R. Goldman* 213

INTRODUCTION

Ellen T. Drake
College of Oceanography
Oregon State University
Corvallis, OR 97331

On 21st June1988, researchers from different disciplines representing various organizations and institutions convened at Corvallis, Oregon, on the campus of Oregon State University, to share knowledge concerning Crater Lake. The Symposium was organized under the aegis of the Pacific Division of the American Association for the Advancement of Science, in the hopes that by pooling our knowledge and understanding the ecosystem of Crater Lake we, as temporary custodians of this planet, might make the right decisions that will ensure the preservation of this most precious national heritage for future generations.

All participants invited to present papers were and are closely involved in scientific research of various aspects of the Crater Lake ecosystem. The articles in this volume represent reports of the results of these scientific investigations and their interpretations, irrelevant to the politico-economical ramifications of such reporting. The geological setting and background of Crater Lake with its violent history is detailed. The ten-year limnological program of the National Park Service in studying Crater Lake is summarized. The chemistry of Crater Lake water and the chemistry and history of the caldera sediments are given. Evidence for possible hydrothermal activity at Crater Lake is presented and an alternative explanation for some of the data observed is hypothesized. The chemistry and ecology of caldera streams and nearby cold springs are discussed, as well as the relationship of lake water level and climate. Secchi disk readings with respect to phytoplankton data and possible long-term sewage influx are presented. Phytoplankton and zooplankton assemblages, including some non-native fish populations, are characterized. The volume concludes with a summary of Crater Lake studies in a comparison of Crater Lake with the early stages of eutrophication of Lake Tahoe.

As much as scientists try to remain immune to societal pressures, however, science can no longer be practised in total isolation within ivory towers. At the same time, scientists cannot allow outside pressures to subvert the scientific process. As a historian of science I have been greatly impressed that out of purely scientific questions, i.e., has the clarity of the lake changed and does hydrothermal venting occur at the bottom of Crater Lake, has emerged a complex situation involving people from industry, government agencies, scientists and other individuals. The crux of the controversy is whether resources surrounding the Crater Lake Park should be tapped and whether such activities could have an impact on the pristine environment of the lake.

Similar situations surrounding other environmental-developmental issues have arisen in many parts of this country, resulting often in bitter debate and even violent confrontations. In such struggles, scientists who are best equipped with studied opinions, unfettered by motives other than scientific, should be allowed to present their results without vigorous opposition from factions with private interests or political jurisdictions at stake.

The articles in this volume are devoid of political taints. If any hint of the latter had appeared during earlier drafts, these have been expunged. This volume, therefore, represents the reporting to-date of what has been done in a concerted attempt to understand the ecosystem of Crater Lake in the best tradition of science.

ACKNOWLEDGMENTS

Many people have made this volume possible.

Besides the help and cooperation I received from my co-editors Gary Larson, Jack Dymond and Bob Collier, and the encouragement we received from Alan E. Leviton, Executive Director of the Pacific Division of AAAS, I wish to thank all those who served as peer reviewers of the articles. Without their conscientious and perceptive reviews, which invariably improved the respective manuscripts, this book could not have become a reality.

The following individuals generously gave of their time and expertise: David E. Armstrong, Peter Bisson, Stephen Carpenter, James P. Cowan, Earl Davis, Stanley Dodson, W. T. Edmondson, Pertti V. Eloranta, Arthur D. Hasler, G. Ross Heath, Lewis Hogan, Thomas Johnson, William J. Liss, Stanford Loeb, Mitchell Lyle, John J. Magnuson, Alexander McBirney, John Melack, James Murray, Peter O. Nelson, Thomas Northcote, Ken Phillips, Harry Phinney, Cynthia Pilskaln, Paul Quay, David Rea, H. James Simpson, Larry Small, Ray Smith, Robert Stallard, John Stockner, Robert Stottlemyer, Ted Strub, Edward Taylor, Todd Thornburg, Max Tilzer, Frank Triska, and Ron Zaneveld.

STATUS OF THE TEN-YEAR LIMNOLOGICAL STUDY OF CRATER LAKE, CRATER LAKE NATIONAL PARK

Gary L. Larson
Cooperative Park Studies Unit, National Park Service
College of Forestry, Oregon State University
Corvallis, Oregon 97331

This paper summarizes the development and program status and planning of a 10-year limnological study of Crater Lake in Crater Lake National Park, Oregon. The program was mandated by Congress in the fall of 1982 after the lake data base (1896-1981) was found to be inadequate to determine whether the lake was decreasing in clarity as had been suggested from independent studies between 1978 and 1981. Goals of the study include establishing a data base for future comparisons, evaluating relationships among physical, chemical, and biological features, and establishing a long-term monitoring program. We want to know whether the lake is changing. If changes are detected, we will develop studies to identify the causes and recommend mitigation measures for anthropogenic changes.

Crater Lake is a primary natural resource of Crater Lake National Park in southern Oregon. The lake covers the floor of Mt. Mazama caldera that formed about 6,800 years ago (Fig. 1; Bacon 1983). It is one of the largest (48 km^2) and highest (1883 m) caldera lakes in the world, and it is the deepest (Larson, 1989). Steep walls form a high rim around the lake, inflowing waters originate within the caldera, and there is no surface outlet (Larson 1988).

Numerous aquatic studies were conducted on the lake from 1896 to 1969, including physical, chemical and biological features (Larson 1987). Many of these studies, however, were short and limited in scope. Enough information was collected to indicate that the lake was very deep, extremely transparent, moderate in alkalinity and conductivity, thermally stratified in summer, well oxygenated, low in phosphorus and nitrate, and containing low densities of phytoplankton and zooplankton. Based on data collected between 1978 and 1981, the transparency of the lake apparently has decreased and the algal community has changed (Table 1).

The limnological data were evaluated in 1982 by a panel of limnologists. They concluded that the lake may be changing, but the data base was too sparse to substantiate their conclusion. On their recommendation, the National Park Service sponsored a limnological study of the lake in the summer of 1982. However, in the fall of 1982, Congress passed a public law (97-250) mandating that the Secretary of the Interior conduct a 10-year limnological study of Crater Lake. The broad goals of the 10-year study include:

1. Develop a reliable data base for use in the future.
2. Develop an understanding of the physical, chemical and biological features of the lake.
3. Establish a long-term monitoring program to examine the limnological characteristics through time. If changes in lake conditions are detected, studies will be developed to identify the cause(s) and if anthropogenic, mitigation measures recommended.

PROGRAM DEVELOPMENT

The project began in the summer of 1983 with emphasis on establishment of the monitoring program (D. W. Larson 1983 and 1984a; and G. L.

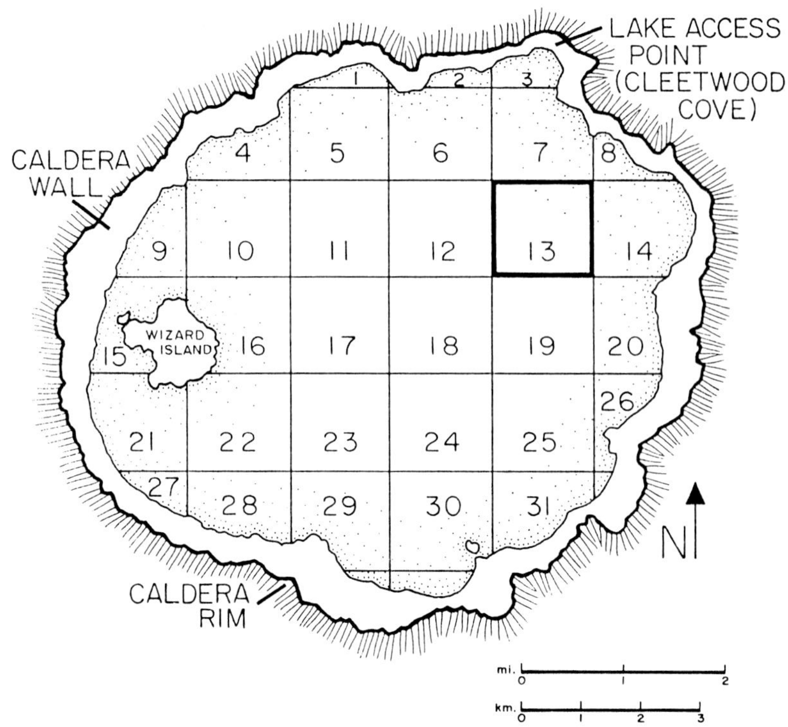

Figure 1. Crater Lake station grid system established by Hoffman (1969). The main trend station (13) is 589 m deep.

TABLE 1. COMPARATIVE DATA FOR SECCHI DISK READINGS AND DOMINANT PHYTOPLANKTON SPECIES FOR SAMPLES TAKEN BETWEEN 1913 AND 1969 AND BETWEEN 1978 AND 1981 (FROM LARSON 1987)

	1913-1969			1978-1981		
	Reference	Method/Comment	Observation	Reference	Method/Comment	Observation
Secchi disk	Larson (1984a)	General lack of environmental conditions when readings taken except for the data collected from 1968-69.	25.6 to 40 m	Larson (1984a)	Environmental conditions documented	21.9 to 36.5 m
Dominant phytoplankton	Kemmerer et al (1924)	Net plankton samples (1913)	*Mougeotia* sp. present throughout water column but maximum at 60 to 150 m *Asterionella* sp. found only below 60 m, maximum at 100 to 200 m	Larson (1984b)	Filtered or settled samples (1978-1980)	*Nitzschia* gracilis, maximum 0 to 20 m *Tribonema* sp., maximum 80 to 120 m
	Utterback et al (1942)	Net and centrifugal water samples (1940)	Few algal cells found in upper 20 m Greatest abundance between 50 and 200 m maximum at 75 m *Anabaena* sp. most abundant alga. *Mougeotia* sp. *Asterionella* sp. *Nitzschia* sp.			*Stephanodiscus hantzschii*, maximum 160 to 200 m

Larson 1985). By 1985, much of the ground work had been done and conceptual models were developed to focus the monitoring and research (Figs. 2 and 3). The first model illustrates the overall components and their interrelationships within the ecosystem, such as the interrelationships among climatic, terrestrial, and limnological characteristics as well as anthropogenic perturbations. The second model focuses on the ecosystem components within the lake. Based on these models, a set of working objectives was developed, and a format for the monitoring program was completed (Tables 2 and 3). The working objectives can be divided into three subdivisions: (1) baseline; (2) lake structure and organization (the parts of the lake system and how they are organized); and (3) examination of changing lake conditions through an analysis of baseline data, paleolimnology, and color and optical properties.

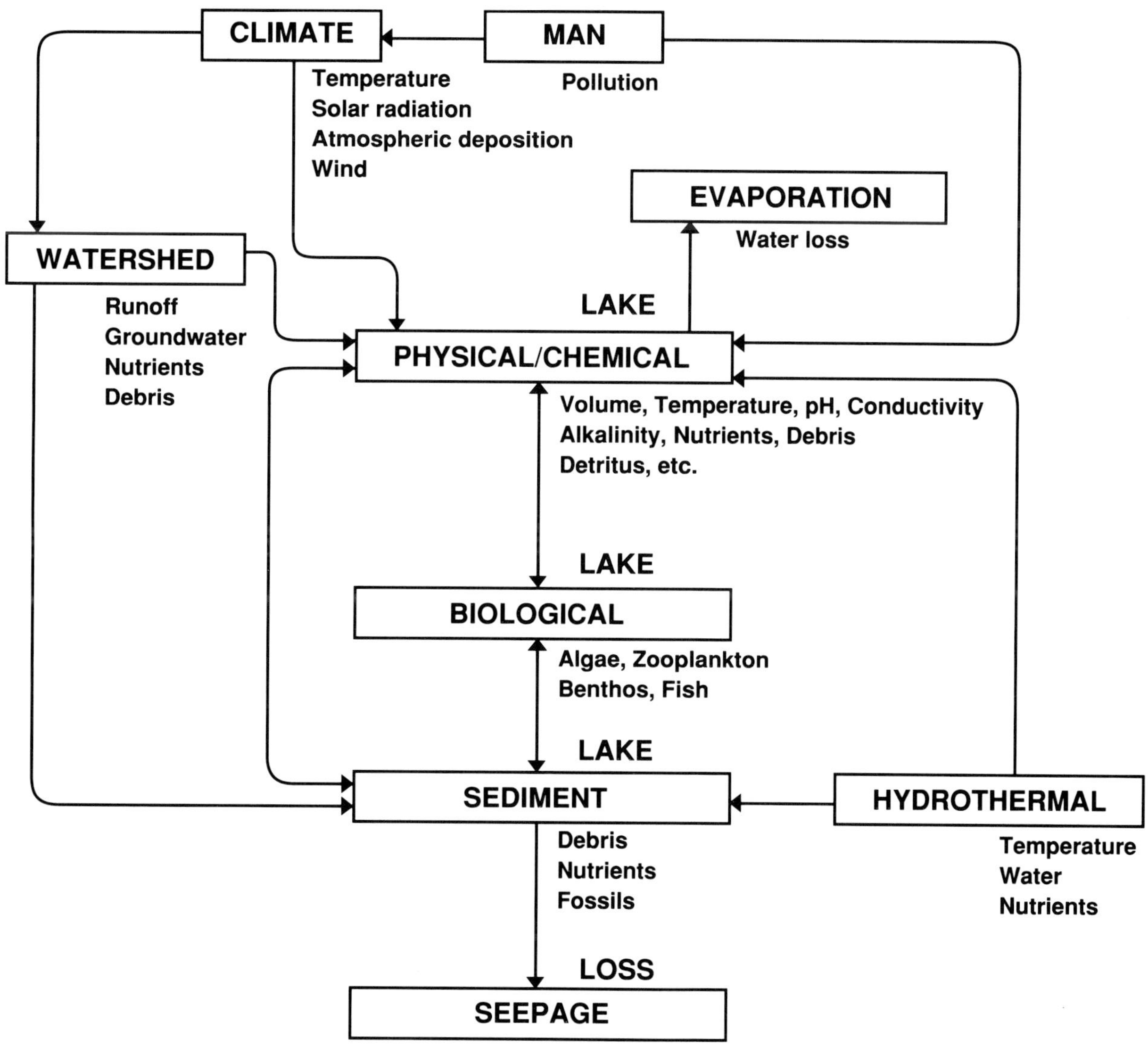

Figure 2. Conceptual model of the Crater Lake ecosystem (after Larson 1988).

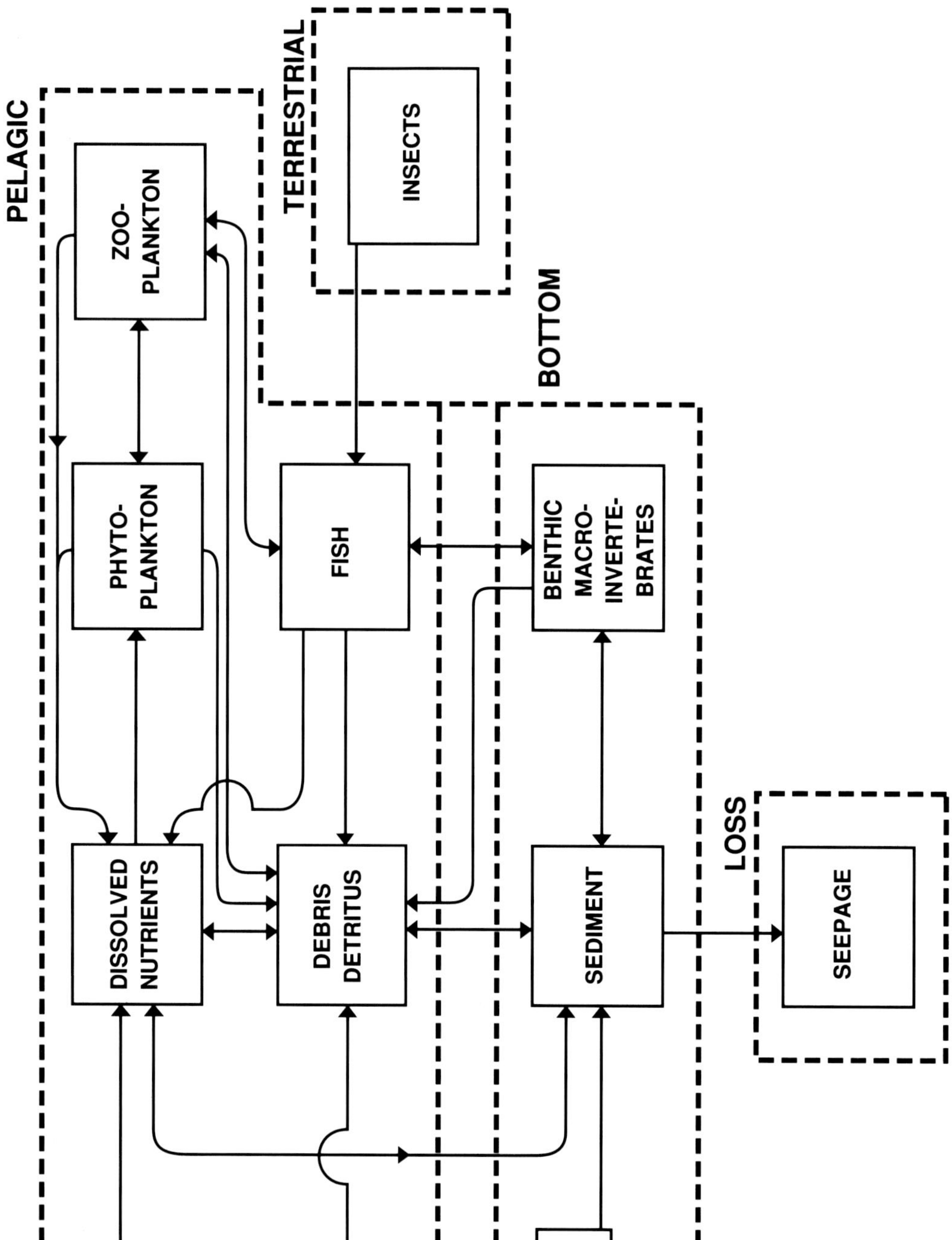

Figure 3. Details of the lake aspects of the conceptual model shown in Fig. 2 (after Larson 1988).

TABLE 2. WORKING OBJECTIVES OF THE CRATER LAKE LIMNOLOGICAL STUDIES (AFTER LARSON, 1988)

1. BASELINE DATA BASE

A. Describe the general physical, chemical, and biological characteristics of the lake for the period 1983-1992.
  1. Determine the amount of seasonal and annual variation of each variable.
  2. Determine how each variable varies with depth.
  3. Determine the amount of spatial (horizontal) variation of each variable.
  4. Evaluate the lake clarity data relative to the physical, chemical, and biological conditions of the lake.

2. LAKE ORGANIZATION AND STRUCTURE

A. Lake volume, stratification and circulation.
  1. Estimate water input into the caldera.
  2. Document relationships between input and changing lake levels.
  3. Evaluate the conditions necessary for the development of thermal stratification.
  4. Describe the circulation patterns and processes of the lake.

B. Nutrients.
  1. Estimate nutrient loading from atmospheric sources and caldera springs into the lake.
  2. Estimate loss of nutrients to sedimentation.
  3. Evaluate nutrient levels in the lake through time relative to inputs, losses, and recycling.

C. Biological features.
  1. Describe the relationships of phytoplankton species, abundance, biovolume, distribution, and production relative to physical and chemical lake features and zooplankton.
  2. Describe the relationships of zooplankton species, abundance, biomass, and distribution relative to physical and chemical lake features, phytoplankton, and fish.
  3. Describe the relationships of benthic macro-invertebrate species, distribution, and abundance relative to physical and chemical lake features.
  4. Describe the relationships of fish species, abundance, biomass, and distribution relative to physical and chemical lake features and zooplankton, benthic macro-invertebrates and terrestrial insects.

3. OPTICAL CHARACTERISTICS, LAKE COLOR, AND PALEOLIMNOLOGY

A. Color and optical properties.
  1. Determine color and optical properties of the lake.
    a. Compare with Pettit's 1935 lake color study.
    b. Compare with Smith et al., 1969 optical study.
  2. If changes are observed, (3.A1), interpret these relative to modern lake clarity conditions (1.A4).

B. Paleolimnology
  1. Evaluate historic lake conditions from analyses of sediment cores from the lake.
    a. Determine selected physical characteristics through time.
    b. Determine selected chemical characteristics through time.
    c. Examine the fossil record through time.
  2 . Determine relationships between the characteristics of surface sediments and settling materials.

4. EVALUATE THE SYSTEM FOR CHANGE (FROM 1, 2, AND 3)

A. Determine if any variable under study shows signs of change greater than would be expected from modern annual variations.

B. Determine if any detected changes could result in a loss of lake clarity.

C. To the extent necessary, identify and conduct special studies to evaluate factors that may be impacting lake water quality.

TABLE 3. COMPONENTS OF THE CRATER LAKE BASELINE LIMNOLOGICAL MONITORING AT THE MAIN TREND STATION (13)AND INTRACALDERA SPRINGS (AFTER LARSON, 1988)

1. LAKE PROGRAM

A. Temperature
 1. Record temperature profiles to 250 m (maximum length of thermister cable) at:
  1 m intervals from 0 to 20 m
  5 m intervals from 20 to 100 m
  20 m intervals from 100 to 200 m, and
  25 m intervals from 200 to 250 m
 2. Conductivity, temperature and depth probe (CTD) to 550 m
B. Optical
 1. Secchi disc (20 cm)
 2. Photometer (to 150 m)
 3. Transmissometer (to 550 m)
C. Chemical.
 Determine pH, total alkalinity, specific conductance, dissolved oxygen, total phosphorus, orthophosphate, nitrate-nitrogen, total Kjeldahl-nitrogen, ammonia-nitrogen, silica, and trace elements at all or selected depths from the following depth sequence: 0, 5, 10, 20, 60, 100, 200, 300, 400, 500, and 550 m.
D. Biological
 1. Chlorophyll-*a*
  Estimate the in vitro chlorophyll at the following depth sequence:
  5 m intervals from 0 to 10 m
  20 m intervals from 20 to 200 m
  25 m intervals from 200 to 300 m
 2. Primary production (carbon^{14} light/dark bottle)
  Estimate primary production at the chlorphyll sampling depths to 180 m.
 3 Phytoplankton
  Determine species, densities, and biovolumes at all chlorophyll sampling depths.
 4. Zooplankton
  Determine species, densities, and biomasses. Samples taken with a vertical haul .5 m diameter number 25 closing net.
 5. Fish
  Determine species, abundances, biomasses, distributions, age, sex, growth and food habits. Samples collected with gill nets, hook and line and down rigger. Pelagic distributions will be estimated using an echo-sounder.

2. SPRINGS

A. Location
  Each spring identified by a numbered tag.
B. Physical and chemical water quality and bacteria.
  Record temperature and take samples for pH, conductivity, alkalinity, nutrients, trace elements, and bacteria (total coliforms, fecal coliforms, and fecal streptococcus).

From 1983 to 1985, studies of the lake were conducted from late June through September because the lake was not accessible during winter. In 1985, a boathouse was constructed on Wizard Island (Fig. 1), which permitted sampling the lake during winter. The first winter trip took place in March, 1986; since then, one or two winter trips have been made each year.

Monitoring is limited to deep-water station 13 (Fig. 1) for two reasons. First, much of the historical data were collected at this site. Second, limnological conditions at station 13 are representative of the other deep-water station (23).

The program has increased in breadth since 1983. Studies of the sources and fates of particles and initial testing for the presence of hydrothermal activity began in 1983. Zooplankton studies began in 1985. Studies on lake color, climate and lake level fluctuations, and fish began in 1986. A conductivity, depth and temperature probe and a transmissometer were purchased in 1987 so that temperature, conductivity and turbidity could be measured from the top to the bottom of the water column. Collection of bulk atmospheric deposition began in 1987, as did initial surveys of periphyton. The current investigators are listed in Table 4.

SUMMARY OF IMPORTANT FINDINGS

Thermal stratification of the lake occurs between August and September. A thermocline can form as early as mid-August or as late as September. The epilimnion extends to greater depths in September than in August due to cooler air temperatures and mixing from fall storms. Surface temperatures from late June/early July to September have ranged from 8.8 to 19.2°C. In summer, water temperature de-

TABLE 4. CURRENT INVESTIGATORS WORKING ON THE CRATER LAKE PROGRAM

Project	*Investigator*	*Affiliation*
Water Quality	NPS[1]	
	Cameron Jones	Forestry Sciences Laboratory (OSU)[2]
Climate/Lake Level (water budget)	Kelly Redmond	State Climatologist Climatic Institute (OSU)
Optical Properties		
Secchi disc	NPS	
Photometer	NPS	
Transmissometer	NPS	
	Robert Collier	Oceanography (OSU)
Color	Peter Fontana	Physics (OSU)
Caldera Springs		
Chemistry	Stan Gregory	Fisheries and Wildlife (OSU)
	NPS	
Bacteria	NPS	
Bulk Deposition	NPS	
(mass balance)	Peter Nelson	Civil Engineering (OSU)
	Joe Reilly[3]	Civil Engineering (OSU)
Phytoplankton		
Chlorophyll	NPS	
Primary production	NPS	
	Mike Conrady	Radiation Center (OSU)
Community analysis	C. David McIntire	Botany (OSU)
	Mary DeBacon[3]	Botany (OSU)
Periphyton	C. David McIntire	Botany (OSU)
	Harry Phinney	Botany (OSU)
	Standford Loeb	Purdue University
Macrophytes	C. David McIntire	Botany (OSU)
	Harry Phinney	Botany (OSU)
Zooplankton	Elena Karnaugh[3] (Thomas)	Fisheries and Wildlife (OSU)
Macrobenthos	Norm Anderson	Entomology (OSU)
	Bob Wisseman	Entomology (OSU)
Fish	Mark Buktenica[3]	Fisheries and Wildlife (OSU)
	NPS	
Particle Flux	Jack Dymond	Oceanography (OSU)
	Albert Collier	Oceanography (OSU)
Hydrothermal processes	Jack Dymond	Oceanography (OSU)
	Robert Collier	Oceanography (OSU)

[1] National Park Service includes Gary Larson, Mark Buktenica, and Mike Hurley
[2] OSU indicates Oregon State University.
[3] Masters of Science project.

creases to a depth of about 300 m, and thereafter increases slightly with increased depth. The ambient temperature at the lake bottom is 3.5 C at Station 13. In the south basin (grids 23 and 24, Fig. 1), small increases in temperature have been recorded near the lake bottom (Collier *et al.* 1990).

Lake pH is slightly basic, total alkalinity ranges from about 25 to 31 mg/1, and conductivity ranges from 100 to 125 µmhos/cm. Alkalinity and conductivity increase in the deep lake, especially near the lake bottom in the south basin. The pH decreases slightly in the deep lake. The water column is well oxygenated, although small decreases in concentration have been detected near the lake bottom in fall. Orthophosphate-P ranges between 9 and 24 µg/1, with the highest concentrations typically occurring in the deep lake. Nitrate-N is near or below detection limits in the upper 200 m of the lake. Below 200 m the concentrations increase and are typically highest at 550 m (11-17 µg/1).

Secchi disk readings vary seasonally and annually. The highest readings usually occurred in June and July and the lowest in August and September. The lowest August readings were recorded in 1982. A reading of 37 m was recorded in early July 1985, but thereafter the readings were in the high 20s. In 1986, the readings were low in mid-August (high 20s), increased in late August to near July levels (low 30s) and then decreased slightly again in mid-September. In 1987, the readings were in the low- to mid-30s in late June and early July, decreased to the high 20s by early August, and then increased to the low- to mid-30s in late August and in September. Secchi disk readings have a positive correlation with the depth of the 1% level of incident light (Larson and Hurley, in press). The 1% depths range from about 80 to 100 m.

Laboratory studies of near-surface lake samples have shown the water to be very blue. This suggests that the water contains a small amount of particles. Considerable seasonal variation has been observed, however (Fontana 1988).

Crater Lake has undergone pronounced variations in lake level since 1900. Recent studies by Redmond (this volume) have modelled these fluctuations relative to the amount of precipitation. The model has been used to revise the lake water budget. One important finding is that the amount of evaporation appears to be greater in winter than in summer.

Results from the particle flux study are developing important insights about recycling of nutrients and organic matter in the lake. Preliminary results suggest that most of the primary production is recycled in the euphotic zone; additional particulate organic matter is recycled between the euphotic zone and the lake bottom, and more than 80% of the organic carbon and nitrogen that reach the sediments is recycled back into the lake (Dymond and Collier 1990).

In addition to changes of water quality noted earlier near the lake bottom in the south basin, the studies of possible hydrothermal processes by Collier *et al.* (1990). have also shown that the normal homogeneous, light-colored, smooth surface sediment is dotted with coarse grained and multicolored areas. These may represent hydrothermal precipitates (Dymond and Collier 1990).

Chlorophyll is in low concentration (<2µg/1) in Crater Lake. During winter and spring there is a fairly uniform concentration from the lake surface to about 200 m, thereafter decreasing in concentration with increased depth. By early summer, a deep water maximum begins to develop between 100 and 120 m (sometimes to 140 m) and is fully developed by July and remains through September (no October samples have been taken). The vertical depth profile for chlorophyll does not correlate well with the vertical profiles for phytoplankton density and biovolume (Debacon and McIntire 1988).

During summer months, primary production maxima occur from 80-100 m. Small peaks in primary production have been observed near the lake surface in August. There is some indication that peak production occurs at shallower depths in winter, spring and fall. The vertical distribution of primary production does not correlate very well with those for chlorophyll and phytoplankton density and biovolume (Debacon and McIntire 1988).

Crater Lake phytoplankton community can be described as a sparse but diverse assemblage of 132 taxa (Debacon and McIntire 1990). During winter, *Stephanodiscus hantzschii* and *Gymnodinium inversum* are the dominant taxa from the lake surface to depths of about 250 m. In spring, *G. fuscum, S. hantzschii, Tribonema affine*, and *Synedra delicatissima* have relatively high cell densities from the lake surface to a depth of 180 m. In summer, *Nitzschia gracilis* dominates the upper 20 m, *Tribonema* sp. is present between 40 and 100 m, *G. inversum* is found throughout the water column to about 140 m, and *S. hantzschii* is rare above 100 m and has a maximum density at about 140 m. The maximum biovolume occurs between April and September.

The zooplankton community includes nine species of rotifer and two crustacean species (Karnaugh 1990). The rotifers include *Keratella cochlearis, Keratella quadrata, Kellicotia longispina, Polyarthra dolichoptera, Philodina* cf. *acuticornis, Filinia terminalis, Synchaeta oblonga, Conochilus unicornis* and *Collotheca pelagica*. The crustacean species include *Daphnia pulicaria* and *Bosmina longirostris*. No invertebrate predators have been collected. The community is dominated numerically by rotifers. Most rotifer populations occurred between 80 and 120 m and the crustaceans between 20 and 100 m, but vertical zonation occurs during summer. Only *Polyarthra* has been found in substantial numbers in the upper 40 m of the lake and *Philodina* is the only species in high density below 160 m.

K. cochlearis was the most abundant species in 1986, but its density decreased in 1987, while that of *Philodina* increased. *Daphnia* has been less abundant than *Bosmina*, but both increased in abundance in 1987 as compared to 1986.

Although naturally barren of fish, the lake was stocked with several salmonid species between 1888 and 1941. Kokanee salmon and rainbow trout are the only known species in the lake at this time. Kokanee are mainly in deep water (maximum depth of capture was about 86 m) and offshore during the day and then migrate into shallower water at night. Rainbow trout appear to be nearshore during the day and night. Growth rates of both species were similar to other northwest populations in oligotrophic lakes. Kokanee fed almost entirely on small-bodied (1.2 mg mean dry weight) aquatic prey; less than 5% (by weight) of ingested prey are terrestrial, and all are presumably eaten at the lake surface. Their primary prey were Chironomidae, Trichoptera, Amphipoda and Cladocera (*Daphnia*). Rainbow trout fed heavily on large-bodied prey (9.8 mg mean dry weight). Primary prey include Trichoptera, Hymenoptera, Chironomidae pupae, terrestrial Coleoptera, Diptera, aquatic Coleoptera, Ephemeroptera, Gastropoda and terrestrial Hemioptera (Buktenica 1989).

PROGRAM PLANNING 1988-1992

Development of a data base and a long-term monitoring program are two of the main goals of the 10-year study. In this context, year-round baseline monitoring of water quality, nutrients, optical properties, chlorophyll, phytoplankton, primary production, zooplankton, and fish studies will continue. Phytoplankton and zooplankton studies, however, will continue to consume considerable processing time in the laboratory. Studies of the periphyton and macrobenthos are scheduled for 1988-90. Studies of caldera spring water quality also will continue and, to the extent possible, will include discharge measurements at selected sites. Based on these results, a long-term monitoring program will be designed by 1992 and implemented in 1993.

The study of the relationships between the climate and fluctuations of the lake level will continue in 1988 and 1989. The objectives are to refine the climate-lake fluctuation model for the period from 1900 to 1988 and evaluate the hydrologic budget each year since 1900 relative to runoff patterns, the amount of precipitation, how the lake level responded to such variations, and nutrient loading from the atmosphere and watershed. This work also will contribute to our evaluation of changes in Secchi disk clarity, especially for the years with high readings (1937 and 1969) and readings in 1954 and since 1978, lake color, spectral sensitivity, light transmission, and turbidity.

Lake color studies will continue to evaluate the small change in color noted since 1934-1935. It is anticipated that samples from lakes of different trophic status will be compared as well.

The transmissometer will be used extensively in 1988-1992 to evaluate spatial and temporal changes in lake turbidity. The quality and quantity of particles in the water column relative to our other optical studies will be emphasized. Another study will be a repeat of the 1969 optical assessment conducted by Smith, Tyler, and Goldman (1970). This project involves an assessment of the downwelling and upwelling spectral characteristics of light in the lake. The project will be conducted when funding is available.

Particle flux studies in 1988 and 1989 will again center on deployment of moorings. These will be retrieved in July, set, retrieved again in September and set for winter. These studies will add to our understanding of nutrient fluxes from the euphotic zone, recycling within the water column, and burial of particles on the lake bottom. Studies of the chemical history of the lake will be initiated on lake cores. Work on hydrothermal activity will be emphasized in 1988-90. In 1988, the primary goals were to: (1) locate and sample hydrothermal fluids using a 1-person submarine; (2) evaluate specific ecosystem responses to hydrothermal inputs in the past from paleolimnological studies of sediment chemistry and phytoplankton and zooplankton fossils, and 3) explore the lake bottom for unusual features such as deep-water moss and periphyton. In 1989, detailed sampling of hydrothermal fluids were carried out in order to evaluate the influences of the hydrothermal system on mixing of the water column and water chemistry. Detailed biological studies and sampling of lake bottom deposits around the sources of the fluids also were conducted. In 1990, data will be synthesized and the project completed.

A study of the internal movement of the water column was conducted in 1988-89. This work will provide important knowledge about how much the lake "turns over" in the classical limnological sense. Understanding the extent of such mixing and the processes involved will provide a much better interpretation of the nutrient cycling and the movement of particles within the lake.

Our studies have shown Crater Lake to be a dynamic and complex ecosystem. The ongoing studies and ones planned as discussed above will add to our understanding of the lake system. Yet several important questions remain, including a main goal of the project—has the lake changed?

Although the basic water quality of the lake does not appear to have changed substantially during the last 75 years (Larson 1988), two important questions need to be resolved. First, why is the conductivity and alkalinity of the lake so high? Is this a consequence of hydrothermal sources? Second, how does the deep water nitrate pool form, what maintains it, does the nitrate circulate into the euphotic zone, and how does it affect the productivity of the lake? Important clues to the answers to these questions will be addressed from the studies of particle flux, lake circulation and hydrothermal processes.

Spatial and temporal patterns of phytoplankton, zooplankton and fish illustrate the diversity of the lake habitats and food web. Studies are needed to evaluate the discontinuity in the vertical distributions of chlorophyll, phytoplankton abundance and biovolume, and primary production, and how these discontinuities relate to the zooplankton community. Additional studies are required to define the long-term relationships between fish and their prey. This is especially important for kokanee salmon because of their interaction in the food web with rainbow trout and their predation on *Daphnia.*

Secchi disk readings have fluctuated from the low 20 m to the high 30 m range since 1982. For the month of August, however, the readings since 1978 have been substantially lower than those recorded in 1937 and 1969. This change has been the basis for the hypothesis that the lake has changed in clarity, but two points must be made about this hypothesis. First, a Secchi reading of about 33 m was made in 1954 in the middle of August using proper techniques under acceptable environmental conditions (mid-day, clear, bright and calm). This reading documents that lake clarity has not always been at 39 - 40 m during the month of August between 1937 and 1969, suggesting that dynamic fluctuations in lake clarity are a natural part of the lake system and not necessarily a result of anthropogenic contamination. Second, disk readings in very clear lakes are subject to large changes because the Secchi disk depth and particle density relationships are hyperbolic, and the relationships for Crater Lake are probably on the upper descending portion of the curve. Small changes in particle density, therefore, can cause large changes in disk readings. Evaluating the reasons for changing disk readings will require identification of the total particle community and an understanding of how it changes in density and optical properties through time. Particle densities, kinds, types and shapes could be influenced by natural environmental conditions, distant and on-site anthropogenic sources and lake processes, such as hydrothermal activity and food web interactions among zooplanktivorous fish, zooplankton, phytoplankton and nutrients.

Decreasing disk readings relative to increased productivity have been demonstrated in Lake Washington (Edmondson 1972) and Lake Tahoe (Goldman 1988). For Crater Lake, however, a weak negative relationship was found between disk readings and primary production (Larson and Hurley, in press). As is the case at Lake Tahoe, many years of monitoring may be required to document a decrease in clarity if lake productivity is actually increasing. Studies of climate, nutrient loading, lake circulation and possible hydrothermal inputs will assist in our evaluation of this possible long-term trend.

Very little is known about the phytoplankton community prior to the late 1970s. Based on the information in Table 1, the present community appears to differ in species structure and distribution as compared to the findings by Kemmerer *et al.* (1924) and Utterback *et al.* (1940). Two important studies are planned to evaluate these changes. First, sampling methodologies will be compared, i.e., net samples, centrifuge, and settled samples. Second, a paleolimnological study will be conducted using dated cores collected in the hydrothermal study. This information should provide considerable insight about how the algal communities have changed through time.

Like the phytoplankton community, very little historical information is available about the Crater Lake zooplankton community, especially rotifers. Part of this problem undoubtedly results from different sampling techniques. But the historical data from Evermann (1897) showed that the community included *Cyclops* in 1896. In 1913, three years after the first large stocking of rainbow trout (50,000), Kemmerer, *et al.* (1924) did not collect *Cyclops*, but *Asplanchna* was present. A paleolimnological study of zooplankton fossils is planned from the same cores used in the diatom study. The results should provide valuable insight about the structure of the zooplankton community in the past. These data may provide information about the periodicity of certain species, such as *Daphnia pulicaria*.

Synthesis of the data collected from 1982-1992 will involve extensive statistical treatments and modeling. Much of this work will be done on an annual basis as individual projects are completed. The major synthesis, however, will begin in 1990. We will emphasize how climate and nutrients affect the ecology of the lake and how the structural components of the lake and their organization are interrelated.

ACKNOWLEDGMENTS

I want to thank the many scientists involved in the ten-year study and the staff of Crater Lake National Park for their continued encouragement and cooperation. The goals for the ten-year program could not be accomplished without this team spirit. I want to especially thank Dr. Stanford Loeb for his many contributions to the program. I thank Dr. Ellen Drake for her assistance and leadership in organizing the Symposium and Proceedings.

LITERATURE CITED

Bacon, C. R. 1983. Eruptive history of Mt. Mazama and Crater Lake Caldera, Cascade Range, U.S.A. Jour. Volcanol. Geotherm. Res. 18:57-115.

Buktenica, M. W. 1989. Ecology of kokanee salmon and rainbow trout in Crater Lake, a deep ultra-oligotrophic caldera lake (Oregon). M.S. Thesis. Oregon State Univ., Corvallis, Ore. 89 pages.

Debacon, M. and C. David McIntire. 1988. Seasonal and vertical distribution and abundance of phytoplankton in Crater Lake. Pages 94-125 *in* G. Larson, ed., Crater Lake Limnological Studies 1987.

Debacon, M. K. and C. David McIntire. 1990. Spatial and temporal patterns in the phytoplankton of Crater Lake (1985-1987). Pages 167-175 *in* E. T. Drake *et al.*, eds, Crater Lake: An Ecosystem Study. Pacific Div., Amer. Assn. Advance. Sci., San Francisco, Calif.

Collier, R., J. Dymond, J. McManus and J. Lupton. 1990. Chemical and physical properties of the water column at Crater Lake, OR. Pages 69-79 *in* E. T. Drake *et al.*, eds., Crater Lake: An Ecosystem Study. Pacific Div., Amer. Assn. Advance. Sci., San Francisco, Calif.

Dymond, J. and R. Collier. 1990. The chemistry of Crater Lake sediments: Definition of sources and implications for hydrothermal activity. Pages 41-60 *in*: E. T. Drake *et al.*, eds, Crater Lake: An Ecosystem Study. Pacific Div., Amer. Assn. Advance. Sci., San Francisco, Calif.

Edmondson, W. T. 1972. Nutrients and phytoplankton in Lake Washington. Amer. Soc. Limnol. Oceanogr., Spec. Symp. 1:172-193.

Evermann, B. W. 1897. United States Fish Commission investigations at Crater Lake. Mazama 1:230-239.

Fontana, P. R. 1988. Lake color. Pages 85-93 *in* G. Larson, ed., Crater Lake Limnological Studies 1987.

Goldman, C. R. 1988. Primary productivity, nutrients, and lake transparency during the early onset of eutrophication in ultra-oligotrophic Lake

Tahoe, California-Nevada. Limnol. Oceanogr. 33:1321-1333.

Hoffman, F. O. 1969. The horizontal distribution and vertical migration of the limnetic zooplankton in Crater Lake, Oregon. M.A. Thesis. Oregon State Univ., Corvallis, Ore. 60 pp.

Karnaugh, E. 1990. Sampling strategy and a preliminary description of the pelagic zooplankton community in Crater Lake. Pages 177-183 *in* E. T. Drake *et al.*, eds., Crater Lake: An Ecosystem Study. Pacific Div., Amer. Assn. Advance. Sci., San Francisco, Calif.

Kemmerer, G. K., J. F. Bovard and W. R. Boorman. 1924. Northwestern lakes of the United States: Biological and chemical studies with reference to possibilities in production of fish. Bull. Bur. Fish. 30(Doc. 944):51-140.

Larson, D. W. 1983. Limnological studies of Crater Lake 1982. Ann. Rep. Nat'l Park Serv.

Larson, D. W. 1984a. Limnological Studies of Crater Lake 1983. Ann. Rep. Nat'l Park Serv.

Larson, D. W. 1984b. The Crater Lake study: Detection of possible optical deterioration of a rare, unusually deep caldera lake in Oregon, U.S.A. Verh. Internat'l Verein. Limnol. 22:513-517.

Larson, G. L. 1985. Limnological studies of Crater Lake 1984. Ann. Rep. Nat'l Park Serv.

Larson, G. L. 1987. A review of the Crater Lake limnological programs. Pages 58-69 *in* T. P. Boyle, ed., New Approaches to Monitoring Aquatic Systems. Amer. Soc. Testing & Materials, STP., Philadelphia, Pa.

Larson, G. L. 1988. Limnological studies of Crater Lake 1987. Ann. Rep. Nat'l Park Serv.

Larson, G. L. 1989. Geographical distribution, morphology and water quality of caldera lakes: A review. Hydrobiologia (171):23-32.

Larson, G. L. and M. D. Hurley. (in press.) Interpreting variations in Secchi disk transparencies of Crater Lake, a deep caldera lake (Oregon). Proc. Internat'l Watershed Symp., Subalpine Processes and Water Quality. June 8-10, 1988.

Redmond, K.T. 1990. Crater Lake climate and lake level variability. Pages 127-141 *in* E. T. Drake *et al.*, eds., Crater Lake: An Ecosystem Study. Pacific Div., Amer. Assn. Advance. Sci., San Francisco, Calif.

Utterback, C. L., L. D. Phifer and J. R. Robinson. 1942. Some chemical, planktonic and optical characteristics of Crater Lake. Ecology 23(1):97-183.

THE GEOLOGIC SETTING OF CRATER LAKE, OREGON

Charles R. Bacon and Marvin A. Lanphere
U.S. Geological Survey
345 Middlefield Road
Menlo Park, California 94025

Crater Lake partly fills the 8-by-10-km diameter caldera that collapsed during the climactic eruption of Mount Mazama about 6850 yr. B.P. This eruption vented 50 km^3 of magma and deposited pumice and ash over a large part of the Pacific Northwest. Mount Mazama was a large composite volcano whose summit elevation apparently was about 3600 m. This volcano was made up of several overlapping shield and stratovolcanoes constructed episodically from at least 420 ka to about 40 ka. Most rocks older than 100 ka are hydrothermally altered and are therefore less permeable than younger lava flows and fragmental deposits. Relatively young, permeable materials are below lake level between Pumice Point and Wineglass, and these may regulate seepage from the lake and limit maximum lake level. Beneath the east half of Mount Mazama is a large field of older rhyodacite lava, itself resting on basaltic andesite on its southwest side and on andesite and dacite on the southeast. The mountain lies at the head of the Klamath graben, a downfaulted basin. Related north-south-trending normal faults cut Mazama lavas on the southwest flank of the volcano. Postcaldera andesitic lava flows and pyroclastic material formed the central platform, Merriam Cone, and Wizard Island within the caldera shortly after it collapsed. The small rhyodacite dome east of Wizard Island was erupted later, about 4000 yr. B.P. All postcaldera lavas save the top of the Wizard Island pile are beneath the lake surface.

Crater Lake partly fills one of the most spectacular collapse calderas of the world. This caldera, a basin at least 1 km deep formed by catastrophic collapse associated with a violent pyroclastic eruption, was first described by Diller and Patton (1902), and later became widely renowned through Howel Williams' (1942) vivid account of its geology. It was Williams who built the accepted theory of caldera formation (Williams 1941) largely on his experience at Crater Lake, and as a result, this caldera has become well known to geologists as a type example. More recent studies have benefited from modern analytical techniques so that the eruptive and glacial history of Mount Mazama, the volcanic mountain in which the caldera formed, is known with more precision. Details of the geology and physical volcanology can be found in Bacon (1983), Druitt and Bacon (1986), and Nelson *et al.* (1988); the chemistry and mineralogy of volcanic rocks in Ritchey (1980), Bacon (1986), Bacon and Druitt (1988), Druitt and Bacon (1989), and Bacon (1990); and a road guide to the geology in Bacon (1987). This paper is an attempt to give an overview of the geology of the Crater Lake area, with comments on how it may affect conditions in the lake.

ANALYTICAL METHODS

The geologic history of the Crater Lake area has been interpreted from detailed geologic mapping by one of the authors (CRB) of approximately the area of Crater Lake National Park and of the caldera walls. Various analytical methods have been employed to quantify the compositions and ages of samples.

Terminology of volcanic rocks has come to be defined on chemical composition. The classification of LeBas *et al.* (1986) is used, slightly modified at the silicic end, in which basalt has <52 wt.% SiO_2, basaltic andesite 52-57%, andesite 57-63%, dacite 63-68%, and rhyodacite >68%. Analytical methods are described in Bacon and Druitt (1988).

Ages of volcanic rocks determined by the K-Ar method are used in constructing the chronology of events described in this paper. The K-Ar data are supplemented by radiocarbon ages determined by S.W. Robinson (Bacon 1983, and unpublished data) and paleomagnetic studies by D. E. Champion (unpublished data, 1988). Radiocarbon ages are given in years before present (yr. B.P. where "present" = 1950 A.D.), which are not strictly equivalent to calendar years; K-Ar ages are given in ka (10^3 years) B.P.

GEOLOGY BENEATH AND AROUND MOUNT MAZAMA

Mapping of the flanks of Mount Mazama and study of rocks brought up in its climactic eruption reveal something of the geology beneath the volcano (Fig. 1). On the southwest, Mazama lies on a lava field composed of Pleistocene basaltic andesite. The heavily glaciated Union Peak volcano (Williams 1942) is a large shield southwest of Mount Mazama that typifies basaltic andesitic volcanism in the High Cascades near Crater Lake. Similar basaltic andesite occurs northwest of the limit of Mazama lavas. On the south at least two thick rhyodacite lava flows are sandwiched between pre-Mazama basaltic andesite and Mazama lavas. More rhyodacite extends beneath Mazama andesites, including the dacite flows of Mount Scott (part of Mount Mazama), counterclockwise around Mount Mazama to its northeast flank. Pre-Mazama rhyodacite flows yield K-Ar ages of 400-500 ka and around 725 ka. Locally, on the southeast in the canyons of Sun and Sand Creeks and at Dry Butte, the rhyodacite lies on older andesite and dacite. The north flank of Mazama is largely buried by ignimbrite of the climactic eruption so that pre-Mazama rocks are not exposed. Accidental lithic (rock) fragments in deposits of the climactic eruption suggest that the pre-Mazama rhyodacite extends beneath the east part of the caldera. The same reasoning suggests that at somewhat greater depth there exists a granodiorite pluton, the origin of which may be related to erupted Mazama dacites or, perhaps, the pre-Mazama rhyodacite. Bits of this rock are found all around the caldera, indicating a body at least 5 km in diameter (the minimum diameter of a subsided block beneath the caldera floor).

The basaltic andesite lava fields described above were fed by "monogenetic" volcanoes—fissures and local vents marked by cinder cones. The petrology of lavas of monogenetic volcanoes near Crater Lake is described by Bacon (1990). Many are younger than Mount Mazama (e.g., Red Cone), and some are older. A few produced andesite (Crater Peak). The largest of the young features is Timber Crater (north of area of Fig. 1), a shield composed of uniform andesite too young to date by K-Ar (c. 20 ka). These lavas represent the sort of magma input to the Mazama system over time and are typical of High Cascade volcanism in Oregon.

Tholeiitic basalt flows in Castle Creek, along the Rogue River from Bybee Creek to Prospect, and east of Bald Crater (north of area of Fig. 1), probably issued from fissures near the Cascade axis. These are the only true basalts in the Crater Lake area (Bacon, 1990). Their ages, except for basalt at Prospect (1250±110 ka, Fiebelkorn *et al.* 1982), and their relation to the calc-alkaline lavas are uncertain.

Monogenetic vents commonly form north-south alignments or are associated with normal faults of like trend. Normal faults, virtually all of which show down-to-the-east displacement, have been mapped from west of Annie Creek to the vicinity of Discovery Point and north of Red Cone as far west as the headwaters of Crater Creek. The faults die out within relatively young lavas of Mount Mazama, or at least are not detectable, probably because displacements are too subtle to be evident in the glaciated, forested, and ash-covered slopes. No tectonic faults have been observed to cut rocks of the caldera walls. Maklaks Crater and Sand Ridge to its south-southeast lie on the northward extension of the down-to-the-west fault that forms the east boundary of the Klamath Graben as it trends into Crater Lake National Park. These observations are consistent with a mildly extensional tectonic regime, which is manifested south and east of Mazama by the Klamath graben and related Basin and Range structures.

MOUNT MAZAMA

For the present purpose we define Mount Mazama as the shield and stratovolcano complex, exclusive of the preclimactic rhyodacite lava flows, that collapsed during the climactic eruption and within

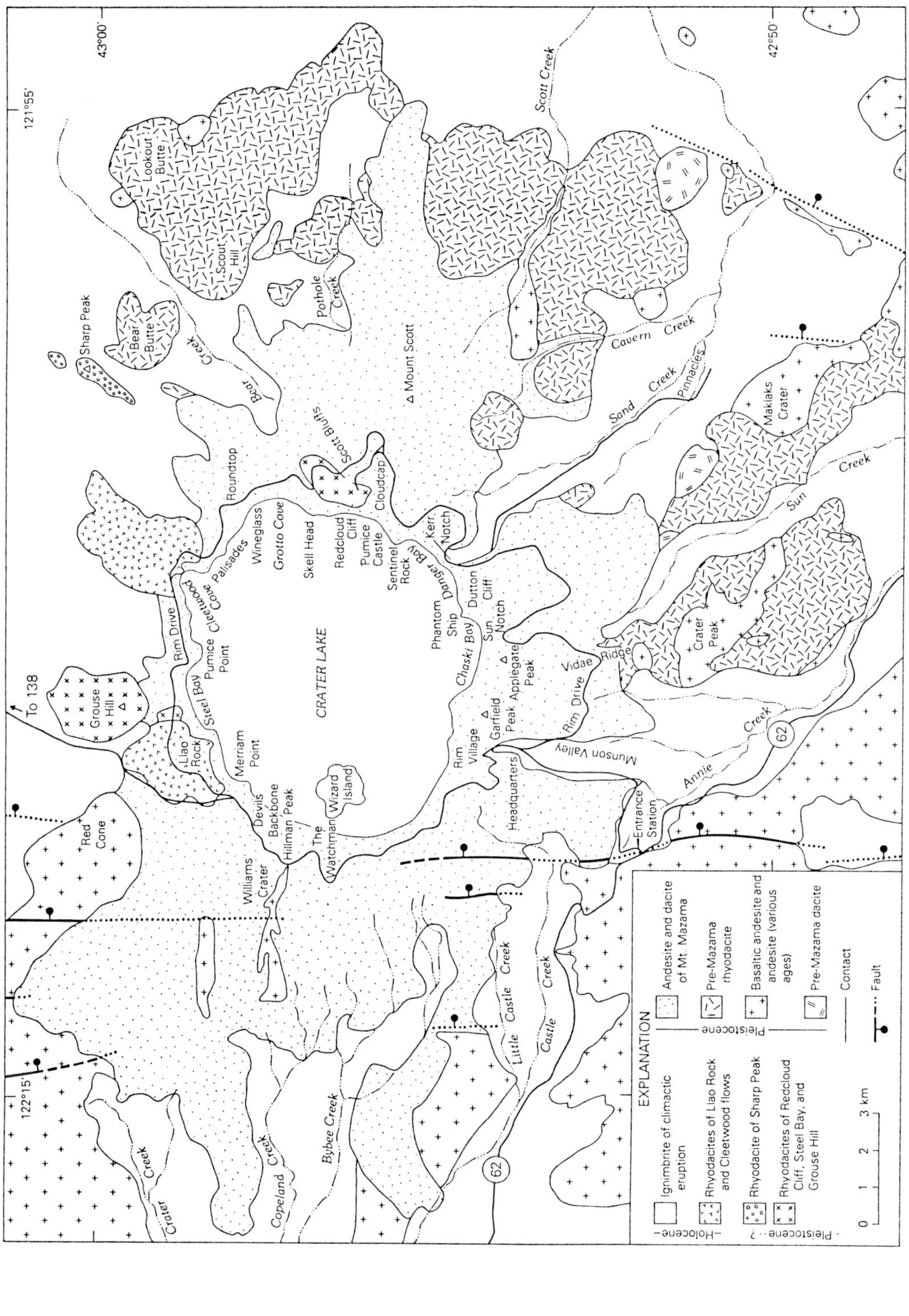

Figure 1. Generalized geologic map of the Crater Lake area.

which the caldera lies (Fig. 1). At its highest, the summit of Mount Mazama apparently was ~3600m. The bulk of Mazama consists of andesitic and low-silica dacitic lavas; basaltic andesitic and dacitic lavas are locally dominant. Except for a few thick flows, most of these lavas apparently were erupted from low fountains because individual flow units can be traced from proximal agglutinated bomb deposits through streaky lava to distal homogeneous lava. Each flow unit consists of a dense lower zone that generally is at least half the thickness of the unit, overlain gradationally by fragmental material. These "rubbly" flow tops were more permeable than fractured dense lava, and hence show more pronounced alteration effects that appear as light bands on the caldera walls, particularly on the south side of the caldera, where hydrothermal fluids within the volcano reacted with them.

Lavas of the low-silica dacitic cone of Mount Scott have yielded the oldest K-Ar ages (~420 ka) of any flows here considered to be part of Mount Mazama. The oldest lavas of Mount Mazama exposed in the caldera (~400 ka) are andesites of the Phantom Cone (Williams 1942) at water level below Dutton Cliff. Andesite and low-silica dacite flows, low on the walls between there and Redcloud Cliff, range up to ~340 ka; similar lavas extend to the west along Chaski Bay. Sun Notch is carved in thin andesite flows dated at ~320 ka that lie on the thicker ≥340 ka flows. Above Sun Notch, Dutton Cliff itself and Applegate Peak are composed of comparatively unaltered silicic andesite flows as young as ~210 ka that rest on the eroded surface of the ~320 ka andesites. Flows at the top of the caldera wall south of Sentinel Rock and just below Garfield Peak and Rim Village date from ~225-245 ka, and somewhat older lava (~275 ka) extends as far clockwise around the caldera as just south of The Watchman. The highest flow at Garfield Peak (~225 ka) is distinctive because, unlike most Mazama lavas, it contains hornblende phenocrysts. Apart from Mount Scott and thin basaltic andesite flows midway up the caldera wall below Redcloud Cliff that probably were erupted from a local source around 220 ka, most of the volcanic activity up to about 200 ka occurred at vents of the main stratovolcano complex centered on what is now the south part of the caldera.

The character of volcanism, as interpreted from the remains of Mount Mazama, changed from andesitic stratocone construction to basaltic andesite shield building around 200 ka. Several monogenetic vents date from around this time, albeit with relatively large analytical uncertainty: Desert Cone and Bald Crater (both north of the area of Fig. 1), Maklaks Crater, and a poorly preserved cinder cone north of Crater Peak. Basaltic andesite flows below Redcloud Cliff (~220 ka) and at lake level below Llao Rock (~185 ka) were erupted from small shield volcanoes. During the same period, however, dacite flowed ≥9 km west from a vent, now cut by a north-south normal fault, 2 km west of Rim Village.

The shield below Llao Rock was built in at least two episodes, apparently ending around 130 ka. Thick andesite flows that rest on this shield midway up the caldera wall between Merriam Point and Pumice Point, the flow above the boat landing at Cleetwood Cove, and the Palisades flow (Williams 1942) were all emplaced between ~130 and ~100 ka. The slightly older (~160 ka) Roundtop flow (Williams 1942) is another thick andesite flow that, like the others, apparently issued from vents well north of the old summit of Mount Mazama. The feeder for one such eruption is exposed where it cuts the shield lavas southwest of Merriam Point. Silicic andesite also formed extensive thick flows on the northwest flank of Mazama (~80 ka) and above Grotto Cove (~70 ka).

By ~70 ka, a new stratocone was forming at Hillman Peak. This volcano consists of three sets of lava flows, the lowest lying on the andesite of Merriam Point. The middle lavas of Hillman Peak are chemically distinctive basaltic andesites with prominent hornblende phenocrysts and have been mapped over several km^2 on the west flank of Mazama. The bottom and top sets are more typical pyroxene andesite but differ sufficiently in appearance that they have been mapped separately. Hillman Peak is the westernmost of the stratovolcanoes that comprise Mount Mazama.

Dacitic eruptions became more common and produced widespread tephra also around 70 ka. Pumice was erupted at a vent near Redcloud Cliff during a Plinian eruption (high convective eruption column) to form the tephra of Pumice Castle, which has been identified 110 km east at Summer Lake (Davis 1985). Related dacite lava forms thick flows south of Redcloud Cliff, at Cloudcap, and Scott Bluffs. The tephra of Pumice Castle can be traced in the caldera wall as far west as Llao Rock (e.g., it forms the lower one-third of the pumice slope at Pumice Point), and commonly is overlain by one or two dacitic lava flows. Above these at Steel Bay is an airfall pumice deposit whose welding indicates a local source. In short, several vents erupted rather similar dacite over a brief period. Mazama then reverted to producing andesitic lava, such as that found near the caldera rim from Discovery Point clockwise to Pumice Point; the Devils Backbone dike fed an eruptive center that has been removed by glacial erosion. These andesites date from ~60 to ~40 ka. The Watchman (dacite) flow (Williams 1942), whose feeder can be seen in the caldera wall, is overlain by such andesite. The Watchman flow has yielded a K-Ar age of ~50 ka. Dacitic ignimbrite in the caldera wall to the south, and locally exposed on the southwest flank of Mount Mazama, probably records the onset of Watchman eruptive activity, although the correlation is uncertain. The caldera rim andesites and the Watchman flow provide the last direct record of the growth of Mount Mazama—its final products are only preserved as fragmental deposits and small flank flows.

At the head of Munson Valley, from Park Headquarters to the caldera rim, are debris-flow and lithic-pyroclastic-flow deposits that apparently are around 40 ka. Their age is based on glacial features and on an unusual paleomagnetic direction they share with Red Cone, a basaltic andesitic cinder cone northwest of the caldera. Most of the material consists of blocks of dacite. Similar deposits occur high on the west side of Garfield Peak, near the headwaters of Castle Creek, 1.5 km south of The Watchman, and immediately south of Devils Backbone. The distribution and character of these deposits suggest that they were formed when a dome, or cluster of domes, was destroyed during their emplacement just west of the summit of Mount Mazama. Indeed, dacitic summit domes are commonplace on stratovolcanoes. An analogy can be made with the familiar Crater Rock dome on Mount Hood and the apron of related pyroclastic debris below it. Should the dome in the crater of Mount St. Helens oversteepen and collapse, it probably would produce deposits like those in Munson Valley.

The youngest dacite preserved on Mount Mazama forms a tiny dome by the rim road west of Hillman Peak and is a major component of the mingled (incompletely mixed) basalt/andesite/dacite lava flows of Williams Crater. Williams Crater, a basaltic cinder cone 1 km west of Hillman Peak and named after Howel Williams, marks one of a series of vents in a west-northwest line that produced first a basalt flow and cinder cone, and then mingled lava (Bacon, 1990). The "basalt" is in fact basaltic andesite contaminated with fragments of gabbro. Besides forming a homogeneous lava flow west of Williams Crater, the basalt occurs as inclusions in the andesite and dacite of the mingled lavas. Tephra from Williams Crater lies on the dacitic pyroclastic-flow deposits south of Devils Backbone. The dacitic end-members of both Williams Crater and Munson Valley products are similar and were erupted at about the same time, suggesting a common source beneath Mount Mazama. The andesite is simply a mixture of dacitic and basaltic components. At Williams Crater gabbro-contaminated basaltic andesitic magma evidently interacted with the margin of this dacitic magma body to produce inclusions and mingled lavas. The dacite of Williams Crater represents the last magma to be erupted from Mount Mazama before onset of rhyodacitic volcanism that we believe can be tied to growth of the climactic magma chamber.

GROWTH AND ERUPTION OF THE CLIMACTIC MAGMA CHAMBER

The first rhyodacite attributed to the climactic chamber formed the flows and related domes and pyroclastic deposits of Grouse Hill, Steel Bay, and Redcloud Cliff (Fig. 1) between 30 and 25 ka (Bacon and Druitt 1988). Eruption of an extensive rhyodacitic pumice fall and the lava flow of Llao Rock is dated at 7015±45 yr. B.P.; related dikes are exposed in the caldera wall. All of these rhyodacites are more differentiated than that of the climactic eruption, dated at 6845±50 yr. B.P. Rhyodacite of the Cleetwood flow (Williams 1942) is identical to that of the climactic eruption and was still hot when the climactic event started. Rhyodacite domes of Sharp Peak also are compositionally identical to climactic rhyodacite, but their precise age is uncertain.

The climactic eruption took place in two phases within and between which there is no evidence for any significant pause (Bacon 1983): (1) a single-vent phase producing a Plinian column that deposited widespread airfall tehpra, followed by valley-hugging welded ignimbrite known as the Wineglass Welded Tuff (Williams 1942) when the column collapsed to a lower height; and (2) a ring-vent phase of highly-mobile pyroclastic flows generated from relatively high columns as the caldera collapsed. The ring-vent-phase deposits consist of lithic breccia near the caldera and on topographically high areas and of nonwelded to partly welded pumiceous ignimbrite in valleys (Bacon 1983; Druitt and Bacon 1986). Magmatic products of the single-vent phase and much of the ring-vent phase consisted of uniform rhyodacite and very minor silicic andesite. Later ring-vent phase products were dominantly crystal-rich andesitic and basaltic andesitic scoria, followed by mafic cumulate blocks, some of which contain olivine. The total volume of magma erupted was ~50 km^3. Partially fused granitoid blocks also were ejected, particularly near the close of the eruption. The climactic ejecta apparently represent the contents of a horizontally stratified magma reservoir (Williams 1942), erupted in a geologic instant. The layering in this chamber was controlled by densities of magmas and crystal cumulates. The petrology of these materials is described in Bacon and Druitt (1988) and Druitt and Bacon (1988, 1989).

THE CALDERA AND POSTCALDERA VOLCANISM WITHIN IT

Crater Lake caldera is a collapse depression enlarged by caving and sliding of the walls to form a scalloped outline. The area of structural subsidence probably is not much larger than 5 km in diameter but the topographic caldera is about 10 km east-west by 8 km north-south. Within the subsided portion there may be up to 2 km of intracaldera tuff and interbedded landslide deposits, virtually all of which were deposited by the end of the climactic eruption, although there is no direct evidence for the nature of rocks beneath "acoustic basement" (Nelson *et al.* 1988). Filling in the irregularities in the surface of the syncollapse deposits are sediments described by Nelson *et al.* (1988).

Postcaldera volcanism has been confined to the caldera (Fig. 2). Andesite was erupted from several vents, apparently in the first few hundred years following the climactic eruption. This forms the central platform, Merriam Cone, and Wizard Island (Bacon and Druitt 1988; Nelson *et al.* 1988). Whether these lavas represent magma related to the climactic chamber or a new influx is uncertain at present. In any case, the lake was filled to nearly its present level before the end of andesitic postcaldera volcanism. A small rhyodacite dome was extruded at ~4000 yr. B.P. on the east flank of the andesite pile of Wizard Island.

GLACIATION OF MOUNT MAZAMA

Evidence for glacial erosion and for ice presence abounds in the caldera walls and on the flanks of Mount Mazama. For example, low-silica dacite flowed into a glacial valley carved on the east flank of Mount Mazama at ~305 ka; these two intracanyon flows now form the cliffs at Sentinel Rock. Other thick massive intracanyon flows now stand as high ground near the heads of Sun Creek and the east fork of Annie Creek. Some lava interacted with ice, such as silicic andesite east of Sun Notch and the dacite that forms Munson Point (~275 ka) west of Munson Valley. Fracturing of lava that is believed to reflect rapid cooling while in contact with ice occurs in younger flows as well, most notably in the caldera wall at the top of the "stem" of the Wineglass and the thick flow at Pumice Point.

Glacial ice occupied valleys and the higher parts of Mount Mazama several times during the volcano's history. The deeper valleys, Munson, Sun, and Kerr (south of Kerr Notch), were carved by repeated advances of glaciers. Ice-polished surfaces and unequivocal glacial striae are present both in caldera wall exposures and on the flanks of Mazama where they commonly were degraded by passage of ring-vent-phase pyroclastic flows. Although there is the possibility that large valley glaciers on stratovolcanoes do not accurately reflect times of continental ice presence, it is tempting to correlate glaciated surfaces and glacial deposits with the marine oxygen isotope record (Bacon 1983). However, such is beyond the scope of this paper.

Lateral moraines are preserved in some drainages, such as Annie, Cavern, Scott, Pothole, and Bear Creeks, all near the park boundary. These presumably date from the last glaciation. Large areas of till have been mapped on the west flank of Mazama and on basaltic andesite farther west, mostly at elevations lower than 6000 feet. Some of these deposits are sufficiently weathered that they may be older than the lateral moraines of the south and east drainages. Glacial deposits in the caldera walls are few and are confined to the north part of the caldera,

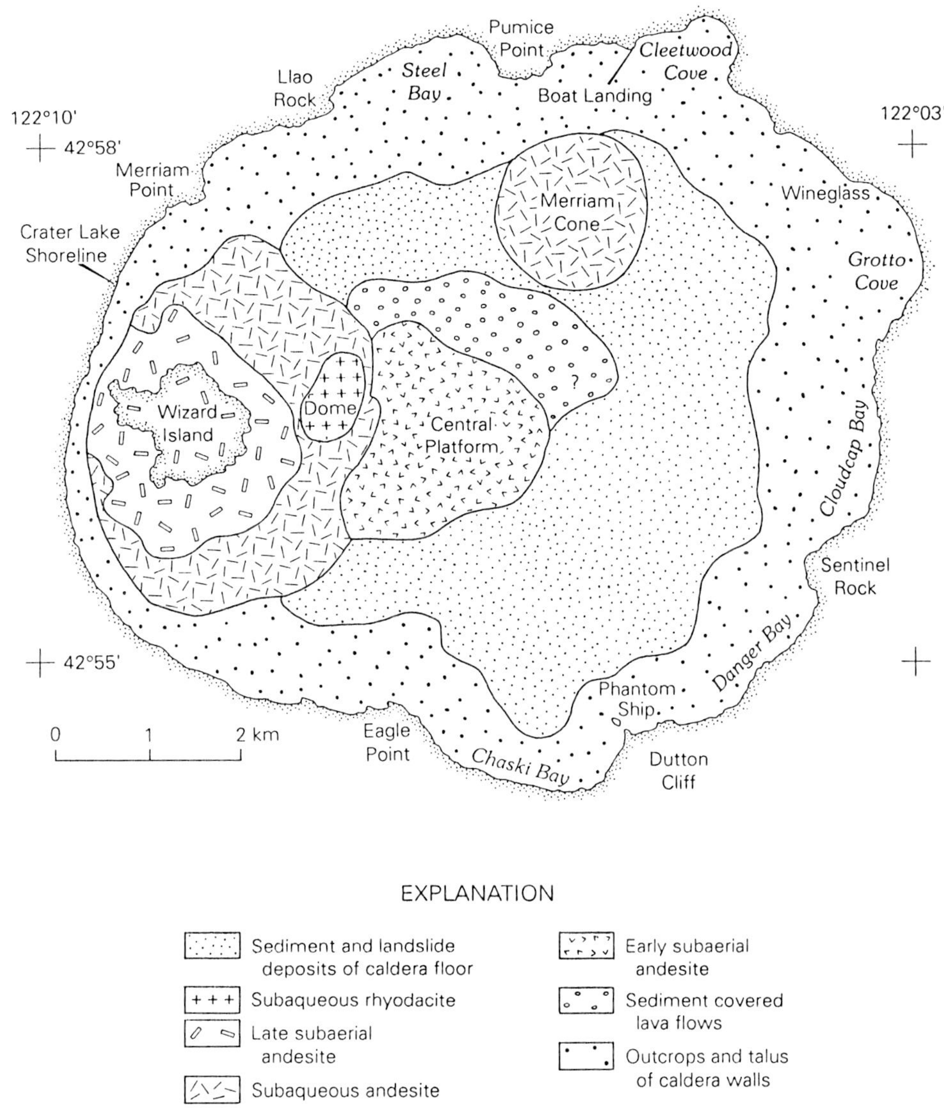

Figure 2. Interpretive geologic map of the floor of Crater Lake caldera.

although Williams (1942) suggested they were abundant. Reinterpretation of deposits in the caldera walls is based on increased knowledge of pyroclastic deposits in the 50 years since Howel Williams was in the field at Crater Lake.

GEOLOGIC FEATURES THAT MAY AFFECT CRATER LAKE

Permeability of rocks, i.e., their capacity for transmitting fluid, is closely related to the formation of Crater Lake. Lacking an outlet, it is the combination of percolation of water through the caldera walls and evaporative loss that balances precipitation to maintain the lake's level (Phillips and Van Denburgh 1968). Most volcanic rocks are not particularly permeable on the scale of a hand specimen. However, their structure and fracture (joint) pattern exert strong influences on permeability, so that dense rock may be highly permeable on the scale of an outcrop. This is the case with the walls of Crater Lake caldera. All of the lava flows are fractured, mostly due to contraction during cooling. Some of the more silicic lavas contain internal breccia zones that formed before flow ceased. Fracture permeability may be significant, but even more important is the high permeability of the rubbly tops of lava flows, and in many cases, of brecciated or columnar basal zones. Fragmental deposits, such as formed by debris avalanches, probably are the most permeable. Glacial till, which typically contains ultrafine matrix material, and ashy pyroclastic-flow deposits may be less permeable. In all of these materials, reaction with fluids, particularly with hot water, causes new minerals to form that seal fractures and clog spaces between clasts. As can be seen at many places in the caldera walls, dikes cutting older deposits may form relatively impermeable barriers. Tectonic faults, not seen above lake level, nevertheless may be present below the lake surface, in which case they could provide sheet-like conduits parallel to their surfaces and barriers to flow across them owing to the presence of gouge. Minor faults parallel to the caldera wall and related to collapse are present near Garfield Peak and Chaski Bay; impermeable gouge in these faults could form barriers to leakage from the lake.

Aspects of the geology of the caldera walls suggest a possible mechanism of lake level regulation. The drowned subaerial lava flows of Wizard Island end in a steeper slope at a depth of ~70 m (Fig. 2), below which lava appears to have flowed into water. Therefore, the lake was within 70 m of its present level around the time andesitic postcaldera volcanism ceased, or by ~6000 yr. B.P. It is possible that lake level has been regulated at approximately this elevation ever since by a change in bulk permeability of the caldera walls in the following way. Everywhere at lake level except between Pumice Point and Wineglass (Fig. 2), the rocks at lake level are subtly to severely hydrothermally altered, and their bulk permeability may be relatively low. However, in the northeast sector there are many fresh lava flows and fragmental deposits at water level where the upper surface of altered rocks dives gently below the lake. This may act as a buried spillway of sorts due to the likely permeability contrast between fresh fragmental deposits and altered lava flows. Kibby *et al.* (1968) found that surface currents from midlake terminate in Cleetwood Cove, also suggesting that unaltered rocks in this area may be responsible for significant leakage.

Other effects of geology on the lake are more speculative still. The detailed nature of the caldera fill is unknown (see Nelson *et al.* 1988), but by analogy with well-exposed ancient caldera-fill deposits (Lipman 1984) it may consist largely of relatively impermeable welded tuff. Nevertheless, the common occurrence of thermal springs in the ring-fracture zones of large calderas indicates that fracture permeability can be significant in caldera-fill deposits. By their nature, hydrothermal systems tend to be self-sealing: fracture permeability can be destroyed by precipitation of minerals. Maintenance of permeability in the caldera fill may require repeated fracturing, either by tectonic movements or by seismicity related to volcanism. Vigorous hydrothermal venting is common in volcanic areas of active tectonic extension (e.g., Yellowstone, mid-ocean ridges). Although Crater Lake lies in an environment where slow extension clearly has been taking place for a long time, events may be sufficiently infrequent that modern hydrothermal circulation (Williams and Von Herzen 1983) has a more diffuse manifestation (Dymond and Collier 1990) than in areas with active seismicity.

ACKNOWLEDGMENTS

This summary is based on a mass of data to which many Geological Survey scientists have contributed. We are indebted to our colleagues D. E. Champion, C. H. Nelson, and T. H. Druitt (now at the

University of Wales), and to chemists, technicians, and field assistants far too numerous to list here. Comments by M. Nathenson, E. M. Taylor, J. Fierstein, and an anonymous reviewer helped to clarify the manuscript. The cooperation of the staff of Crater Lake National Park during field aspects of this study is appreciated.

LITERATURE CITED

Bacon, C. R. 1983. Eruptive history of Mount Mazama and Crater Lake caldera, Cascade Range, U.S.A. Jour. Volcanol. Geotherm. Res. 18:57-115.

Bacon, C. R. 1986. Magmatic inclusions in silicic and intermediate volcanic rocks. Jour. Geophys. Res. 91:6091-6112.

Bacon, C. R. 1987. Mount Mazama and Crater Lake caldera, Oregon. Geol. Soc. Amer. Centennial Field Guide 1:301-306.

Bacon, C. R. 1990. Calc-alkaline, sho-shonitic, and primitive tholeiitic lavas from monogenetic volcanoes near Crater Lake, Oregon. Jour. Petrol. 31. (in press)

Bacon, C. R., and T. H. Druitt. 1988. Compositional evolution of the zoned calcalkaline magma chamber of Mount Mazama, Crater Lake, Oregon. Contrib. Mineral. Petrol. 98:224-256.

Davis, J. O. 1985. Correlation of late Quaternary tephra layers in a long pluvial sequence near Summer Lake, Oregon. Quat. Res. 23:38-53.

Diller, J. S., and H. B. Patton. 1902. The geology and petrography of Crater Lake National Park. U. S. Geol. Survey Prof. Pap. 3. 167 pp.

Druitt, T. H., and C. R. Bacon. 1986. Lithic breccia and ignimbrite erupted during the collapse of Crater Lake caldera, Oregon. Jour. Volcanol. Geotherm. Res. 29:1-32.

Druitt, T. H., and C. R. Bacon. 1988. Compositional zonation and cumulus processes in the Mount Mazama magma chamber, Crater Lake, Oregon. Trans. Roy. Soc. Edinburgh: Earth Sci. 79:289-297.

Druitt, T. H., and C. R. Bacon. 1989. Petrology of the zoned calcalkaline magma chamber of Mount Mazama, Crater Lake, Oregon. Contrib. Mineral. Petrol. 101:245-259.

Dymond, J., and R. Collier. 1990. The chemistry of Crater Lake sediments: Definition of sources and implications for hydrothermal activity. Pages 41-60 *in* E. T. Drake *et al.*, eds., Crater Lake: An Ecosystem Study. Pacific Div., Amer. Assn. Adv. Sci., San Francisco, Calif.

Fiebelkorn, R. B., G. W. Walker, N. S. MacLeod, E. H. McKee, and J. G. Smith. 1982. Index to K-Ar age determinations for the state of Oregon. U. S. Geol. Survey Open-File Rep. 82-596.

Kibby, H. V., J. R. Donaldson, and C. E. Bond. 1968. Temperature and current observations in Crater Lake, Oregon. Limnol. Oceanogr. 13:363-366.

LeBas, M. J., R. W. LeMaitre, A. Streckeisen, and B. Zanettin. 1986. A chemical classification of volcanic rocks based on the total alkali-silica diagram. Jour. Petrol. 27:745-750.

Lipman, P. W. 1984. The roots of ash flow calderas in western North America: Windows into the tops of granitic batholiths. Jour. Geophys. Res., 89: 8801-8841.

Nelson, C. H., P. R. Carlson, and C. R. Bacon. 1988. The Mount Mazama climactic eruption (~6900 yr B.P.) and resulting convulsive sedimentation on the Crater Lake caldera floor, continent, and ocean basin. Geol. Soc. Amer. Spec. Pap. 229:37-57.

Phillips, K. N., and A. S. Van Denburgh. 1968. Hydrology of Crater, East, and Davis lakes, Oregon. U. S. Geol. Surv. Water Supply Pap. 1859-E. 160 pp.

Ritchey, J. L. 1980. Divergent magmas at Crater Lake, Oregon: Products of fractional crystallization and vertical zoning in a shallow, water-undersaturated chamber. Jour. Volcanol. Geotherm. Res. 7:373-386.

Williams, D. L., and R. P. Von Herzen. 1983. On the terrestrial heat flow and physical limnology of Crater Lake, Oregon. Jour. Geophys. Res. 88: 1094-1104.

Williams, H. 1941. Calderas and their origin. Univ. Calif. Publ. Geol. Sci. 25:239-346.

Williams, H. 1942. The geology of Crater Lake National Park, Oregon. Carnegie Inst. Washington. Pub. 540. 162 pp.

SEDIMENTARY HISTORY OF CRATER LAKE CALDERA, OREGON

John H. Barber, Jr. and C. Hans Nelson
U.S. Geological Survey, M/S 999
345 Middlefield Rd., Menlo Park, CA 94025

Reflection seismic and sedimentologic studies reveal an evolution of depositional styles in Crater Lake caldera that reflects waning volcanic and seismic activity during a 7000-year history. Rapid construction of volcanic edifices (central platform, Wizard Island, and Merriam Cone) divided the caldera floor into three basins. A hummocky acoustic basement is recognized in seismic records and probably represents the surface of syncollapse deposits. This irregular surface forms the caldera floor upon which subaerial and lacustrine sediment has accumulated. The deepest acoustic unit contains both flat-lying and chaotic reflectors that may reflect subaerial sheet-wash and mass-wasting. It is followed by subaerial debris flows that are recognized by their chaotic acoustic signature and wedge-shaped geometry and appear to be a response to seismicity associated with post-caldera volcanism. Among these is a major landslide wedge off the southern margin of the lake. Subaerial conditions may have prevailed only a short time (hundreds of years?) following the formation of the caldera, because hemipelagic lake sediment on the central platform dates almost as old as the caldera itself, indicating subaqueous deposition over most of the caldera history.

Basin-margin sediment wedges evolving to flat-lying, parallel, basin-wide acoustic reflectors, a decrease in mean grain size from basin edge to basin center, and graded sand beds indicate that lacustrine deposits consist of base-of-slope aprons and basin-filling turbidites. A change from high- to low-amplitude acoustic reflectors may be related to a reduction in turbidite sedimentation about 4,000 years ago that resulted from waning volcanism and seismicity.

There has been a history of deposition in Crater Lake caldera following its formation about 6,900 yr B.P. (Bacon 1983). In this paper, we characterize stratigraphy and sedimentary processes within the caldera that suggest the timing of the onset of lake filling as well as other post-caldera volcanic and sedimentary events. We also speculate on aspects of the lake-floor geology that may influence present-day sedimentation patterns.

Initial studies of the sediment on the floor of Crater Lake focused on the surface sediment distribution, near-surface sediment lithology from shallow (50-cm) cores, and sedimentary processes (Nelson 1967). Later studies have described seismic stratigraphy and its relationship to sedimentary history and the caldera collapse (Nelson *et al.* 1986b and 1988). This paper summarizes prior work, and provides an overview of sedimentary history on the caldera floor and its relationship to present-day processes.

METHODS OF STUDY

Boomer and air-gun seismic-reflection surveys were conducted in 1979 and cores were collected in 1980. The seismic surveys provide a means of looking at cross-sections of the lake-floor stratigraphy, revealing strata up to 100 m below the lake floor. Seismic profiles were spaced approximately 200 to 500 m apart. Techniques used to assign acoustic velocities and determine thickness of the acoustic units described below are described in greater detail in Nelson *et al.* (1988). Radiocarbon ages of lake sediment were provided by S.W. Robinson (U. S. Geol. Surv., written comm., 1986) and are reported in yr B.P. (years before present, where present = 1950 A.D.).

MORPHOLOGY OF THE LAKE FLOOR

The present lake floor is dominated by volcanic edifices (emergent Wizard Island, submerged Merriam Cone and the central platform) and three basins (Fig. 1). Other important morphological features include (1) a small rhyodacite dome between Wizard Island and the central platform, and (2) a hummocky ramp extending out from Chaski Bay off the southern margin of the lake.

SEISMIC STRATIGRAPHY

Seismic profiles over the caldera walls and volcanic edifices show acoustic basement at the surface, which indicates that there is little sediment cover on these features. Bottom photographs, however, show a thin drape (on the order of mm to cm) of fine-grained sediment covering outcrops. Cores recovered as much as 2 m of sediment in local depressions on the central platform (Nelson 1967; Dymond and Collier 1990). This sediment on the platform provided a condensed stratigraphic section that was critical for establishing a chronology for depositional events in the caldera. The thick deposits underlying the basin floors required seismic stratigraphy to differentiate and characterize the sedimentary units.

A classification of the seismic units in Crater Lake was developed by Nelson *et al.* (1986b, 1988) and is summarized here. Five acoustic units have been identified on seismic-reflection profiles in the basins (Figs. 2 and 3). The five units are: (I) acoustic basement; (II) a heterogeneous unit that typically contains high-amplitude, parallel reflectors that are locally discontinuous or inclined; (III) a well-defined unit containing disorganized, chaotic reflections, that generally thickens toward the caldera walls, forming a wedge; (IV) a well-layered sequence of flat-lying, high-amplitude (strong), parallel reflectors; and (V) a well-layered sequence of flat-lying, low-amplitude (weak), parallel reflectors.

Unit I

Unit I consists of incoherent and attenuated acoustic returns. The top surface of unit I can be identified on most airgun profiles and is the limit of acoustic penetration, or acoustic basement (Figs. 2 and 3). A contour map showing depth to this surface gives an indication of how the caldera looked before any post-collapse sedimentation took place (Fig. 4). This surface shows two main features underlying the sediment in the basins: (1) small depressions (~50-100 m diameter) that are possibly phreatic-explosion craters (see Nelson *et al.* 1988 for a more thorough discussion of these features), and (2) an irregular surface that may outline coherent blocks topping the syncollapse deposits of unit I created during caldera formation (see Bacon and Lanphere 1990). The depressions, together with presently known areas of high heat flow (Williams and Von Herzen 1968) and the summit craters of Wizard Island and Merriam Cone outline an elliptical zone that may define a ring-fracture zone along which Mt. Mazama collapsed when the caldera was formed (Fig. 4) (Bacon 1983; Nelson *et al.* 1988).

Unit II

The next unit (II) has covered and filled in the depressions and irregularities of the acoustic basement, and is as much as 50 m thick in the east basin, 30 m thick in the northwest basin, and 20 m thick in the southwest basin (Nelson *et al.* 1988). Unit II has a complex and heterogeneous character. The reflectors are only locally continuous, strong, and parallel. They are sometimes flat-lying and sometimes inclined, and change character across the basins (Figs. 2 and 3). We interpret this unit generally as bedded volcaniclastics that were deposited shortly after the caldera collapse: the flat-lying reflectors may outline ejecta from phreatic-explosion craters and/or sheetwash deposits similar to those noted as the first post-eruption deposits at Mount St. Helens (Brantley and Power 1985; Collins and Dunne 1986).

Unit III

Because of the ponding of unit II, unit III has been deposited on a surface of low relief compared to the acoustic basement (Figs. 2 and 3). The thickness of the unit is a relatively constant 10 m in all basins, increasing to 20 m in wedges at the base of the caldera wall in the east basin. Reflections within the unit are chaotic and poorly bedded. Because of the lack of internal organization and the shape of unit III, we think that it is composed of single or multiple debris flows and avalanches that make up debris fans (Nelson *et al.* 1988).

Unit III also appears to crop out at the lake floor, where a major landslide under Chaski Bay is out-

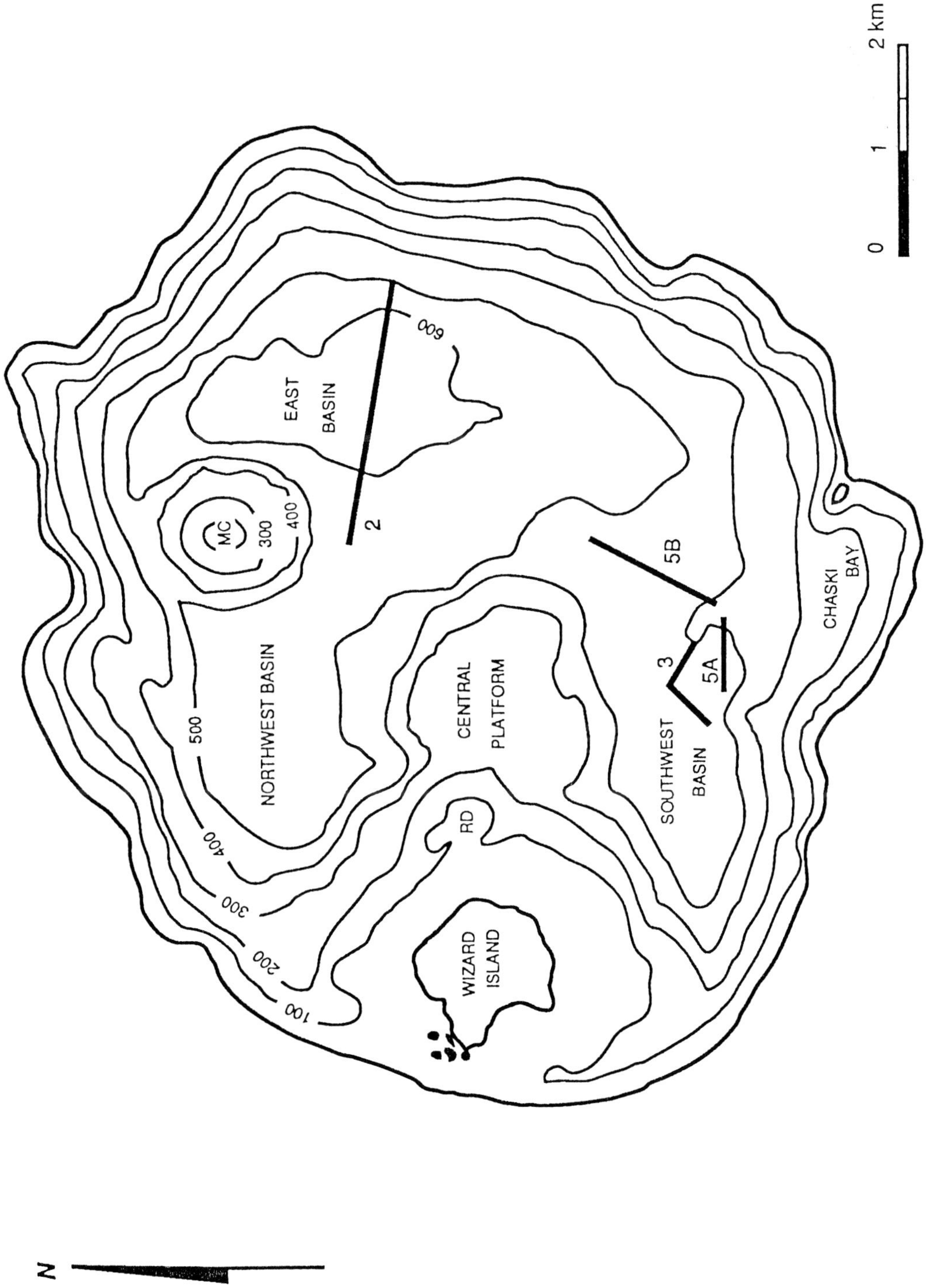

Figure 1. Bathymetry and important morphological features of Crater Lake. Contour interval is 100 m. MC = Merriam Cone. RD = rhyodacite dome. Numbers indicate seismic profiles shown in Figs. 2, 3, and 5.

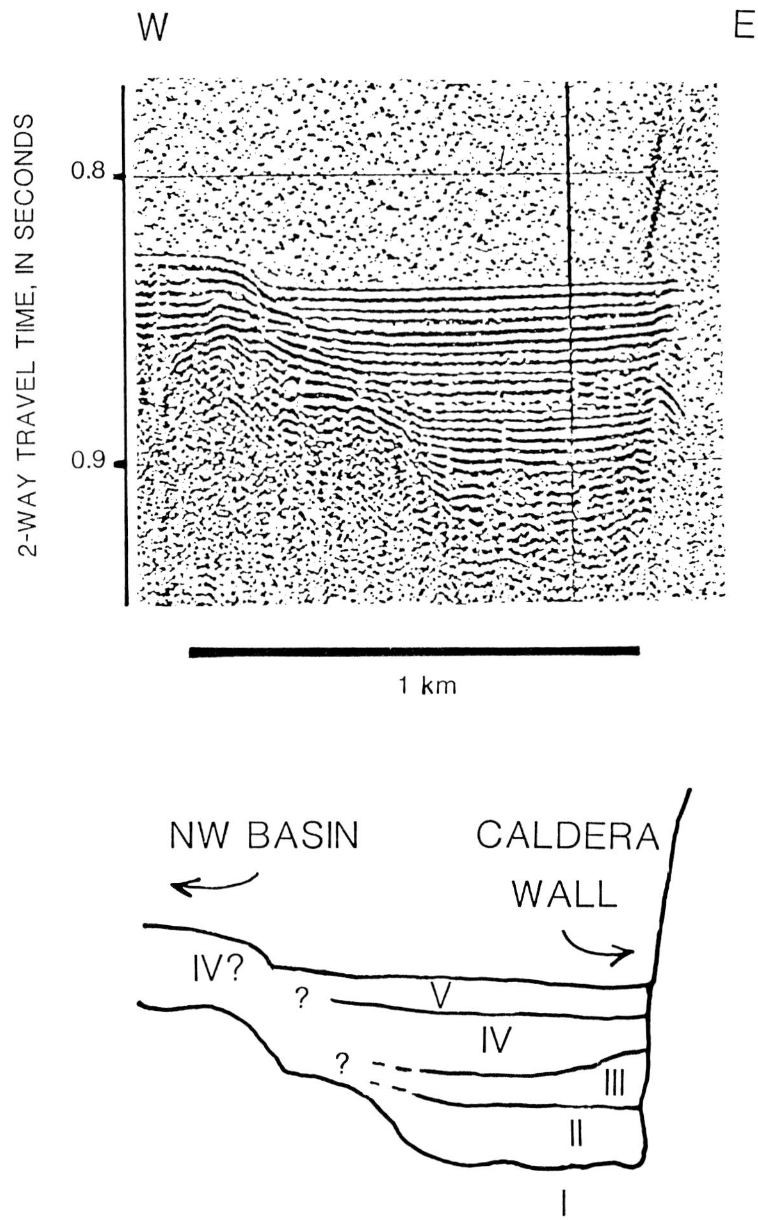

Figure 2. Deep-penetration (1-in^3-airgun) profile across the east basin and interpretive drawing. I-V are acoustic units discussed in the text (see Nelson *et al.* 1988, for another example of the fully developed 5-unit stratigraphy). Vertical exaggeration varies from about 9x in the water column to about 5x in unit II. Location of the profile is shown in Fig. 1.

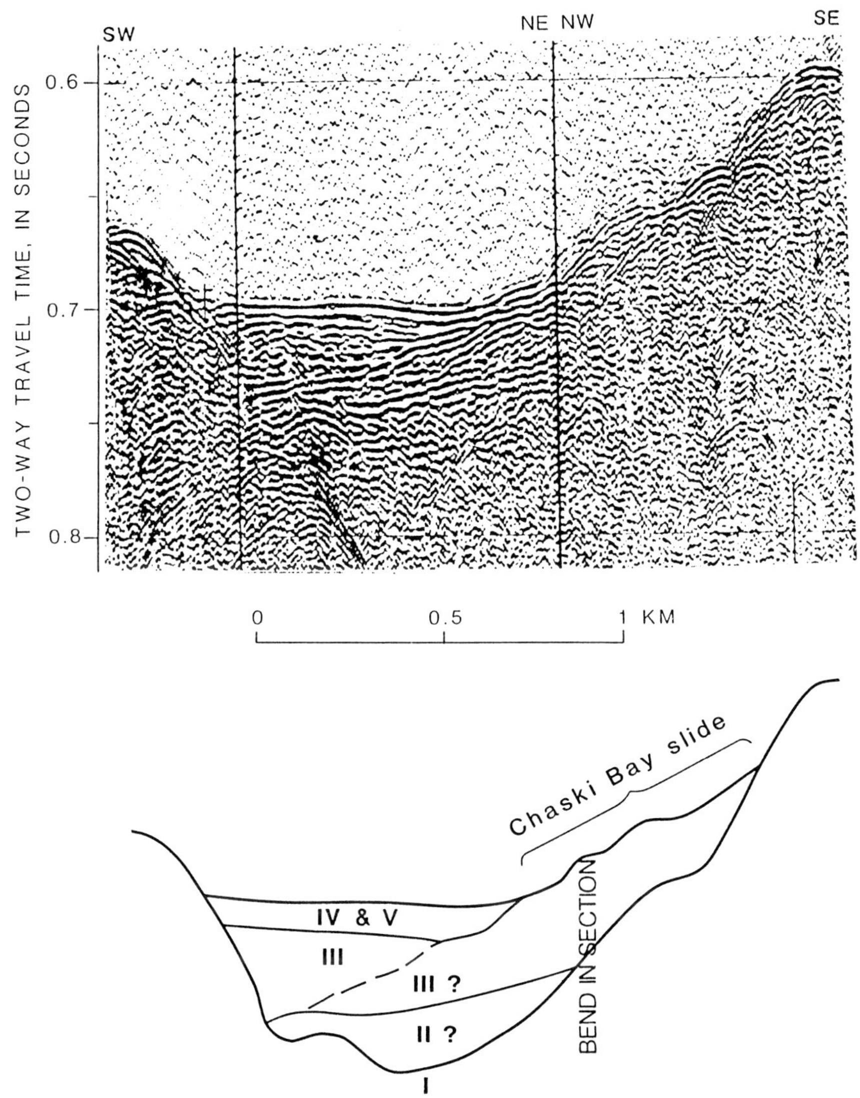

Figure 3. Deep-penetration (1-in^3-airgun) profile across the southwest basin and interpretive line drawing (modified from Nelson *et al.* 1988). Acoustic units are discussed in the text. Slope on the right side of the profile is over Chaski Bay slide. Vertical exaggeration as in Fig. 2. Location of the profile is shown in Fig. 1.

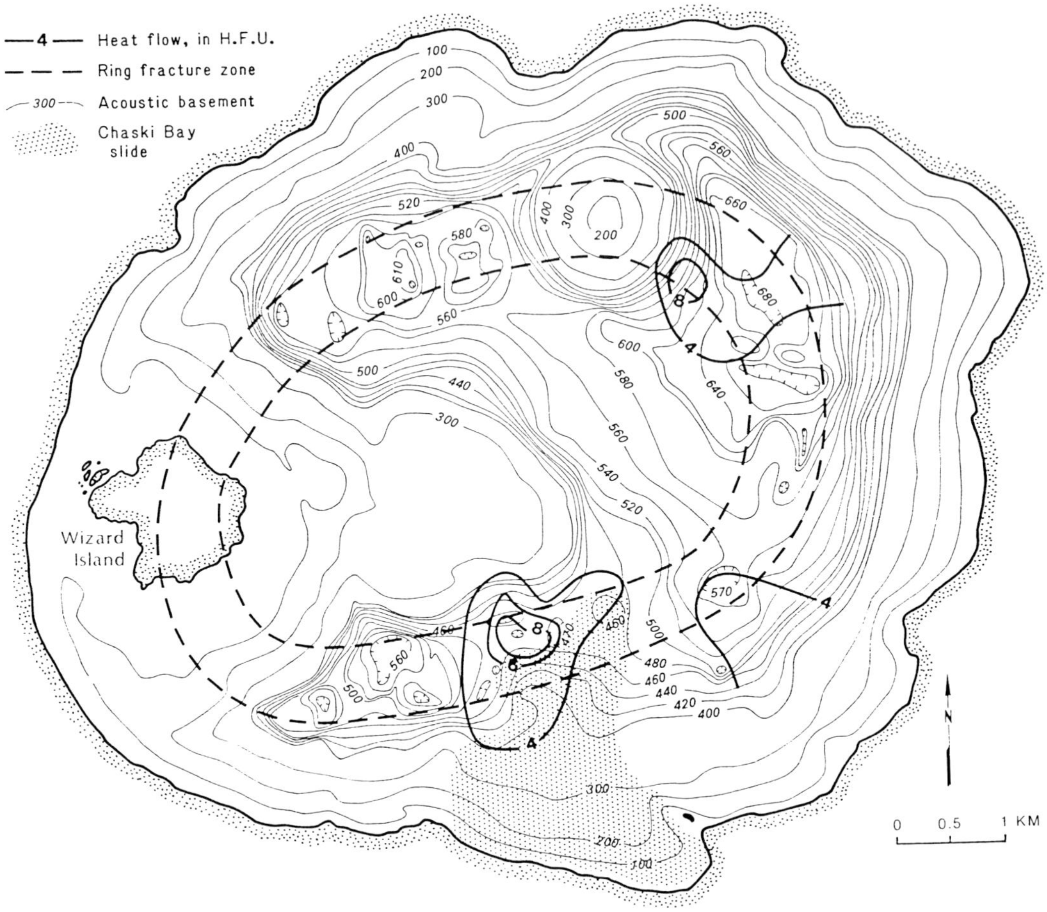

Figure 4. Structure-contour map of the acoustic-basement surface (unit I), and heat-flow contours (1 H.F.U.=10^{-6} cal/cm^2s) from Williams and Von Herzen (1983) (modified from Nelson *et al.* 1988). Note the proposed ring-fracture zone and Chaski Bay slide. Note also the small depressions (phreatic-explosion craters?) associated with the ring-fracture zone. Contour interval on the acoustic basement is 100 m (above 400 m) and 20 m (below 400 m), with 10-m contours added to help define the craters. Modified from Nelson *et al.* (1988).

lined by surface-reflector characteristics, low slope gradient, and contours bowed downslope (Nelson *et al.* 1988). We further distinguish the landslide by the pronounced hyperbolic reflections seen on high-resolution seismic profiles, indicative of an irregular, hummocky, blocky surface sloping to the north and plunging beneath the basin floor (Fig. 5). This large landslide extends from the base of the caldera wall to the southeastern edge of the central platform (Fig. 4), and includes a 'dome' that has been dredged. C. R. Bacon (written comm. 1989) describes the dredge sample as containing rocks of heterogeneous lithology that indicate a rubble mound rather than a volcanic dome or cone.

Units IV and V

Overlying unit III is a 20-to-40-m-thick sequence of flat-lying, parallel reflectors that are continuous across each of the basins (Fig. 6). The seismic characteristics of this unit are typical of turbidites in other lacustrine and marine environments (Hyne *et al.* 1972; Poppe *et al.* 1985, Newhall *et al.* 1987). We believe this set of parallel reflectors represents subaqueous sediment that was deposited after significant accumulation of lake water in the caldera (Nelson *et al.* 1988). High-amplitude (strong) reflectors, apparently related to coarser-grained, thicker sand beds (Nelson *et al.* 1986b) are characteristic of the entire sequence above unit III in the small

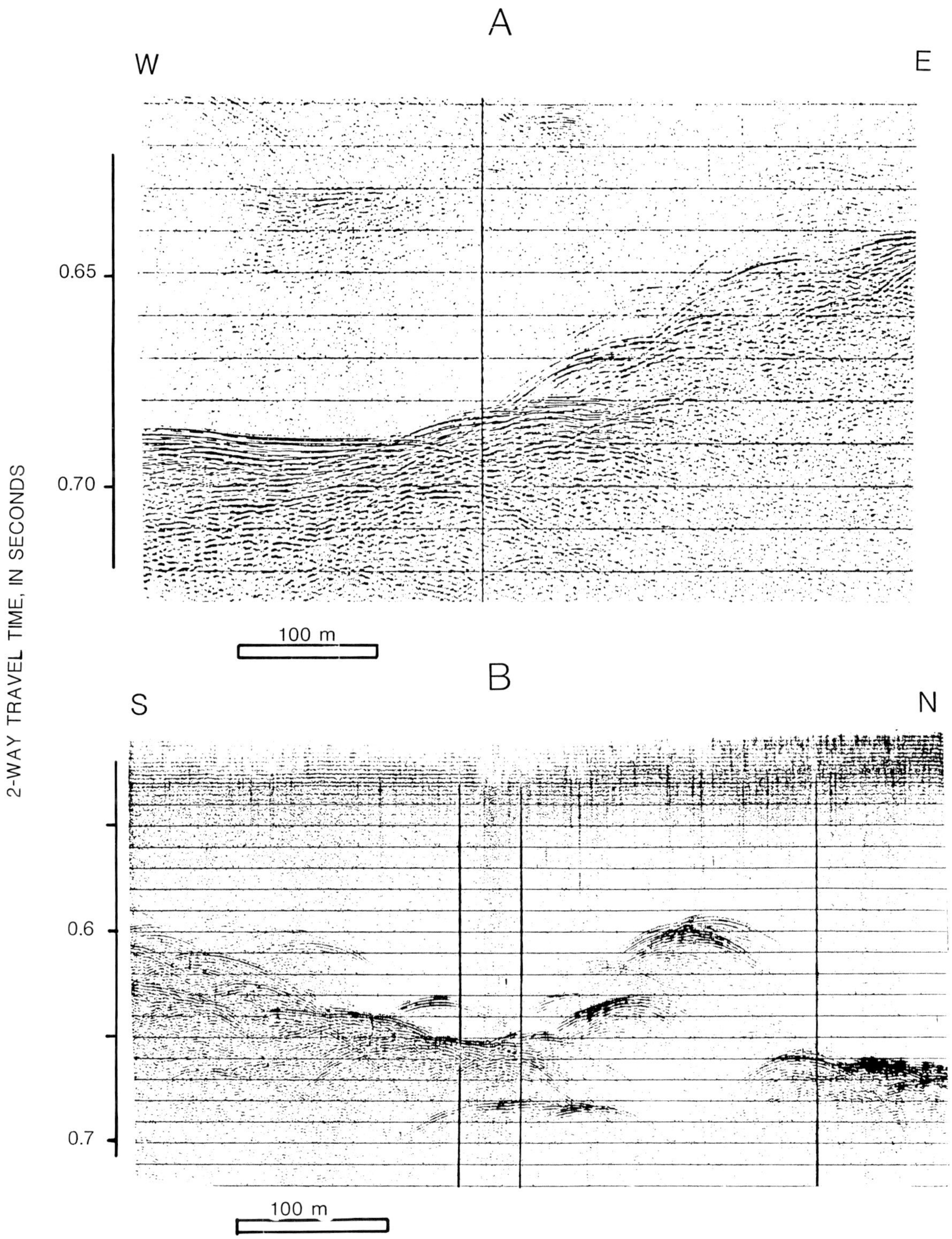

Figure 5. High-resolution (boomer) profiles over Chaski Bay slide, showing hummocky, hyperbolic acoustic returns indicative of the slide surface. Vertical exaggeration is about 5x in A and about 2x in B. Locations of the profiles are shown on Figure 1.

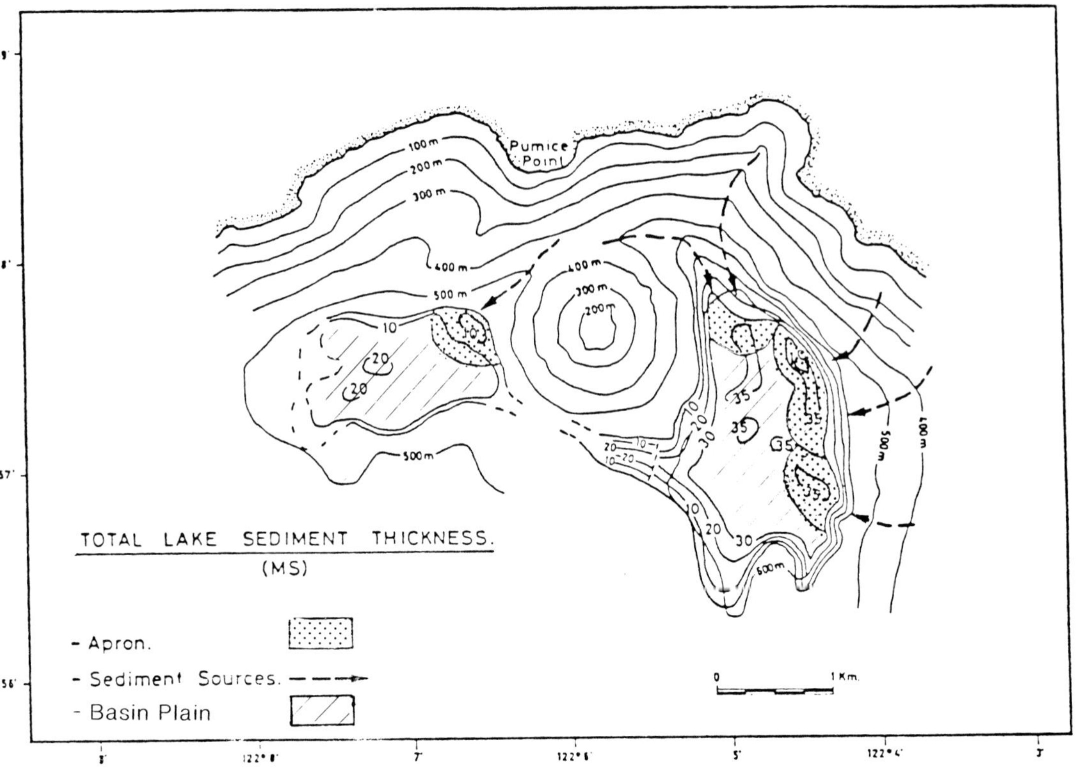

Figure 6. Isopach map of units IV and V combined, showing base-of-slope aprons in the east and northwest basins, with some inferred caldera-wall debris-chute sources (modified from Nelson *et al.*, in press).

northwestern and southwestern basins (Fig. 3). In the large eastern basin, however, the reflections can be separated into a lower set of high-amplitude reflections (unit IV), and an upper set of low-amplitude (weak) reflectors (unit V) that probably indicate fine-grained and thin-bedded turbidites similar to those noted in near-surface cores (Fig. 2) (Nelson 1967; Nelson *et al.* 1986b).

SEDIMENT OF THE BASIN FLOORS

Basin fill over the collapse debris of unit I reaches a total thickness of 90 to 100 m in the east basin and 70 to 80 m in the smaller western basins (Fig. 4). Sediment at these depths is beyond the reach of present sampling techniques, so the acoustic units cannot be characterized and dated directly.

We have sampled three types of near-surface sedimentary deposits from the basin floors: hemipelagic interbeds, basin-plain turbidites, and base-of-slope aprons with sediment-gravity-flow deposits. Basin-plain sediment has an alternation of hemipelagic mud, and silty to sandy beds characterized by vertical grading of (a) texture, (b) Bouma sequences of sedimentary structures, and (c) composition that are typical of turbidites (Nelson 1967; Nelson *et al.* 1986b). Base-of-slope aprons exhibit higher sand/shale ratios, thicker sand and gravel beds, and coarser-grained turbidites. The aprons form single or coalescing cones that are observed in the seismic records and can be outlined in isopach maps of the lacustrine sequence (units IV and V combined) (Fig. 6) (Nelson *et al.*, in press).

SEDIMENT OF THE CENTRAL PLATFORM

Sediment deposited on the central platform 200 to 300 m above the basin floors is much thinner (<2 m) than in the basins, as it is sheltered from basin-floor turbidite deposition and thus is formed largely by hemipelagic deposition. Consequently, the central-platform sediment provides a condensed stratigraphic section that probably encompasses the entire sedimentary history of the lake. The age of the oldest lacustrine sediment recovered from the platform is ~6900 yr B.P., approximately as old as the

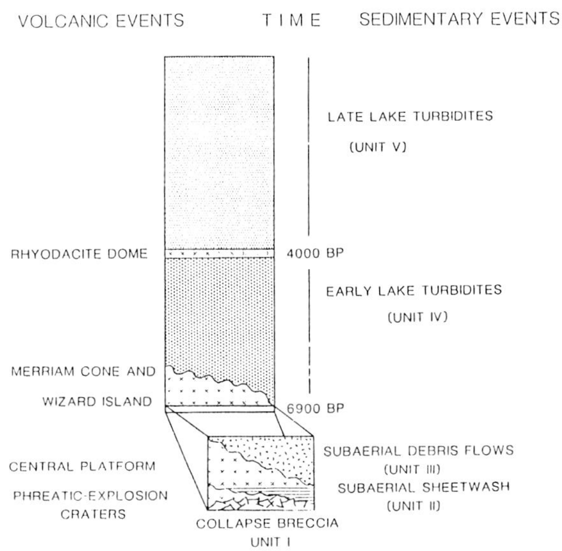

Figure 7. Schematic stratigraphic column showing the development of volcanic and sedimentary deposits on the floor of Crater Lake.

formation time of the caldera (Nelson *et al.* 1986a). The middle of the sequence of hemipelagic sediment is interrupted by an ash layer that correlates with the rhyodacite dome at the northwestern edge of the central platform (Bacon and Lanphere 1990). This ash is the last record of volcanic eruptions in the caldera (Nelson *et al.* 1986a; Bacon and Lanphere 1990).

SEQUENCE OF EVENTS

We have developed a history of volcanic and sedimentary events in the caldera by combining interpretation of the acoustic profiles with the constraints provided by dating and analyses of sediment from cores (Fig. 7). Immediately following the formation of the caldera, the region remained unstable and volcanic activity continued on the caldera floor. Evidence for phreatic-explosion craters is suggested by depressions of the acoustic basement surface in the ring-fracture zone (Nelson *et al.* 1988). The central platform formed early in the history of the caldera, prior to formation of the lake (Bacon and Lanphere, 1990). This is confirmed by the old age of lacustrine sediment covering the platform (Nelson *et al.* 1986a). Merriam Cone and much of Wizard Island formed after there was standing water in the caldera (Bacon and Lanphere 1990).

Sheetwash and/or phreatic-explosion debris, together with debris flows, (unit II) initially filled in the crater depressions of the ring-fracture zone (Nelson *et al.* 1988). This was followed by deposition of major landslides (Figs. 3-5), debris avalanches, and debris flows (unit III) that covered all the basin floors and probably resulted from seismic activity associated with the intracaldera eruptions that formed the subaerial volcanic edifices such as the central platform.

Deposition of units II and III was probably very closely spaced in time and occurred shortly after the caldera collapse. This is similar to the rapid post-eruption sheet-flow and debris-flow deposition of 12-80 m of subaerial sediment on the caldera floor of Mount St. Helens in the years since the 1980 eruption (Beach 1985; Brantley and Power 1985; Collins and Dunne 1986).

We associate the onset of lake sedimentation with the sharp change in style of seismic stratigraphy that begins with the nearly flat-lying, continuous and parallel reflections of unit IV in seismic profiles (Figs. 2 and 3). These characteristics are similar to those of subaqueous turbidites in other lakes (Hyne *et al.* 1972; Poppe *et al.* 1985; Newhall *et al.* 1987) as well as the marine environment (see Nelson and Nilsen 1984, for many examples). Graded turbidite beds have been sampled in near-surface deposits of unit V, which has seismic reflections similar to unit IV (Nelson *et al.* 1986b).

In contrast, we attribute unit III to *subaerial* mass-wasting processes, because of the chaotic nature and lack of continuity of reflections and the inclined and wedge-shaped geometry of the deposits (Nelson *et al.* 1988) and, in particular, the general absence of uniform, flat seismic reflectors and persistence of chaotic reflectors *across entire basin floors* (Figs. 2 and 3). If large, basin-filling mass flows (Chaski Bay slide, for example) had occurred underwater, they would have deposited gradational turbidite sheets similar to those that result from present subaqueous mass flows of coarse-grained debris from the caldera walls (Nelson *et al.* 1986b). Sand-rich subaqueous apron systems like those seen in the near-surface of Crater Lake have an evolution of sediment-gravity flows that result in nearly flat-lying, gradational and continuous reflectors across the entire basin (Nelson *et al.*, in press).

An age of ~6900 yr B.P. in hemipelagic mud on the central platform (Nelson *et al.* 1986a) shows that subaqueous deposition started very early in the history of the caldera, and that the east basin was at least 370 m below the lake surface at this time (depth of the central platform – depth of the top of unit III in the east basin) (Figs. 1 and 7). Therefore, subaerial deposition is constrained to within hundreds of years of the caldera collapse (Nelson *et al.* 1986a). Assuming present conditions of precipitation and evaporation, but no leakage, the lake would take a minimum of 225 years to accumulate (Nelson 1961, and in preparation).

The high-amplitude reflections of the earlier turbidites (unit IV) probably represent coarser-grained, thicker beds that occurred more frequently because of seismicity related to postcaldera subaqueous formation of Merriam Cone and much of Wizard Island. The lack of volcanic and associated seismic activity since the formation of the rhyodacite dome about 4000 years ago has resulted in the lower-amplitude reflections of unit V that represent thinner and finer-grained turbidites of the central east basin floor (Fig. 2) (Nelson *et al.* 1986b).

PRESENT-DAY PROCESSES AND INTER-BASIN RELATIONSHIPS

At present, the floor of the east basin is approximately 100 m lower than the other two basins (Fig. 1). There is a connection between the east and northwest basins, and the present-day floor of the northwest basin dips toward the east basin. Consequently, sediment from the northwest basin may funnel to the east basin. A possible fault and channelling related to sediment funneling at the northwest edge of the east basin warrant further investigation (Fig. 2).

Sediment from the saddle between the southwest basin and the east basin has distinctive semi-lithified and oxidized layers. In this same area Williams and Von Herzen (1983) recorded relatively high heat flow values. Recent work has suggested the possibility of hydrothermal vents associated with elevated water temperatures in this area (Dymond and Collier 1990). The saddle, where there is evidence of a large landslide on the southern margin of the lake, seems to isolate the southwest basin from the east basin (Fig. 4).

The landslide may provide a more permeable pathway for heated water from ring-fracture sites to reach the lake-floor surface. This is suggested by the coarse, hummocky debris seen on seismic records (Fig. 5) and confirmed by diving operations (Bacon and Lanphere 1990; Dymond and Collier 1990). Elsewhere on the lake floor, the thick sediment fill and the fine-grained uppermost lake sediments of unit V may serve as a cap inhibiting such fluid migration.

SUMMARY

An interplay of volcanic and sedimentary processes has produced the deposits on the floor of Crater Lake. Major volcanic and sedimentary events occurred rapidly during the early subaerial history of the caldera floor. Phreatic-explosion craters developed on the ring fracture and were rapidly filled by explosion debris, sheetwash and mass-flow deposits. Subaerial volcanic flows built the central platform (Bacon and Lanphere 1990); apparently the associated volcanic and seismic activity resulted in deposition of debris fans that covered the basin floors and included the large Chaski Bay landslide. Water quickly accumulated in the caldera and covered the central platform, where hemipelagic deposits began accumulating slowly, creating a condensed section that eventually recorded the entire lacustrine history. Coarser and thicker turbidites covered the basin floors in the early lake history and apparently were associated with subaqueous volcanic and seismic activity related to the formation of Wizard Island and Merriam Cone. Extrusion of the rhyodacite dome about 4,000 years ago signaled the end of significant post-caldera volcanic eruptions (Bacon and Lanphere 1990). Thereafter, reduced turbidity-current activity continued to build proximal base-of-slope aprons with sand and gravel layers and distal basin plains with interbedded sandy turbidites and hemipelagic mud. These flat-lying lake deposits show no indication of deformation by recent volcanic activity, although there is the possibility that local permeability could allow heated fluids to reach the lake floor (Fig. 2).

ACKNOWLEDGMENTS

The cooperation of the National Park Service is appreciated. Many colleagues have contributed to various aspects of the work summarized here, in particular, C. R. Bacon, P. R. Carlson, J. Dymond, B. Larsen, M. Larsen, A. W. Meyer, S. W. Robinson,

D. Schwartz, and D. Thor. The manuscript has benefited from reviews by C. R. Bacon, T. C. Johnson, T. Wiley, and an anonymous reviewer.

LITERATURE CITED

Bacon, C. R. 1983. Eruptive history of Mount Mazama and Crater Lake caldera, Cascade Range, U.S.A. Jour. Volcanol. Geotherm. Res. 18:57-115.

Bacon, C. R., and M. A. Lanphere. 1990. The geologic setting of Crater Lake, Oregon. Pages 19-28 *in* E. T. Drake *et al.*, eds., Crater Lake: An Ecosystem Study. Pacific Div., Amer. Assn. Adv. Sci., San Francisco, Calif.

Beach, G. L. 1985. Comparative geomorphic analysis of the Mount St. Helens crater area: An assessment of elevational change, 1980-1983. Corvallis. Oregon State Univ., Dep. Geogr., Unpublished rep. to U. S. Geol. Surv., 25 pp.

Brantley, S., and J. Power. 1985. Reports from the U. S. Geological Survey's Cascades Volcano Observatory at Vancouver, Washington. U. S. Geol. Surv. Earthquake Information Bull. 17:20-32.

Collins, B. D., and T. Dunne. 1986. Erosion of tephra from the 1980 eruption of Mount St. Helens. Geol. Soc. Amer. Bull. 97:896-905.

Dymond, J., and R. Collier. 1990. The chemistry of Crater Lake sediments: Definition of sources and implications for hydrothermal activity. Pages 41-60 *in* E. T. Drake *et al.*, eds., Crater Lake: An Ecosystem Study. Pacific Div., Amer. Assn. Adv. Sci., San Francisco, Calif.

Hyne, N. J., Paul Chelminski, J. E. Court, D. S. Gorsline, and C. R. Goldman. 1972. Quaternary history of Lake Tahoe, California-Nevada. Geol. Soc. Amer. Bull. 83:1435-1448.

Nelson, C. H. 1961. Geological limnology of Crater Lake, Oregon. M. S. Thesis. Univ. Minnesota, Minneapolis, Minn. 179 pp.

Nelson, C. H. 1967. Sediments of Crater Lake. Geol. Soc. Amer. Bull. 78:833-848.

Nelson, C. H., C. R. Bacon, and S. W. Robinson. 1986a. The caldera floor sedimentary history of Crater Lake, Oregon [abstract]. Internat'l Assn. Sedimentologists, 12th Internat'l Sedimentological Congr., Canberra, Australia, Abstracts vol., p. 226.

Nelson, C. H., P. R. Carlson, and C. R. Bacon 1988. The Mount Mazama climactic eruption (~6900 yr B.P.) and resulting convulsive sedimentation on the Crater Lake caldera floor, continent, and ocean basin. Pages 37-57 *in* H. E. Clifton, ed., Sedimentologic Consequences of Convulsive Events. Geol. Soc. Amer. Spec. Pap. 229.

Nelson, C. H., A. Maldonado, J. H. Barber, Jr., and Belen Alonso. (in press) Modern sand-rich and mud-rich siliciclastic aprons, alternative base-of-slope turbidite systems to submarine fans and implications for hydrocarbon exploration. *In* Paul Weimer and M. H. Link, eds., Seismic Facies and Sedimentary Processes of Modern and Ancient Submarine Fans. Springer-Verlag, New York.

Nelson, C. H., A. W. Meyer, D. Thor, and M. Larsen. 1986b. Crater Lake, Oregon: a restricted basin with base-of-slope aprons of non-channelized turbidites. Geology 14:238-241.

Nelson, C. H., and T. H. Nilsen. 1984. Modern and ancient deep sea-fan sedimentation. Soc. Economic Paleont. Mineral. Short Course 14. 404 pp.

Newhall, C. G., C. K. Paul, J. P. Bradbury, A. Higuera-Gundy, L. F. Poppe, S. Self, N. B. Sharpless, and J. Ziagos. 1987. Recent geological history of Lake Atitlan, a caldera lake in western Guatemala. Jour. Volcanol. Geotherm. Res. 33 (1-3):81-108.

Poppe, L. J., C. K. Paul, C. G. Newhall, J. P. Bradbury, and J. Ziagos. 1985. A geophysical and geological study of Laguna de Ayarza, a Guatemalan caldera lake. Jour. Volcanol. Geotherm. Res. 25:125-144.

Williams, D. L., and R. P. Von Herzen 1983. On the terrestrial heat flow and physical limnology of Crater Lake, Oregon. Jour. Geophys. Res. 88: 1094-1104.

THE CHEMISTRY OF CRATER LAKE SEDIMENTS: DEFINITION OF SOURCES AND IMPLICATIONS FOR HYDROTHERMAL ACTIVITY

Jack Dymond and Robert Collier
College of Oceanography
Oregon State University
Corvallis, Oregon 97331

Crater Lake sediments exhibit variations in composition that can largely be explained by mixtures of three distinct sources: aluminosilicate debris from the caldera walls, biogenic materials which settle from the euphotic zone, and an iron-rich precipitate. The origin of the precipitate component is uncertain. Redox mobilization of iron and other metals driven by decomposition of organic matter is unlikely, because the spatial variability of metal enrichments are incompatible with this mechanism. Precipitation of iron from anoxic springs is a feasible explanation. Since the metal deposits have only been observed in a region of the lake which has thermal and chemical anomalies in the water column, the precipitation may be from thermal springs rather than cold springs. Precipitation from hydrothermal fluids is the favored hypothesis because recent submersible observation have demonstrated that iron-rich precipitates are currently forming in the lake in association with warmer waters, anomalous concentrations of tracers of hydrothermal venting, and bacterial communities.

Comparison between the composition of surficial sediments and materials collected with sediment traps moored in the mid-water column indicates that most of the recycling of biogenic debris occurs at the sediment-water interface. Less than 10 % of the organic carbon and nitrogen that reaches the bottom as particles is preserved in the sediments. Approximately 20 % of biogenic opal is preserved. The sediment trap data also reveal that resuspension of sediments and downslope transport of material are important mechanisms for introducing sediments to the deep basins of the lake.

Downcore analyses of lake sediments indicate the presence of manganese reduction and mobilization. Manganese mobilization is most pronounced in sediments from the portion of the lake that exhibits surficial sediment enrichment of iron and water column anomalies. We propose that the strong manganese mobilization, iron precipitates, and observed iron-rich deposition on the lake bottom are a consequence of slow and dispersed advection of thermal fluids through the lake sediments and rock outcrops. Downcore variations in opal content may indicate variations in biological productivity in the lake in the past. These variations are positively correlated with variations in the abundance of the hydrothermally-associated element, lithium; this observation suggests that hydrothermal inputs may contribute to the biological productivity of the lake.

Sediments covering much of the bottom of Crater Lake provide a record of the lake's biological and chemical status. Interpretation of this record can enhance our understanding of major lake processes. For example, Nelson *et al.*, (1986) demonstrated the importance of sediment-gravity-flow (turbidite) processes in Crater Lake. Studies of sediments from other lakes have documented the importance of nutrient recycling at the lake floor and redox reactions within the sediments. Sediments are the ultimate sink for dissolved and particulate materials which are removed within the lake. It is probable that the lake approximates a steady-state in which inputs and outputs of elements are in balance. Therefore, changes in the burial rates of elements in lake sediments may be used to define changes in the input fluxes.

In this paper we will examine the chemical composition of Crater Lake sediments in order to understand the sources of particulate and dissolved material to the lake. In addition we will address the question of temporal variability in these sources by examining the composition of sediments through time.

Data

Figure 1 shows the location of five gravity cores from the lake. Three of the cores, 79-1, 80-7, and 80-14, were collected by the U.S. Geological Survey. Cores 85-3 and 85-4, were collected by Oregon State University. The cores represent three distinct areas of the lake. Core 80-14 is from the Eastern Basin, which has the greatest water depths in the lake. The relatively flat basin floor appears to be a consequence of the catastrophic slumps and gravity flow deposits which are believed to have been common during the first few hundred years of lake history (Nelson *et al.* 1986; Barber and Nelson 1990). Currently, individual particle settling may be the dominant mechanism of sedimentation, and the sedimentation rates are probably much lower than during the early part of lake history. Cores 79-1 and

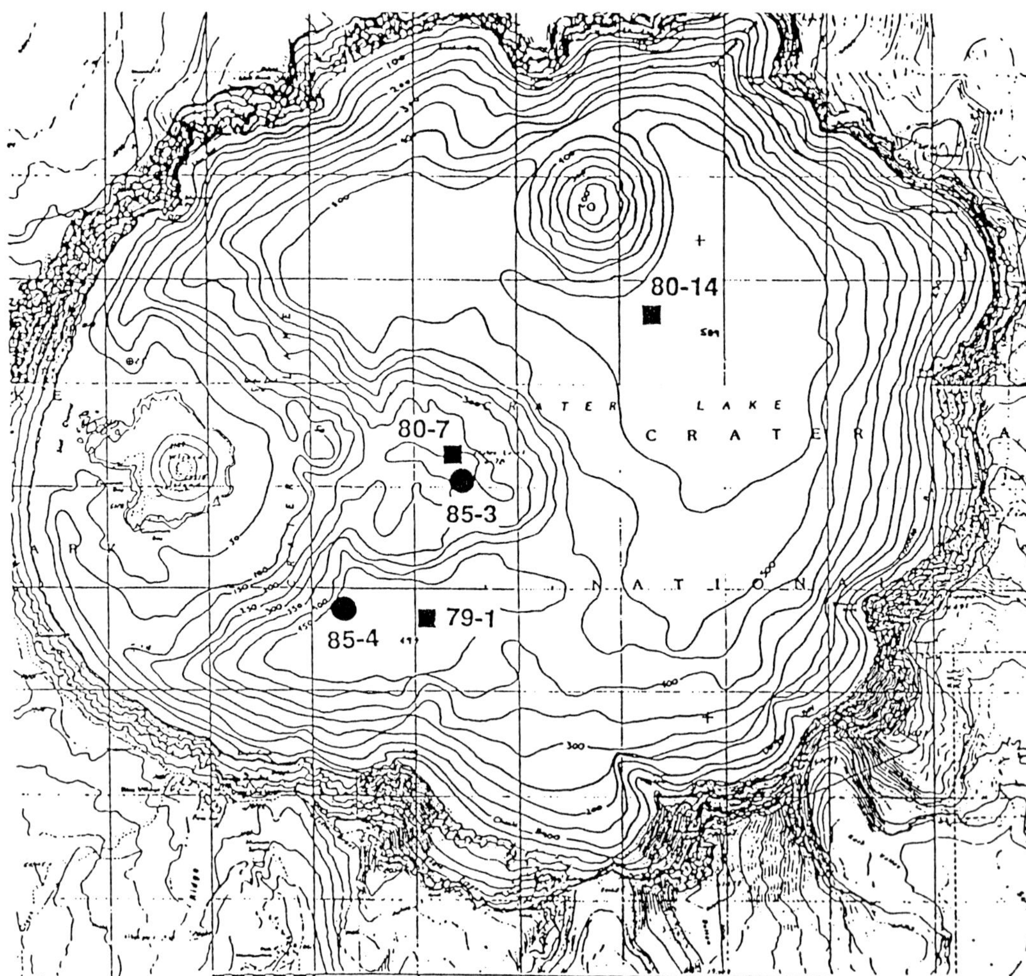

Figure 1. The location map for sediment cores discussed in this report. 79-1, 80-7, and 80-14 are cores recovered by the U. S. Geological Survey. 85-3 and 85-4 are cores recovered by Oregon State University.

85-4 are from the South Basin. This basin contains localized regions of exceptional heat flow and anomalous bottom water temperatures (Williams and von Herzen 1983; Collier *et al.* 1990). The South basin also experienced landslide and gravity flow inputs during early lake history (Barber and Nelson 1990). Cores 80-7 and 85-3 are from the Wizard Island Platform, a relatively shallow area to the east of Wizard Island. This shallow platform appears to have been formed by late stage volcanism which followed the catastrophic eruption and caldera collapse that took place 6850 years ago (Bacon and Lanphere 1990). Because of the relatively shallow depths, sedimentation in this area is due to particle settling rather than debris or turbidity flow from the caldera walls. Thus, the sedimentation rates are probably much slower and the area has only thin sediments (Barber and Nelson 1990).

The sediment samples were analyzed by atomic absorption spectrometry (AAS) and instrumental neutron activation analysis (INAA) following procedures reported in Dymond and Lyle (1985) and Fischer (1984). The samples were analyzed in duplicate, and the averages of the two analyses are shown in Table 1. The precision of the measurements varies between 2 and 10%, with the poorer precision appropriate for minor elements such as Li, Zn, and Ba. The accuracy, determined by analyses of USGS standards and comparisons between INAA and AAS for Al, Ca, Mn, and Ba, is similar to the precision. Organic carbon and nitrogen were measured by combustion using a CHN analyzer.

Compositional data from sediment traps which were deployed at various depths on moorings are reported as well (Table 2). Samples from CL-1 and CL-2 were from in the South Basin. The other trap deployments were in the central portion of the Eastern Basin. Material collected with sediment traps can define the composition of settling particles prior to undergoing alteration, dissolution, and decomposition which may occur at the sediment-water interface.

Sediment Compositional Types (End-Members)

Generalized sources of Crater Lake sediments can be defined from their elemental compositions. For example, iron and aluminum concentrations (Fig. 2) define three end-members: (1) aluminosilicate debris from the caldera wall which typically has high aluminum content and Fe/Al values between 0.3 and 0.4; (2) biogenic debris, an important component in sediment trap samples, is depleted in both Al and Fe; (3) material with relatively low aluminum and high

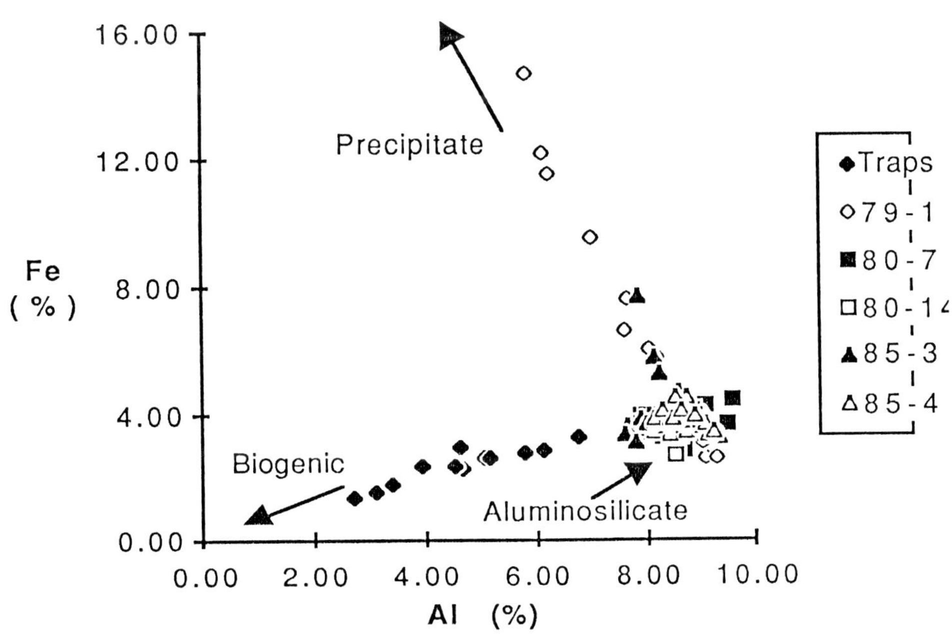

Figure 2. A plot of the concentration of Fe against the concentration of Al for all sediment core and trap samples listed in Tables 1 and 2. Arrows indicate the inferred three end-members.

TABLE 1. CHEMICAL COMPOSITION OF CRATER LAKE LOCATIONS OF SAMPLES ARE SHOWN IN FIGURE 1

Sample	Water Depth	Depth in Core (cm)	N	P	Corg	Si	Al	Opal	Mg	K	Ca	Fe	Li	Ti	Mn	Cu	Zn	Sr	Ba	As	Sb
			<--- ppm	ppm-->	<---			%				--->	<---				ppm				--->
79-1	463 m	0		3745	1.65	26.4	7.60	9.4		1.00	2.44	6.65	24		1545	96	74	511	659		
		2		5183	0.80	25.4	7.67	6.3		1.01	2.58	7.57	22		1490	84	102	527	706		
		5		6651	0.83	24.9	7.01	10.1		0.93	2.44	9.55	23		1580	83	67	510	858		
		6				22.2	5.84	12.2	1.17	0.78	2.1	14.70	20		1780	78	88	524	1300	860	2.6
		9				22.2	6.13	9.9	1.16	0.83	2.17	12.20	23		1340	173	105	669	2500		
		10		19833	1.38	22.5	6.26	9.6		0.87	2.23	11.55	22		1040	190	86	664	2480		
		20		530	0.45	28.0	8.82	4.0		1.10	2.72	3.87	26		530	95	60	555	450		
		30		511	0.68	28.6	8.58	7.3		1.08	2.38	4.13	28		643	120	79	483	455		
		32				27.3	8.20	7.1	1.31	0.93	2.53	5.73	15		475	75	62	559	530		
		33		384	0.15	28.3	8.03	10.9		0.94	2.55	6.01	15		559	84	54	517	457		
		36		436	0.18	28.3	8.60	6.4		1.00	2.33	4.56	22		505	81	52	524	485		
		45		535		28.6	9.07	3.5		1.20	2.88	2.57	27		331	152	73	594	516		
		58		556	0.20	29.1	9.27	3.2		1.25	2.06	2.61	26		276	158	58	380	416		
		66		446	0.32	29.8	8.71	9.5		1.13	2.52	4.16	29		359	112	68	541	467		
		74		620	0.08	29.0	9.05	4.8		1.07	2.81	3.10	14		203	82	55	668	611		
80-7	256 m	0	4000		2.19	28.4	8.27	9.4	1.51		2.47	3.57	36	3950	1145	135	84	420	490		
		2	3100		2.23	30.3	8.14	15.2	1.3		2.31	3.29	38	3700	1190	130	80	410	545		
		5	4100		2.70	27.4	7.89	9.8	1.37		2.38	3.93	34	3850	717	145	80	435	515		
		9				27.4	8.52	4.8	1.54		2.66	3.67	36	4080	814	150	83	460	505		
		20	3200		1.49	29.3	7.95	14.2	1.42		2.64	3.37	38	3950	1150	150	79	480	505		
		30	3200		1.15	28.9	8.60	8.1	1.37		2.74	3.41	44	4000	836	155	75	470	565		
		40				29.6	9.01	6.7	1.45		2.98	3.52	42	4150	860	185	80	505	520		
		50				28.6	8.55	7.6	1.44		2.82	3.53	40	4050	557	205	78	520	525		
		60				29.1	8.56	8.9	1.46		2.75	4.01	36	4000	671	290	87	460	515		
		67				28.6	9.07	3.7	1.79		3.6	4.29	26	4500	654	120	75	645	490		
		70				28.0	9.02	2.3	1.44		3.06	3.69	35	4300	520	150	77	540	550		
		95				29.2	8.98	5.9	1.43		2.96	3.67	38	4200	617	285	83	500	510		
		120				29.6	8.70	9.0	1.28		2.74	2.88	41	4050	440	260	74	465	530		
		140				30.8	9.47	6.2	1.34		3.11	3.68	40	4350	1455	275	80	500	560		
		145				27.9	9.59	0.3	2.32		4.59	4.42	16	4850	945	505	72	765	410		
80-14	590 m	0	4200		2.27	27.5	7.98	9.2		1.15	2.35	3.94	37		848	130	78	435	505		
		2	2600		1.08	28.3	8.58	6.7		1.14	2.72	3.89	29		832	106	72	500	521		
		5	2200		1.00	28.0	8.68	5.1		1.16	2.86	3.90	29		830	92	68	515	531		
		9	3200		1.07	28.5	8.57	7.3		1.20	2.83	3.83	28		877	97	69	520	532		
		17	2700		0.80	28.1	8.82	4.3		1.22	2.06	4.08	29		1015	116	71	420	485		
		30	3800		1.82	28.5	8.34	9.0		1.15	2.47	3.65	33		1260	141	72	460	491		
		45				29.3	8.15	12.7		1.29	2.38	3.47	37		844	144	69	450	539		
		60				27.9	7.88	11.1		1.14	2.47	3.72	32		812	160	69	425	516		
		69				28.1	8.37	7.7		1.17	2.66	3.71	31		818	138	70	470	494		
		71				28.0	8.42	7.2		1.18	2.47	4.04	29		875	130	85	460	492		
		72				28.0	8.63	5.5	1.62		2.7	3.75	26	4200	845	120	86	565	495		
		85				29.0	8.12	12.2	1.42		2.62	3.65	35	3800	892	182	81	440	490		
		100				27.2	8.59	3.7	1.54		2.98	3.37	28	3850	1375	161	72	510	530		
		130				29.4	8.85	7.3	1.57		3.25	3.50	30	4050	1200	133	68	505	525		
		162				30.5	8.52	12.9	0.83		2.34	2.65	34	2940	565	105	57	445	655		
		170				29.0	8.75	7.2	1.3		3.01	3.43	35	4250	839	140	71	450	530		
		185				28.6	8.46	8.3	1.31		2.5	3.37	36	3800	938	179	75	435	540		
85-3	265 m	0	8800		3.49	27.4	7.80	10.4	1.54		2.32	3.71	34	3635	1470	124	77	420	485		
		1	8700		2.74	28.2	8.16	9.7	1.53		2.34	3.62	36	3785	1070	122	77	410	455		
		2	8500		3.17	27.5	8.18	7.7	1.54		2.35	3.74	37	3830	1030	127	78	420	465		
		3	6100		3.26	27.6	8.04	9.1	1.6		2.36	3.89	36	3830	1130	135	81	440	500		
		4	9300		3.75	27.4	7.66	11.5	1.54		2.32	3.64	36	3735	999	133	79	405	455		

TABLE 1 (continued): CHEMICAL COMPOSITION OF CRATER LAKE
LOCATIONS OF SAMPLES ARE SHOWN IN FIGURE 1

		5	7300	3.50	27.2	7.80	9.9	1.5	2.23	3.54	36	3690	914	136	80	405	485		
		6	5000	3.03	28.2	7.62	13.9	1.37	2.16	3.31	39	3560	808	126	75	390	545		
		7	5800	3.14	28.6	7.80	13.6	1.22	2.01	3.10	42	3355	758	116	70	350	550		
		8	6100	3.17	27.9	7.90	11.0	1.48	2.24	3.77	37	3645		142	79	410	505		
		9	6400	3.38	28.0	8.03	10.2	1.53	2.37	3.69	37	3760	1045	147	81	440	495		
		14			27.8	7.88	10.9	1.54	2.2	3.88	37	3680	1150	155	84	450	505		
		25			27.9	8.01	10.1	1.53	2.22	3.66	37	3660	747	181	83	440	520		
		35			28.4	8.17	10.2	1.48	2.3	3.79	38	3905	583	186	83	487	494		
		40			27.5	8.16	7.9	1.5	2.36	5.73	36	3835	934	161	81	509	756		
		45			27.5	8.22	7.4	1.44	2.3	5.30	38	3850	1050	148	78	460	500		
		50			26.4	7.84	7.5	1.42	2.54	7.72	37	3745	1370	148	73	554	876	105	0.9
		55			28.4	8.60	6.8	1.45	2.43	4.49	40	4075	5020	208	81	498	665		
		65			29.0	8.91	5.9	1.63	3.1	3.75	26	4725	494	132	69	606	496		
		75			29.0	8.58	8.5	1.4	2.52	4.47	44	3990	2260	251	76	492	578		
		79			29.2	8.70	8.1	1.38	2.3	3.58	39	3965	2880	214	73	480	718		
85-4	492 m	0	2800	1.44	30.0	8.76	9.7	1.48	3.15	4.21	20	4110	3155	77	68	650	500		
		1	3000	1.31	28.5	8.73	5.9	1.46	2.83	4.16	17	3935	1565	72	70	639	500	78	0.7
		2	4100	1.14	28.9	9.14	3.7	1.52	3.31	3.52	16	4265	1210	62	62	722	515		
		3	4300	1.35	29.2	8.71	8.0	1.55	2.97	3.77	18	4120	1120	79	68	639	540		
		4	3100	1.06	29.2	9.07	5.2	1.54	3.04	3.59	19	4170	930	65	64	658	595		
		5	2900	0.26	29.2	8.74	7.7	1.47	2.7	3.40	17	4160	1085	68	67	615	470		
		6	5300	1.46	28.3	8.55	6.9	1.48	2.73	4.07	19	3990	1360	83	66	614	510		
		7	3600	1.04	29.0	9.04	4.8	1.48	2.92	3.60	15	4125	1080	63	64	650	500		
		8	3000	1.05	29.1	8.74	7.4	1.52	2.71	3.69	18	4140	1115	82	65	600	489		
		9	3100	1.30	28.9	8.96	5.1	1.46	2.85	3.58	16	4175	1060	63	62	640	510		
		10			28.5	8.87	4.9	1.53	2.84	3.69	17	4255	1015	74	68	636	485		
		12			29.2	8.98	5.9	1.46	2.95	3.80	16	4175	1235	62	63	667	495		
		14			28.7	8.66	7.0	1.55	2.75	3.78	20	4035	1225	91	68	614	500		
		16			28.9	8.16	11.4	1.46	2.54	3.86	26	3700	1100	110	68	545	475		
		18			28.5	8.59	7.0	1.58	2.92	4.69	24	3895	1359	96	66	606	450		
		20			28.3	8.46	7.6	1.52	2.86	3.83	21	3990	1130	83	64	618	515		
		22			28.3	8.28	8.9	1.56	2.66	4.06	22	3910	1190	102	68	573	455		
		24			29.2	8.78	7.3	1.58	2.54	3.73	17	4170	1085	71	64	584	520		
		26			28.4	9.01	3.5	1.46	2.88	3.90	14	4170	905	53	60	656	490		
		28			28.9	9.04	4.7	1.54	2.71	3.62	18	4435	910	71	64	602	490		
		30			28.6	8.74	6.1	1.8	1.97	4.48	30	3915	1215	124	75	428	470		
		40			28.3	8.80	5.0	1.56	2.73	3.67	21	4050	966	79	66	610	505		
		45			29.7	9.26	4.9	1.4	3.22	3.34	15	4585	752	38	59	602	475		
		47			29.2	9.14	4.5	1.42	3.16	3.38	15	4485	719	38	58	692	498		
		49			29.7	9.06	6.5	1.39	2.86	3.34	16	4780		48	60	660	575		
		51			29.0	8.70	7.4	1.45	2.88	4.06	18	4465	888	67	63	619	539		
		53			29.5	9.01	6.3	1.45	2.79	3.62	17	4730	846	57	63	640	503		
		55			29.2	9.06	5.3	1.44	3.04	3.68	15	4715	806	44	61	667	552		
		57			29.0	8.72	7.3	1.42	2.92	3.78	14	4695		42	58	632	536		
		59			28.5	8.56	7.2	1.6	2.74	3.88	19	4060	1135	82	67	569	504		
		61			28.8	8.47	8.7	1.6	2.65	3.86	21	4110	890	88	69	570	433		
		63			28.1	8.53	6.4	1.86	2.44	4.48	18	4250	1020	83	79	532	409		
		65			28.3	8.96	3.7	1.64	2.84	3.89	17	4365	838	72	72	600	406		
		67			29.2	9.34	3.1	1.32	3.24	3.24	12	4365		39	58	698	519		
		69			28.9	9.12	4.0	1.33	3.08	3.34	11	4650		36	64	715	554		
		70			28.8	9.20	3.1	1.34	3.24	3.39	12	4565	738	39	59	704	543		
		77			28.6	8.96	4.5	1.39	2.76	3.78	15	4430	1050	61	60	610	485		
		84			28.9	8.64	7.7	1.63	2.06	4.06	28	3895	1000	125	70	465	508		
		91			29.0	9.06	4.8	1.49	3	3.72	18	4220	829	62	63	664	518		
		98			29.5	9.22	4.7	1.36	3.09	3.40	16	4335	792	43	56	699	556		
		105			28.9	8.88	5.9	1.52	3.06	3.90	17	4690		61	64	686	617		

TABLE 2. CHEMICAL COMPOSITION OF SEDIMENT TRAP SAMPLES

Time of Year	Water Depth	Trap Depth	Corg	N	Si	Al*	Al	Opal	Mg	Ca*	Ca	Fe	Ti	Mn*	Mn	Cu	Zn	Sr	Ba	Li
			<-----	-----	-----	-----	-----	%	-----	-----	-----	----->	<-----	-----	-----	-----	ppm	-----	-----	----->
7/12/83	463m	457m	2.2	0.35	27.9	8.48	8.61	5.5		2.83	2.90	3.80		812	783	82	89		496	
9/13/83	"	200	8.4	0.96	26.0	6.65	6.54	16.6	1.19	2.06	2.08	2.99	3225	506	444	74	136	359	353	22
7/11/84	590	200	16.2	1.74	22.5		4.60	22.7	1.05		1.59	2.3	2500		690	94	207	250	300	23
"		583	8.1	0.76	26.3		6.78	15.6	1.29		1.97	3.29	3500		1470	138	160	350	500	31
9/18/84		200	12.2	1.93	23.8		4.69	25.5	0.95		1.42	2.24	2297		439	67	83	252	280	18
"		583	10.0	1.30	25.1		5.80	20.2	1.13		1.71	2.80	2938		988	105	124	304	397	25
7/11/85		200	27.1	3.82	15.9		2.71	20.2	0.62		1.38	1.36			370	61	189		130	11
"		390	25.0	3.02	16.6		3.42	16.6	0.82		1.47	1.76			817	96	102		160	16
"		580	29.6	3.12																
9/6/85		200	9.4	1.32		5.00				1.60			2600	607						
"		390	9.5	1.13		5.16				1.79			2600	776						
"		580	7.2	0.87		6.02				2.06			2700	1343						
7/9/86		200	26.7	2.85		3.41				1.52				548						
"		390	20.6	2.11		4.01				1.46				822						
"		580	18.2	1.72		4.78				1.76				1271						

* These analyses are by INAA, equivalent elements are by AAS

iron concentrations. We believe this component is dominated by iron oxyhydroxide which forms by precipitation from solution.

End-member mixing occurs primarily between the aluminosilicate component and the other two components (Fig. 2). In other words, there are mixing lines both between the biogenic end-member and the aluminosilicate end-member and between the iron precipitate and the aluminosilicate end-member. There is no clear mixing line between biogenic and the precipitate end- member, and there is little filling of the field which would result from significant contributions of all three components in any given sample. The probable explanation of this observation is that aluminosilicate material dominates the sediment at most sites. Only in sediment traps is biogenic debris a major component. Upon reaching the bottom much of this component decomposes and dissolves (i.e., it is recycled at the sediment-water interface). Also, only at certain sites is the precipitate component significant. The samples which define the trend to high iron in Fig. 2 are from the near surface portions of 79-1 and near the bottom of 85-3. The most iron-enriched samples in 79-1 is a lithified, ochre-colored crust from a depth of 6 cm depth in the core. The iron enrichments observed in both 79-1 and 85-3 are not present in nearby cores (Fig. 3a and 3c). The spatial variability in the abundance of the precipitate end-member is relevant for clarifying the source of this component.

The detrital aluminosilicate end-member is primarily debris weathered and eroded from the caldera walls. This material with Al values between 8 and 9% is typical of Cascade volcanics. For example, the rhyodacite expelled by the climactic eruption has an Al content of 8.1% (McBirney 1968). Our analyses of fine debris from seven scree slopes at various locations around the caldera walls reveal an average aluminum content of 8.9 + 0.5% and Fe/Al values ranging between 0.3 to 0.5.

The biogenic end-member represents remains of flora and fauna which settle through the water column and are preserved in the sediments. This material is depleted in Fe and Al and composed of organic matter and siliceous shell material from diatoms and other phytoplankton. In general, materials from sediment traps deployed at 200 m water depths are more enriched in the biogenic end-member than samples from deeper traps. This is a consequence of two

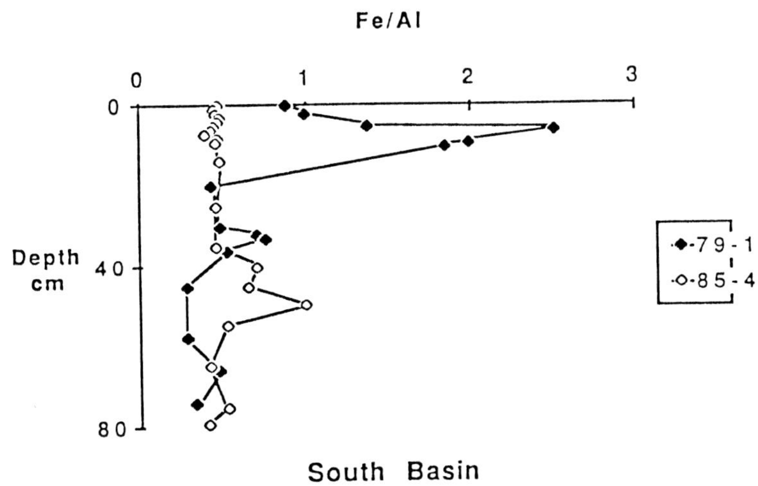

Figure 3a. Iron to aluminum ratio as a function of depth in the sediment core.

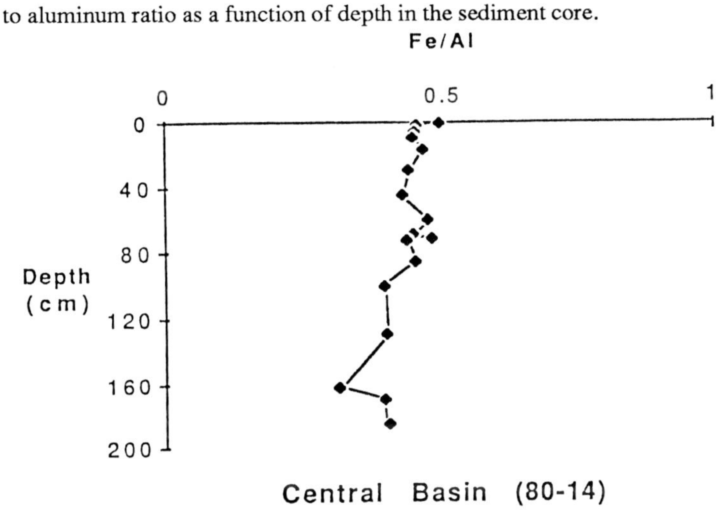

Figure 3b. Iron to aluminum ratio as a function of depth in the sediment core from the Central Basin (Fig. 1). Note the changes in scales compared to Figure 3a.

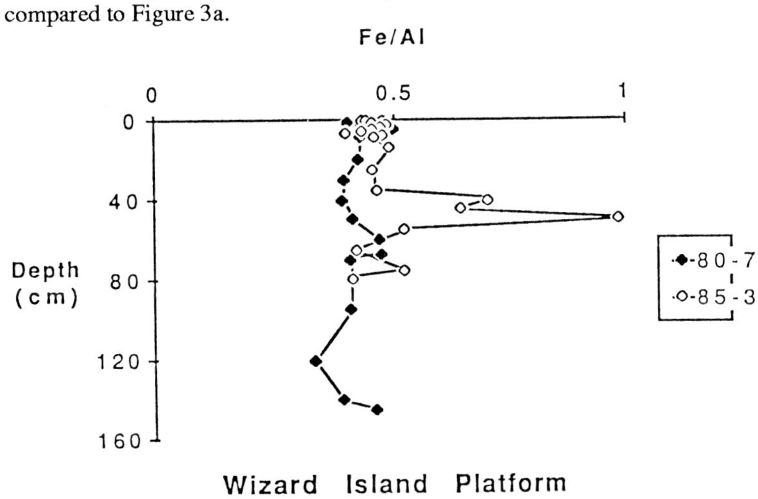

Figure 3c. Iron to aluminum ratio as a function of depth in two sediment cores from Wizard Platform (see Fig. 2).

processes. First, dissolution and decomposition of labile biogenic material during settling produces a relative enrichment of refractory sediment components. Second, resuspension of bottom sediments results in the downslope transport of material that has undergone more loss of labile biogenic material. The downslope inputs predominantly affect the deepest traps.

If the third end-member is a hydrated iron oxide precipitate we can estimate its iron content by extrapolating the Fe/Al relationship shown in Fig. 2 to zero aluminum. This is a reasonable assumption since Al is relatively immobile in aqueous solution, and natural hydrated iron oxide precipitates have only trace quantities of Al. This extrapolation suggests the iron content of the pure precipitate is 33%, similar to values reported for iron oxide precipitates associated with marine hydrothermal systems (Corliss *et al.* 1978; Bostrom and Widenfalk 1984). Examination of the covariation between Fe and other elements in the 79-1 core suggests this component is depleted relative to volcanogenic debris in Si, K, and Ca, all of which are major elements in Mt. Mazama volcanics. In contrast, the precipitate component appears to be enriched in P, Mn, and Ba. Precipitates with enrichments of these elements could be formed by several processes, and we will discuss the possible mechanisms in the next section.

Origin of Iron-rich Precipitates

We will consider three mechanisms for the formation of the precipitate end-member: (1) low temperature redox-driven mobilization within the sediment column; (2) precipitation from cold springs; and (3) precipitation from thermal springs.

The burial of organic matter in sediments can result in low temperature redox mobilization of iron and manganese within the sediment column. This process occurs in some lake and marine sediments and has been described in detail by many researchers (e.g., Froelich *et al.* 1979; Berner 1980). Free oxygen and oxygen-bearing compounds within sediment pore waters serve as the oxidants for organic matter, and solid-phase manganese and iron oxyhydroxides are reduced to soluble +2 valence forms which diffuse upward along concentration gradients. Precipitation of manganese and iron oxyhydroxides in oxygenated sediments near the sediment-water interface can produce surficial enrichments. Figure 3a shows that iron enrichment in the 79-1 core is surficial as would be expected by a redox mobilization process. In contrast, however, neither the nearby core, 85-4, nor any of the other cores exhibit surface iron enrichments (Fig. 3b and 3c). This suggests that the process which forms the precipitates is localized. Since redox mobilization is driven by organic carbon burial, localized mobilization would require heterogeneities in surface primary productivity and/or sedimentation rates on an unreasonably small horizontal scale. Also of significance is the fact that both 85-3 and 85-4 exhibit iron enrichments at depth (Fig. 3a and 3c), an observation which is incompatible with active iron reduction and mobilization.

Precipitation from cool springs is an alternative explanation for the iron-rich component. Since iron and manganese can be mobilized in anoxic waters, these elements are enriched in certain cold springs. Because both iron and manganese would be highly insoluble in the oxygenated lake waters, mixing of spring waters with lake water could form hydrated metal oxide deposits. The enrichments of some elements, particularly barium and phosphorus noted in the Crater Lake precipitate end-member could be accounted for by coprecipitation or scavenging of these elements from normal bottom waters by the newly formed precipitates. This mechanism can explain the localized distribution of the precipitate material in our sediment samples. It requires, however, that cold springs occur in a portion of the lake which has anomalous bottom water temperatures and compositions (Williams and von Herzen 1983; Collier *et al.* 1990). Such an association would be a surprising coincidence; however, it cannot be ruled out.

Anoxic thermal springs are similarly conducive to leaching and transport of iron and manganese. Although reports of thermal springs in deep lakes are rare, research in the oceans during the past decade has revealed a variety of hydrothermal vents and associated deposits. These marine counterparts, which are a consequence of seawater circulation through cooling igneous rocks, occur both in rocky and in sedimented areas of the ocean floor. Some springs are extensively mixed with cool formation waters prior to venting into the bottom; other springs vent into the oceans at temperatures above 300°C. Venting at some sites is by diffuse advection through

TABLE 3. ESTIMATED PERCIPITATE CONCENTRATIONS (WT %) AND COMPARISONS WITH OTHER SAMPLES

	Precipitate Component[1]	Bacterial Mat[3]	Deep Rover Crusts[4]	Hydrothermal Oxides Galapagos[5]	Santorini[6]
Al	0[2]	0.21	0.46	0.32	0.87
Si	9.2	7.2	11.7	13.3	6.8
P	5.3	3.7	2.0	–	–
Ca	1.2	0.80	0.87	1.4	1.3
Fe	32.7	37.9	35.5	27.5	27.8
Mn	–	0.01	0.45	7.9	0.045
Ba	0.3	0.02	0.05	0.07	0.005

[1]The estimate of this component was based on fitting the Fe-Al relationship in 79-1 and extrapolating to zero Al to produce the value of 32.7% Fe shown above. Then, the relationships between Fe and the other elements were determined and values for these other elements at an iron concentration of 32.7% were computed.
[2]Zero by assumption.
[3]Bacterial mat sample collected by Deep Rover, Dive 179 (Dymond *et al.*, submitted).
[4]Crust data for sample collected during Deep Rover Dive 187 (Collier and Dymond 1989).
[5]Data from Corliss *et al.* 1978
[6]Data from Bostrom and Widenfalk 1984

the sediments, and manganese and iron oxides and silicate are deposited as near surface crusts (Corliss *et al.* 1978). The more than 300,000 years of volcanic history of Mt. Mazama (Bacon and Lanphere 1990) and the explosive volcanic origin of Crater Lake are compatible with a consistent and relatively young magmatic source that could drive hydrothermal systems and produce localized thermal springs within the lake.

Supporting a thermal spring origin for the precipitate component is the fact that the 79-1 core was recovered from a region of high sediment heat flow and bottom waters which have both anomalous temperatures and dissolved contents (Collier *et al.* 1987; Collier and Dymond 1988a; Collier *et al.* 1990). These observations include enrichments in some tracers which have been used with great success to locate thermal springs in the deep sea and correlation of the dissolved components with water temperature.

In addition, remotely operated vehicle (ROV) and manned-submersible observations (Collier and Dymond 1988a; Collier and Dymond 1988b) have revealed deposits in the South Basin which may be examples of presently forming precipitates. Perhaps the most striking submersible observations were bacterial mats found within the South Basin (Collier and Dymond 1988b; Dymond *et al.*, submitted). Some of these mats are 2-3 m across and blanket cliff faces; others are small patches a few tens of centimeters across covering sediment surfaces. These mats appear to be the result of bacterially mediated iron and manganese oxidation. They have high (nearly 38%) iron concentrations and relatively large amounts of phosphorus and silicon. In Table 3 we compare the mat compositions with an estimated composition of pure precipitate component and find that the two materials have sufficiently similar compositions to suggest they are products of the same process. Although no shimmering waters or other evidence of outflow could be observed, measured temperatures within the mat were three to six degrees above ambient, and a water sample taken at a mat site was strongly enriched in a number of elements (e.g., Li, Si, Na, Ca, K, Mn, Cl and SO_4). The mats appear to mark sites of diffuse venting of warm fluids into the lake.

At other sites lithified crusts of what appears to be iron and manganese enriched material were observed with the submersibles (Fig. 4), and at some locations slumping revealed lithified iron-rich bands (Fig. 5). Multicolored, pebble-sized material covered the bottom at a number of locations (Fig. 6). The reddish-brown to black color of these pebbles

Figure 4. Lithified crusts observed in the South Basin area, near 79-1 location. The coloration suggests Mn enrichments overlie iron enrichments.

Figure 5. Lithified crust exposed by slumping at a South Basin site. The coloration suggests strong iron enrichment.

Figure 6. Pebble field from the South Basin area. The various colors are very similar to those in the crust observed in Figure 5, suggesting the pebble field formed by break-up of intact crusts.

also suggests iron and manganese enrichment. These pebbles appear to be broken-up examples of the lithified crusts shown in Fig. 4. These crusts and pebbles have a range of compositions; however, all exhibit strong iron enrichments and relatively low aluminum values (Collier and Dymond 1989). In Table 3 we show the composition of a representative crust sample which appears to have little contamination from typical lake sediments. The compositional range exhibited by other crusts appears to reflect the degree of sediment contamination of the crust and the ease with which manganese is separated from iron by redox processes (Collier and Dymond 1989). X-ray diffraction analysis of the crusts samples demonstrate that the most Al-poor, pure iron oxyhydroxide precipitates are amorphous rather than crystalline iron phases.

We suggest that the bacterial mats, lithified crusts, and pebbled surfaces observed in some parts of the lake and the iron enrichments noted in the upper part of core 79-1 represent different stages in the evolution of precipitates forming from hydrothermal fluids. Precipitation from hydrothermal fluids is indicated by the fact that waters with anomalous concentrations of ^{3}He, Mn, ^{222}Rn, Li, Si, and Cl have been observed at and near these iron-rich deposits, (Collier *el al.* 1987; Lupton *et al.* 1987; Collier and Dymond 1989). In addition, these enrichments exhibit positive correlations with temperature, and overall the data suggest the fluids are the result of hydrothermal circulation, meaning they result from interaction between local formation waters and a cooling igneous body. Iron-oxidizing, chemolithotrophic bacteria precipitate large amounts of iron in mats at interfaces between the anoxic advecting fluids and the oxic lake waters. Since these organisms require a source of ferrous (soluble, reduced) iron, they thrive only as long as the advection of fluids continues. Precipitation within the conduits of hydrothermal systems, however, eventually blocks the flow and shuts down venting at any given site. When this occurs the bacterial mat will be transformed from a gelatinous, living structure to a lithified crust. If the thermal fluid advection is through sediments, rather than through rock, excess pore pressure and doming of the sediments similar to that observed in sediment-hosted hydrothermal systems in the oceans will result (Williams *et al.* 1979; Maris *et al.* 1984; Wheat and McDuff 1988). Cessation of venting would result in dome collapse and breaking of the lithified crusts would account for the pebble fields of mixed crust varieties. These pebbles eventually become buried by aluminosilicate and biogenic debris which comprise the typical sedimentation in the lake. The lithified piece found at a depth of 6 cm in core 79-1 is probably an example of a buried iron-rich crust.

In addition to iron and manganese, other elements are enriched in the precipitate component. For example, barium, zinc, and phosphorus are enriched in the near surface sediments of 79-1 and could be used to define these three end-members as well. Figure 7 exhibits the same three end-members using Ba and Al compositions.

A few measurements of arsenic and antimony demonstrate enrichments of these elements in the iron enriched sediments from the South Basin (Table 1). For example, the sample from 6 cm depth in core 79-1 has: 14.7% iron, 860 ppm arsenic, and 2.6 ppm of antimony. In contrast, the sample from 1 cm depth in core 85-4 has: 4.2% iron (a typical value for Crater Lake sediments), 78 ppm arsenic, and 0.7 ppm antimony. Assuming the 79-1 sample represents a mixture of iron-rich precipitate and typical Crater Lake sediments with iron and As values similar to the sample from 85-4, we can estimate the As content in the pure precipitate. This calculation suggests an arsenic content of the Fe-rich precipitate of over 2000 ppm.

Since arsenic and antimony are anions in aqueous solution, they are easily coprecipitated with iron. Both elements are similar to phosphorus in this respect. Thus, high concentrations in the iron-rich sediments may not indicate enrichment in the spring solution. The observed phosphorus enrichments in iron-rich crusts probably can be explained by coprecipitation of phosphorus derived from the lake. For arsenic and antimony, however, the case is not so clear. Neither element has been analyzed in Crater Lake waters, but in general, arsenic and antimony concentrations in surface waters from volcanic terrains are very low compared to the concentrations in thermal springs (Onishi 1969) and hydrothermal fluids (Ellis and Mahon 1977). In addition, As concentrations in the bacterial mat and several Crater Lake crust samples collected by submersible (Dymond *et al.*, submitted) demonstrate very strong enrichments in the iron- rich precipitates (over 4200

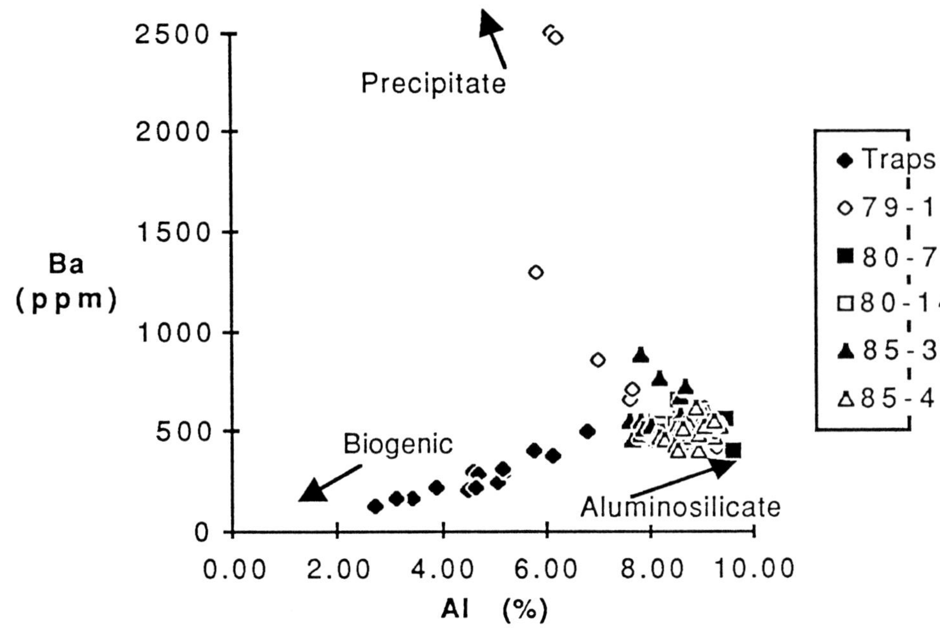

Figure 7. Barium vs aluminum concentration in sediment cores and trap samples. Arrows are directed toward the compositions of the three hypothesized end-members.

ppm in crusts and 3600 ppm in the bacterial mat). These high values support precipitation from hydrothermally-derived waters.

SEDIMENT RECORD

The variations in composition with depth in the sediment cores reflect the depositional history of sediments in the lake as well as post-depositional processes. In this section we will consider some of the recycling and diagenetic processes that occur within the sediments and explore the possibility that the sediment record indicates past episodes of hydrothermal inputs to the lake. This latter topic is a preliminary discussion of a subject that will require a more thorough paleolimnological study to resolve.

The dissolution and decomposition of biogenic debris in the water column and at the sediment-water interface are important processes by which nutrients are recycled within the lake system. The extent of this recycling is one of the major controls on the productivity in the lake. For example, more efficient burial of biogenic particles results in the release of fewer nutrients to the deep lake, and upwelling from this diminished pool of nutrients in the deep lake would support reduced productivity in surface waters. Thus, variations in the efficiency of organic matter burial during the lake's history could produce fluctuations in surface water productivity.

Some recycling of biogenic elements may occur during the settling of the material to the bottom. The observation that deeper sediment trap samples are relatively depleted in biogenic elements and relatively enriched in lithogenic material support this concept (Fig. 8). We suspect, however, that most of these compositional differences are a result of input of resuspended bottom sediments to the settling particle load. Resuspension is indicated by the fact that particulate fluxes of both labile and refractory elements increase with depth (Dymond and Collier, unpublished data). While increases in the concentration of a given component or element can be accounted for by a decrease in another component, flux increases can only be accounted for by lateral (resuspended) input from the sides of the lake. For Crater Lake the flux of biogenic components increases by approximately a factor of two between 200 and 390 m, while the flux aluminosilicate-associated elements increases by more than a factor of three, thus producing an enrichment in refractory elements in particles collected at greater depths.

Comparison of the composition of sediment trap material and lake floor sediments (Fig. 9) can constrain the extent to which carbon, nitrogen, and opal are recycled at the sediment-water interface. Assuming that no aluminum is recycled at the bottom, we can define the extent of recycling from changes in the ratios of biogenic components to aluminum. Since nearly all Al is nearly all associated with refractory aluminosilicate material, this is a reasonable assumption. Table 4 lists the biogenic component to aluminum ratio for both a sediment trap located 10 m above the bottom and for sediments. The fraction recycled within the sediments is computed for each biogenic component assuming aluminum is conservative, and the decrease in the ratios is due solely to a loss of the biogenic components. The data indicate that the fraction of carbon, nitrogen, and opal lost within the upper centimeter of sediments is respectively 86%, 79%, and 60%. At a depth of 5 centimeters the fraction lost is 94%, 90%, and 80%. As discussed by Suess and Muller (1980), a significant portion of the nitrogen and a lesser fraction of the organic carbon in the upper few centimeters of sediments is in the form of living biomass from micro and macroorganisms. Consequently, the extent of carbon and nitrogen recycling may be underestimated by this comparison method. The lower degree of opal recycling suggests that this biogenic component may be more suitable as a paleoproductivity indicator than either carbon or nitrogen.

TABLE 4. A COMPARISON OF THE ORGANIC CARBON, NITROGEN, AND OPAL CONCENTRATIONS (NORMALIZED TO ALUMINUM) OBSERVED IN SEDIMENT TRAP AND SEDIMENT SAMPLES FROM CRATER LAKE

	Sediment Trap[1]	Sediment (0-1)[2]	Sediment (5-6)
Corg/Al	2±1	0.28	0.12
%Recycled		86	94
N/Al	0.25	0.053	0.025
%Recycled		79	90
Opal	2.9	1.15	0.59
%Recycled		60	80

[1]The sediment trap data are from traps placed at a depth of 584 m (10 m above bottom) in the Eastern Basin.

[2]The sediment samples are from the core 80-14, in the Eastern Basin.

As discussed above, the burial of carbon and oxidative recycling of organic carbon and organic nitrogen within the sediment column can result in mobilization of iron and precipitation at a redox boundary near the sediment-water interface. We suggested that the surficial iron enrichments observed in core 79-1 were not due to redox mobilization alone since similar enrichments were not observed in a nearby core (Fig. 3) or any other cores shown in Table 1. Manganese, however, is more readily mobilized than iron, and downcore data suggest that core 79-1 has experienced extensive manganese mobilization (Fig. 10). In 79-1 the Mn/Al values are greatest at the sediment-water interface and decrease with depth to ratios as low as $20x10^{-4}$. The nearby core, 85-4, also exhibits Mn enrichment at the surface; however, the Mn/Al values below a depth of 25 cm are relatively constant (100+17 $x10^{-4}$). This ratio is similar to that observed both in the cores from other parts of the lake and in aluminosilicate debris from the caldera walls (Mn/Al in caldera wall debris = $96+26x10^{-4}$, unpublished data).

The very low Mn/Al ratios from the deeper sections of 79-1 could be explained if 60 to 80 % of the manganese has been mobilized. However, the missing (i.e., "mobilized") manganese from the bottom 60-70 cm of the core cannot be accounted for by the excess manganese in the upper 10-15 cm of the core. A mass balance comparison between the stock of Mn in the deeper parts of 79-1 and 85-4 indicates that less than 5% of the mobilized Mn can be accounted for. This suggests that the manganese at the 79-1 site has been lost to the water column. In highly reducing sediments (abundant carbon burial) manganese and iron can diffuse out of the sediments and precipitate within the water column. But as we suggested earlier, there is no reason to think that sediments at the 79-1 site are more reduced due to carbon burial than those at the 85-4 site since the two sites are less than one kilometer apart and have similar depths.

We wish to emphasize that the depth profile of iron for the 79-1 core contrasts with that of manganese. Although there is a similar surficial enrichment of Fe, there is no depletion of Fe in the deeper parts of the core (Fig. 3a). The Fe/Al values are similar or higher than those measured in the nearby 85-4 core. Moreover, the Fe/Al values in the deeper parts of

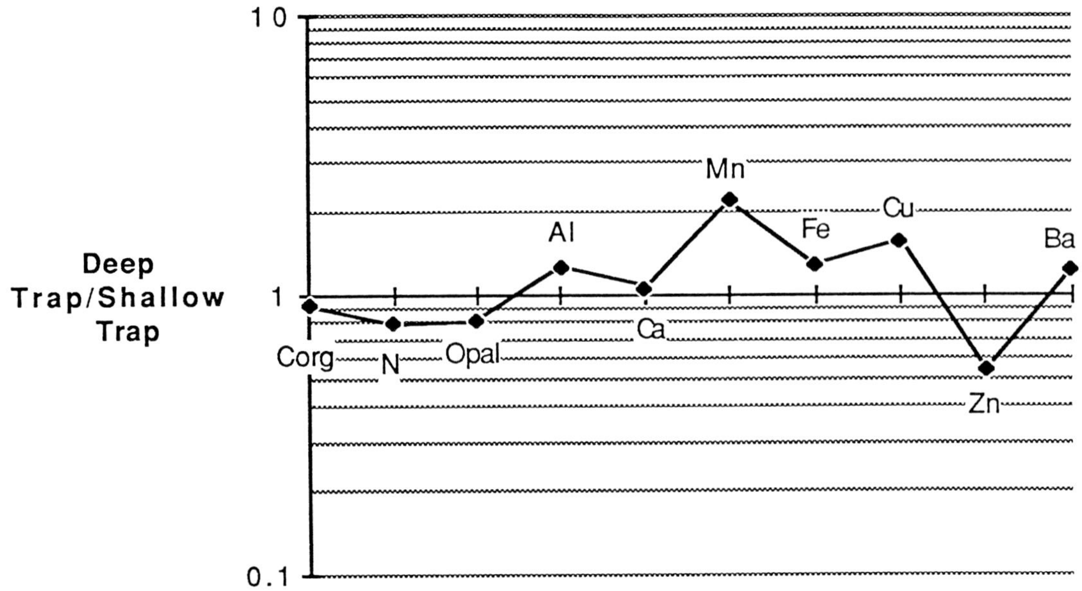

Figure 8. Concentrations of materials in the deep (584 m) trap normalized to the concentrations measured in the shallow (200 m) trap. Values greater than 1.0 indicate enrichments in the deep trap; values less than 1.0 indicate relative enrichments in the shallow trap.

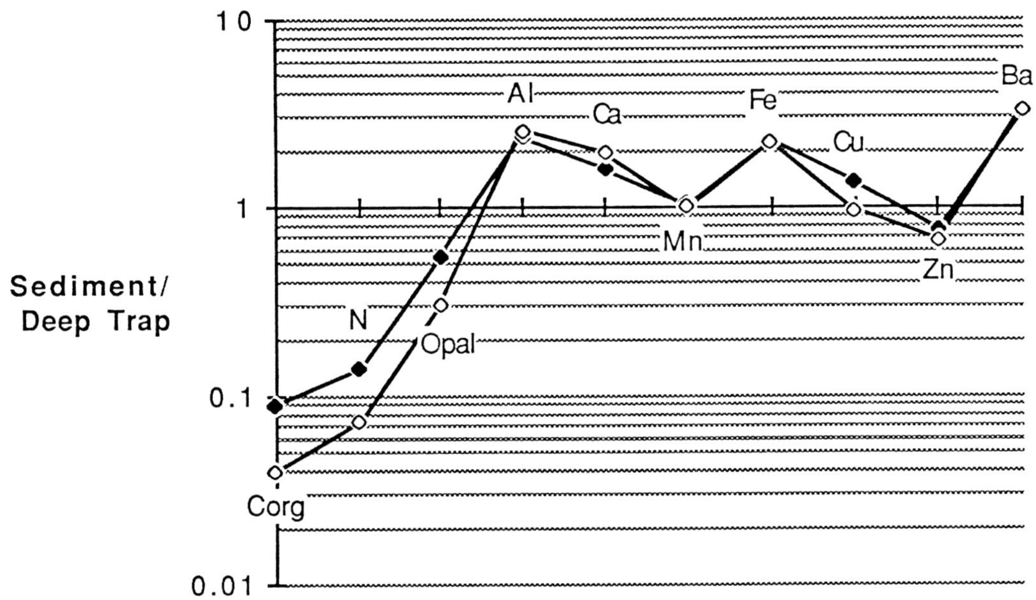

Figure 9. The concentrations of sediments normalized to the concentrations of the deep (584 m) trap. The open symbols represent sediment from a depth of 5 cm in the core; the closed symbols represent sediments from a depth of 0-1 cm in the core. The core is 80-14 from the Central Basin very close to the location of the sediment traps.

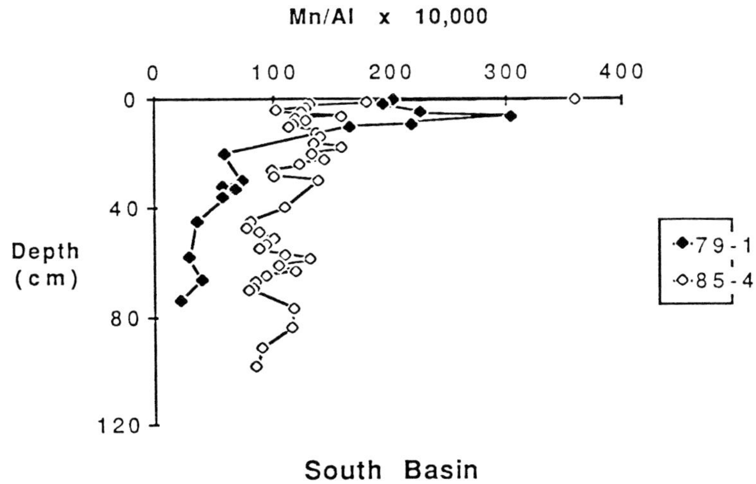

Figure 10. Manganese to aluminum ratios as a function of depth in sediment cores from the South Basin.

79-1 are similar to values measured in debris from the walls, indicating that iron has not been leached from the sediment column as appears to have been the case for manganese.

The sediment trap data also indicates manganese mobilization into the water column. Trap materials from the nearbottom traps (Table 2) exhibit Mn/Al values that range between 170 and 460, generally higher than that observed in aluminosilicate debris or at depth in the cores. Assuming the materials collected by these deep traps are a good representation of the composition of particles reaching the lake floor, some mobilization of Mn must occur in most of the cores, and in 79-1 approximately 90% of the manganese raining to the bottom appears to be recycled to the water column. Thus, the manganese cycle in the lake involves: (1) particulate input of aluminosilicate debris into the upper water column; (2) settling of these particles with biogenic debris (probably as fecal pellets of zooplankton); (3) scavenging of dissolved Mn and aggregation of fine particulate Mn in the water column by settling particles; (4) reduction and dissolution of particle-associated Mn within the sediments as a consequence of organic matter oxidation; (5) diffusion of dissolved Mn^{+2} back into the water column where it is readily oxidized to particulate Mn^{+4}. These processes are compatible with the observed high concentrations of particulate manganese in the deep parts of Crater Lake (Collier *et al.* 1990).

This cycle, however, cannot account for the low Mn/Al values in core 79-1 since the five-step process described above should operate similarly in different parts of the lake. Consequently, we suggest the enhanced manganese leaching of the sediments at the 79-1 site and the enrichments of iron and other elements are the result of advection of pore fluids through the sediment column. Such advection has been observed in marine sediments (Sayles and Jenkins 1982; Becker and von Herzen 1983; Maris *et al.* 1984) and has been attributed to geothermally heated pore fluids. In the case of the 79-1 core the advecting pore fluids are sufficiently reduced to leach manganese from the sediment. Because of the relatively slow kinetics of precipitation and higher solubility of manganese (Krauskopf 1957), only minor amounts precipitate within the more oxidized surface sediments, and the rest escapes into the water column. Since the surficial sediments are enriched in iron, the advecting fluids appear to transport iron. The similarity of Fe/Al values in the deeper parts of both core 79-1 and all other cores, however, suggests iron leaching of the sediments is significant. The lack of iron leaching of the sediments is probably because the iron in most Crater Lake sediments is dominantly in volcanic debris which is relatively unaffected by the advecting pore fluids. The more rapid kinetics of iron precipitation results in sufficient deposition to produce strong enrichment and some lithification of the surficial sediment.

HISTORY OF HYDROTHERMAL ACTIVITY IN THE LAKE

The iron concentration in 79-1 exhibits a subsurface peak at 6 cm depth (Fig. 3). As was noted above, this peak is the result of analyses on a lithified, ochre-colored crust material. Sedimentation rates for Crater Lake are not well constrained; however, the U. S. Geological Survey has made some ^{14}C measurements of organic material in the sediments (S.W. Robinson, pers. comm.). A ^{14}C measurement on moss in 79-1 indicates a sedimentation rate of approximately 30 cm/1000 years and implies the time of maximum hydrothermal influence recorded by 79-1 is about 300 years ago. A smaller peak in the Fe/Al ratio at approximately 35 cm suggests enhanced iron precipitation at this site approximately 1000 years ago. Core 85-4, also from the south basin, exhibits an enrichment of iron at approximately 50 cm depth. In the section below we explore the possibility that measurements of hydrothermally mobilized elements in sediment cores from other parts of the lake may allow evaluation of the history of hydrothermal venting into the lake. The fact that the sediments are sufficiently oxidizing in most parts of the lake to prevent extensive manganese mobilization, indicates that the less mobile Fe oxide phases, once formed through hydrothermal activity, would not be remobilized by sediment redox process that are controlled by organic carbon burial.

In Fig. 3c we compared the Fe/Al data for two cores recovered from the Wizard Island Platform . The most striking feature of these data is the strong enrichment in the ratio in core 85-3 at depths 40-60 cm. Comparison with the nearby core, 80-7, demonstrates that the iron enrichment is very localized and not a ubiquitous redox feature of the sediment record. The ^{14}C age of a pine cone recovered at a depth of 21 cm in 85-3 is 2220±80 years, indicating a sedimentation rate of 9.5 cm/1000 years (S. W. Robinson, pers. comm.). The factor of three lower sedimentation rate for this part of the lake compared to that suggested above for the South Basin is reasonable, since the Wizard Island platform has a depth of only 250 meters and would not accumulate turbidity-flow material and resuspended sediments transported down slope. If we assume a constant sedimentation rate of 9.5 cm/1000 years in the core, the peak iron deposition appears to have occurred approximately 5000 years ago. This event may be related to the period of volcanism which formed the dacite dome (Bacon and Lanphere 1990) just to the east of Wizard Island. A U. S. Geological Survey estimate based on additional ^{14}C data suggests that this volcanism occurred 4000 years ago (C. R. Bacon, pers. comm.), but there may be sufficient error in both dates to allow the possibility that they are synchronous events.

It is possible that these metalliferous inputs may have occurred at a number of sites throughout the history of the lake. From the observations on the two cores which exhibit the iron enrichments it appears that the activity at any one site has limited duration. This suggestion is consistent with the concept that conduits for hydrothermal flow become clogged as a result of precipitation of the dissolved load carried by the fluids. Thus, there would be eventual blockage of any venting site; however, new conduits may form as a consequence of faulting and fracture formation. It is also possible that crustal injection of magma and associated hydrothermal activity has occurred episodically during the history of Mt. Mazama. Consequently, hydrothermal inputs may have waxed and waned throughout the history of the lake.

IMPLICATIONS OF HYDROTHERMAL ACTIVITY FOR BIOLOGICAL PRODUCTIVITY IN THE LAKE

Since we know little about the variability of hydrothermal activity in the lake and can only speculate about the implications of hydrothermal venting on biology, the discussion which follows is speculative. However, we wish to include a preliminary discussion of this topic because we believe the concept has important implications for understanding the paleolimnology of the lake and natural variability in biological productivity. Some relevant examples of this are found in current discussions concerning possible decreases in clarity of the lake (see Dahm *et al.* 1990; G. Larson 1990). There may be natural changes in nutrient inputs and cycling which influence biological productivity and clarity. Hydrothermal activity may influence these processes and the sediments are the recorders of both past biological productivity and hydrothermal activity.

Variability in the sediment-accumulation rates of elements enriched in hydrothermal fluids are an obvious consequence of variable hydrothermal inputs. The enhanced iron contents of certain cores have been noted above, and we suggested that these were a consequence of localized advection of thermal fluids through the sediment column. Because of the insolubility of iron in oxic waters, however, this element does not serve as an indicator of the whole-lake hydrothermal input. A chemical species which is enriched in hydrothermal fluid but is sufficiently soluble to be distributed throughout the lake could better indicate the whole-lake hydrothermal influence. Manganese is one possibility, although it may also precipitate locally, and redox mobilization resulting from organic matter burial could modify the original depositional patterns. Alternatively, barium and lithium, both elements strongly leached from rocks by hydrothermal fluids and relatively soluble in oxic waters, could provide a measure of whole-lake hydrothermal input. In effect, what is needed for such an analysis is a measure of distal hydrothermal deposition. Such a deposition may occur through scavenging of the hydrothermal elements by organic matter settling from the upper water column.

Monitoring biological productivity of the lake is possibly best accomplished by analysis of the opal burial flux or study of diatom assemblages. As we discussed earlier, opal is the biogenic component of the sediment that is best preserved. Certain species of diatoms and diatom assemblages are also indicative of nutrient availability and thus productivity. Biological productivity could be directly correlated with hydrothermal inputs if limiting nutrients are carried by the hydrothermal fluids. Nitrogen and trace metals necessary for nitrogen fixation are possible candidates in this regard. Also, the input of buoyant hydrothermal waters may enhance the mixing of deep lake waters with the surface waters. Since the deep lake is the major reservoir of the biologically limiting nutrient, nitrogen, more rapid cycling of deep lake waters could increase primary productivity in the upper lake. Although this effect could produce a positive relationship between productivity and intensity of venting, enhanced burial could not be maintained without depleting the nitrogen in the deep waters. Thus, any enhanced productivity due to this effect would be relatively short-lived. It is possible, however, that the saltier geothermal waters are denser than typical lake water, and as a result, would produce stable layers in the deep basins. If this were the case, these basins could be a trap for nutrients released from decomposing organic matter that settles to the bottom. Episodes of enhanced hydrothermal activity, therefore, could be times of low biological productivity. Without more information on the composition and volume of hydrothermal input to the lake we can neither define the effects of hydrothermal venting on the biology of the lake nor interpret its influence on the sediment record. Nevertheless, there are reasons to suspect that hydrothermal venting is important to the chemical and biological evolution of the lake and its ecology.

As a preliminary entry to this question we examine the downcore variations in opal and lithium in core 85-4 (Fig. 11a and b). We have chosen this core because we have closely spaced analyses downcore, and the site appears to be distal to venting sources. We infer the lack of local hydrothermal influence by the absence of distinctive iron enrichment. We have normalized Li concentrations to Al because this ratio is a better indicator of enrichment over the aluminosilicate debris from the caldera walls. The variations in opal concentration are more than a factor of three at this site. There appear to be times of greater opal burial which are observed over intervals of 10 to 20 centimeters in the core. As noted above the best estimate of sedimentation rate at this South Basin site is 30 cm/1000 years. Thus, the episodes of enhanced productivity appear to have durations of 300 to 600 years. A factor of two variability in Li/Al is present. In addition, the two measurements appear to correlate. The estimated R2 for the correlation is 0.57, which is significant at the 99% level.

Taken at face value this relationship indicates that increased hydrothermal activity is accompanied by enhanced primary productivity in the lake. As stated above, the most likely cause of such a relationship would be due to enhanced nutrient inputs which accompany venting, since enhanced mixing could not sustain the implied increases in productivity for 102 to 103 years. We wish to emphasize, however, the weaknesses in this analysis and suggest that the data from core 85-4 be considered illustrative of an approach rather than a definitive statement on the relationships between biological and hydrothermal activity. The major weakness in the data is that we are using concentration data for opal and Li; concentration of opal and Li could be controlled by increases and decreases in the aluminosilicate fraction

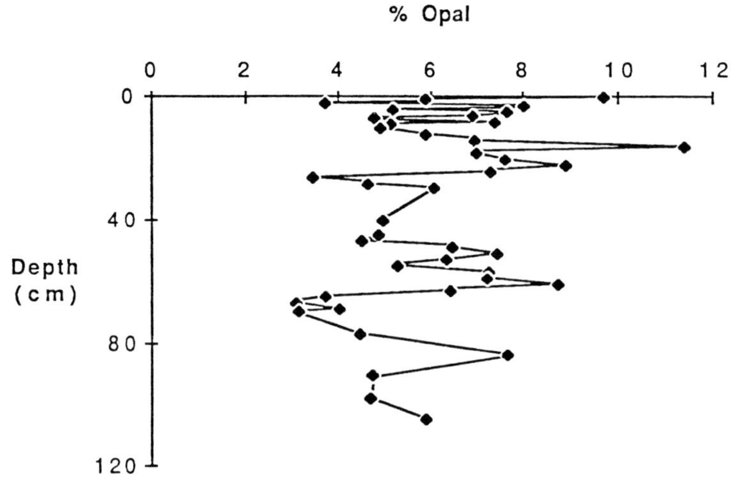

Figure 11a. Variations in opal concentration with depth in the South Basin core 85-4.

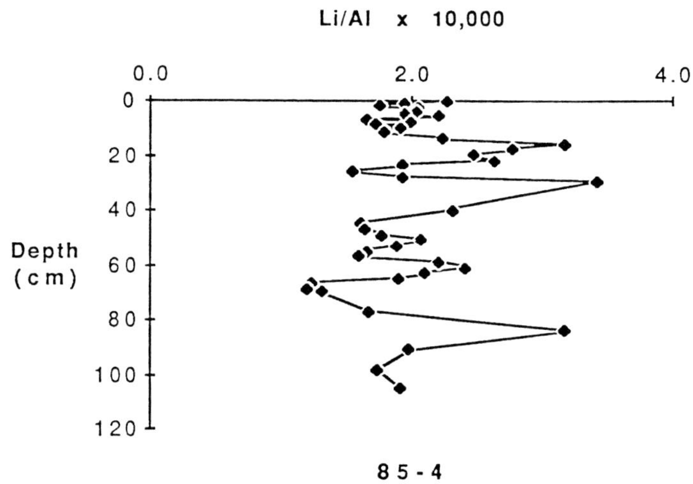

Figure 11b. Lithium/aluminum variations with depth in 85-4.

rather than true increases in the inputs of these two variables. What is needed is burial flux of opal and Li rather than concentration, but examination of variability in burial fluxes requires much better definition of the sedimentation rate than is currently available. Alternatively, diatom assemblages, if they could be related to primary productivity, would not be affected by aluminosilicate dilution. Thus, we can not rule out the possibility that the variability in opal and Li abundances is the result of dilution. We have, however, examined the core for turbidite deposition and find the normal fine sediments interrupted by silt-to-sand-sized layers of a centimeter or two thickness. These layers are probably turbidites, but their location and thickness cannot account for the observed variability in opal and Li abundances.

SUMMARY

The major compositional variability observed in Crater Lake sediments can be accounted for by variations in the relative contributions of volcanic debris from the caldera walls, biogenic debris settling from the euphotic zone, and precipitates which are predominantly hydrated iron oxides. Biogenic debris decomposes and dissolves primarily at the lake floor resulting in preservation of less than 20% of the biogenic particles which reach the bottom. The iron-rich component appears to precipitate from hydrothermal fluids that vent through sediments and rock outcrops. These precipitates may be more evolved analogs to mats of iron oxidizing bacteria that are associated with anomalously warm water and currently exist on the lake floor. Present day sites of hydrothermal iron deposition have only been found in the South Basin, but additional deposition appears to have occurred in the Wizard Island Platform 4-5,000 years ago. Possibly the hydrothermal input to the lake has waxed and waned throughout the history of the basin. This input may influence the ecology of the lake and the biological productivity. Temporal variations in opal accumulation rate and the assemblage of diatoms recorded in sediment cores may provide evidence of this effect.

ACKNOWLEDGMENTS

This research was supported by a grant from the National Park Service. Analyses represent many hours of work by Cydne Perhats and Roberta Conard. Chris Moser built and helped deploy the sediment traps. The cooperation and help from many National Park Service personnel has been especially appreciated. The efforts of Mark Buktenica and John Salinas were especially important for obtaining the sediment cores. The USGS core samples, provided by Hans Nelson, have been especially important in developing the story revealed by the sediments. Brad Beeson, Mitch Lyle, Jim McManus, and Sharon Roth read the manuscript and provided many helpful comments.

LITERATURE CITED

Bacon, C. R. and M. A. Lanphere. 1990. The geological setting of Crater Lake Oregon. Pages 19-28 *in* E. T. Drake *et al.*, eds., Crater Lake: An Ecosystem Study. Pacific Div., Amer. Assn. Advance. Sci., San Francisco, Calif.

Barber J. H. and C. H. Nelson. 1990. Sedimentary history of Crater Lake Caldera, Oregon. Pages 29-39 *in* E. T. Drake *et al.*, eds., Crater Lake: An Ecosystem Study. Pacific Div., Amer. Assn. Advance. Sci., San Francisco, Calif.

Becker, K. and R. P. Von Herzen, (1983). Heat transfer through the Sediments of the Mounds Hydrothermal area, Galapagos Spreading Center at 86°W. Jour. Geophys. Res. 88:995-1008.

Berner, R. A.,1980. Early Diagenesis: A Theoretical Approach. Princeton Univ. Press., Princeton, N.J. 237 pp.

Bostrom, K. and L. Widenfalk. 1984. The origin of iron-rich muds at the Kameni Islands, Santorini, Greece. Chem. Geol. 42:203-218.

Collier, R., J. Dymond, J. Lupton, A. Chen, M. Lilley, and M. Thompson. 1987. Effects of hydrothermal inputs on the chemistry and physics of Crater Lake, OR. EOS 68(50):1721. (p.8)

Collier, R. W. and J. Dymond. 1988a. Studies of hydrothermal processes in Crater Lake: A report of field studies conducted in 1987 for Crater Lake National Park. CPSU, College of Forestry, Oregon State Univ., Corvallis, Ore. 49 pp.

Collier, R. W. and J. Dymond. 1988b. Observations of Bacterial Mats Associated with Thermal Springs at 450 Meters Depth in Crater Lake, OR. EOS 1138.

Collier, R. W. and J. Dymond. 1989. Studies of hydrothermal processes in Crater Lake: A report of field studies conducted in 1988 for the National Park Service. CPSU, College of Forestry, Oregon State Univ., Corvallis, Ore. 79 pp.

Collier, R. W., J. Dymond, J. McManus, and J. Lupton. 1990. Chemical and Physical Properties of the Water Column at Crater Lake, OR. Pages 69-79 *in* E. T. Drake *et al.*, eds., Crater Lake: An Ecosystem Study. Pacific Div., Amer. Assn. Advance. Sci., San Francisco, Calif.

Corliss, J. B., M. Lyle, J. Dymond, and K. Crane. 1978. The chemistry of hydrothermal mounds near the Galapagos Rift. Earth Planet. Sci. Lett., 40:12-24.

Dahm, C. N., D. W. Larson, N. S. Geiger, and L. K. Herrera. 1990. Secchi disk, photometry, and phytoplankton data from Crater Lake: long-term trends and relationships. Pages 143-151 *in* E. T. Drake *et al.*, eds., Crater Lake: An Ecosystem Study. Pacific Div., Amer. Assn. Advance. Sci., San Francisco, Calif.

Dymond, J., and M. Lyle. 1985. Flux comparisons between sediments and sediment traps in the eastern tropical Pacific: Implications for atmospheric CO_2 variations during the Pleistocene. Limnol. Oceanogr. 30:699-712.

Dymond, J., R. W. Collier, and M. E. Watwood. (submitted) Bacterial mats from Crater Lake, Oregon and their relationship to possible deep lake hydrothermal venting. (submitted to Nature)

Ellis, A. J. and W. A. J. Mahon. 1977. Chemistry and geothermal systems. Academic Press, New York. 392 pp.

Fischer, K. 1984. Particle Fluxes in the Eastern Tropical Pacific Ocean—Sources and Processes.

PhD. Thesis. Oregon State Univ., Corvallis, Ore. 225 pp.

Froelich, P. N., G. P. Klinkhammer, M. L. Bender, N. A. Luedtke, G. R. Heath, D. Cullen, P. Dauphin, B. Hartman, and V. Maynard. 1979. Early oxidation of organic matter in pelagic sediments of the Eastern Equatorial Atlantic: Suboxic diagenesis. Geochim. Cosmochim. Acta 43:1075-1090.

Krauskopf, K. B. 1957. Separation of manganese from iron in sedimentary processes. Geochim. Cosmochim. Acta 12:61-84.

Larson, G. L. 1990. Status of the ten-year limnological study of Crater Lake, Crater Lake National Park. Pages 7-18 *in* E. T. Drake *et al.*, eds., Crater Lake: An Ecosystem Study. Pacific Div., Amer. Assn. Advance. Sci., San Francisco, Calif.

Lupton, J. E., R. Collier, and J. Dymond. 1987. Excess ^{3}He in Crater Lake, Oregon: Evidence for geothermal input. EOS 68:1722 (p. 8).

McBirney, A.R. 1968. Compositional variations of the climactic eruption of Mount Mazama. Pages 53-56 *in* H. M. Dole, ed., Andesite Conference Guidebook. Oregon Dep. Geol. Mineral. Ind. Bull., 62:53-56 (p. 4).

Maris, C. R. P., M. L. Bender, P. N. Froelich, P. N. Barnes, and N. A. Luedtke. 1984. Chemical evidence for advection of hydrothermal solutions in the sediments of the Galapagos Mounds hydrothermal field. Geochim. Cosmochim. Acta. 48: 2331-2346.

Nelson, C. H., A. W. Meyer, D. Thor, and M. Larsen. 1986b. Crater Lake, Oregon: a restricted basin with base-of-slope aprons of nonchannelized turbidites. Geology 14:238-241.

Onishi, H. 1969. Arsenic. *In*: K. H. Wedepohl, ed., Handbook of Geochemistry, Section 33-I.

Sayles, F. L., and W. J. Jenkins. 1982. Advection of pore fluids through sediments in the Equatorial east Pacific. Science 217:245-248.

Suess, E., and P. J. Muller. 1980. Productivity, sedimentation rate and sedimentary organic matter in the oceans II. Elemental fractionation. Biogéochimie de la matière organique à l'interface eau-sédiment marin. Marseille, 25-27 Avril 1979. Colloques Internationaux du C.N.R.S., no. 293:17-26.

Wheat, C. G. and R. E. McDuff. 1988. Fluid Circulation and Water- Rock Interaction in the Mariana Mounds Hydrothermal Region. EOS 1499.

Williams, D. L., and R. P. Von Herzen. 1983. On the terrestrial heat flow and physical limnology of Crater Lake, Oregon. Jour. Geophys. Res., 88(B2):1094-1104.

AN ALTERNATIVE HYPOTHESIS TO EXPLAIN TEMPERATURE ANOMALY AND OTHER DATA OBSERVED AT CRATER LAKE, OREGON

Joseph G. La Fleur
California Energy Company, Inc.
2455 Bennett Valley Road, Suite 214-B
Santa Rosa,California 95404

Various data have led some researchers to hypothesize that hydrothermal venting is active on the floor of Crater Lake and that it contributes significantly to the chemical and physical properties of the Lake. Searches in the southern basin using a remote-controlled submersible and a one-person submarine failed to identify any hydrothermal flows, and elevated temperatures measured are consistent with normal groundwater temperatures. An alternative explanation of the data is the possibility that cold, mineralized groundwater seasonally seeps into the south basin of the Lake. This seepage could be facilitated by the Chaski Bay slide which extends from the south caldera wall into the south basin. This alternative hypothesis is proposed in view of the apparent absence of some features normally found associated with hydrothermal systems.

Active hydrothermal input into the south basin of Crater Lake was first hypothesized by Williams and Von Herzen (1983), on the basis of elevated temperature gradients measured in the upper two meters of the lake floor sediments. Subsequently, different authors have attributed various chemical constituents of the lake as supportive of the hydrothermal vent hypothesis. White *et al.* (1985) contended that the concentrations of Na, K, Li, Cl, $SO_4^{=}$ and B measured throughout the Lake evidenced thermal water discharges. Lupton *et al.* (1987) interpreted measurements of ^{3}He concentrations as indicative of hydrothermal inflow. La Fleur (1987) discussed lines of negative evidence which dispute the hydrothermal hypothesis. These include: the absence of elevated bicarbonate concentrations in the Lake water; the absence of vertical hydrothermal plumes; heat flow measurements that can be intrepreted as conductive rather than convective heat transfer; the apparent absence of hydrothermal sinter deposits, i.e., secondary silica and/or carbonates; and the absence of recurrent seismicity that would be necessary for maintaining open hydrothermal conduits (La Fleur 1987).

No thermal inflows were observed by Collier and Dymond (1989), either directly in a one-person submarine or from data collected by a remote-controlled submersible. Temperatures measured could be interpreted as consistent with normal groundwater temperatures. Precipitates of iron oxide and discrete colonies of iron-oxidizing bacteria were observed in the south basin of the Lake (Collier and Dymond 1989). This area of lake floor (Fig. 1) is coincident with the terminus of the Chaski Bay slide (Nelson *et al.* 1988) and with elevated temperature gradients measured in the sediments (Williams and Von Herzen 1983) and is transversed by the inferred location of the caldera ring fracture system (Nelson *et al.* 1988). The observations that have been made in the south basin are consistent with the concept of meteoric waters being channeled into the south basin from the Chaski Bay slide and issuing from the landslide terminus as cold mineral seeps. Cold mineral seeps at the terminus of the Chaski Bay slide probably have no relevance to the bulk chemistry of

Crater Lake. Hypotheses from the modeling of Lake water chemistry are highly speculative because levels of contribution from diagenetic reactions, long-dead hydrothermal systems and ascending non-condensible gases are unknown. Since the lake contains approximately 4.5 trillion gallons of water (Phillips and Van Denburgh 1968), any inflows of mineralized water (no matter the origin) would require very large flow rates to affect the bulk chemistry of the Lake.

The hypothesis of subhorizontal inflow of mineralized meteoric water may offer a more practical explanation of the south basin observations than does the hypothesis of ascending hydrothermal fluids, because several lines of evidence inherent to hydrothermal systems are apparently absent.

ALTERNATIVE EXPLANATION OF VARIOUS DATA SETS

Calculated "Heat Flow" Value

Williams and Von Herzen (1983) measured temperature gradients in the Lake floor sediments by inserting a thermal probe to less than 2 meters depth. They calculated "Heat flow" values by multiplying all measured gradients by a single determined thermal conductivity value. An area of inferred elevated heat flow was thus identified in the south basin (Fig. 1). Williams and Von Herzen hypothesized that hydrothermal venting caused the elevated thermal gradients measured in the south basin, and this area became the "detailed study area" for the subsequent search for hot springs (Collier and Dymond 1988 and 1989). The hydrothermal vent hypothesis is not a unique solution to the temperature gradient data, and interpretation of the data to depths beyond the two meters of penetration is subjective. The steep gradients measured could reflect conductive heat transfer and an anomalously cold lake bottom. Any relatively warmer water or rock mass residing beneath the Lake floor would give rise to steeper temperature gradients. As shown below, because the bottom of Crater Lake is exceptionally cold (3.6°C), ordinary ground waters entering the lake could account for the steep thermal gradients (Fig. 2; Thompson *et al.* 1987; Illian 1970).

Because the south basin thermal gradient anomaly is coincident with the terminus of the Chaski Bay slide, the possibility that the slide feature produces the anomaly must be considered. Meteoric waters percolating through the slide may attain sufficient density through mineral dissolution to descend into the depths of the south basin. This scenario, however, would require unrealistically high salinities in the mineralized, warmer descending water. More reasonably, the slide base might function as an aquitard, creating a temporary hydraulic head which would facilitate the transport of relatively warmer meteoric waters into the south basin during periods of high runoff. Alternatively, the rubbly slide mass may act as an aquifer with the overlying fine-grained sediments acting as a less permeable capping horizon. Either of these scenarios, or a combination of the two, are reasonable explanations of how warmer, less dense meteoric waters could be transmitted into the deep, cold Lake. The proposed geohydrologic setting is shown in Fig. 3. The reader is referred to Nelson (1967) and Nelson *et al.* (1988) for description of the various deposits on the floor of Crater Lake.

Bacterial Mats and Iron and Manganese Precipitates

In the area where the Chaski Bay slide terminates in the south basin, approximately thirteen colonies of iron-oxidizing bacteria were observed by Collier and Dymond (1989). These bacterial mats were all within 300m of each other and are either on or near an abrupt change in slope. The colonies are composed principally of genera *Gallionella* and *Leptothrix* which require an extraneous source of reduced iron to sustain their existence. These types of bacteria are found in low temperature springs, wells, drainages and occasionally in thermal springs. They are not thermophyllic organisms. Flows of water through the flocculent masses of filiform bacteria were neither observed nor measured and no mats were observed on any dives away from the terminus of the Chaski Slide (*ibid.*). Crusts composed of precipitates of iron and manganese oxides were also identified, but these do not necessarily indicate thermal water sources.

The bacterial mats and iron precipitates mark the specific locations where sources of reduced iron contact the oxygenated lake water. Because no flows were observed at any of the locations, the possibility of seasonally episodic flow must be considered. The observations were made in the summer dry season (August 1988), following an unusually dry winter. A seasonal influx of mineralized water appears to be

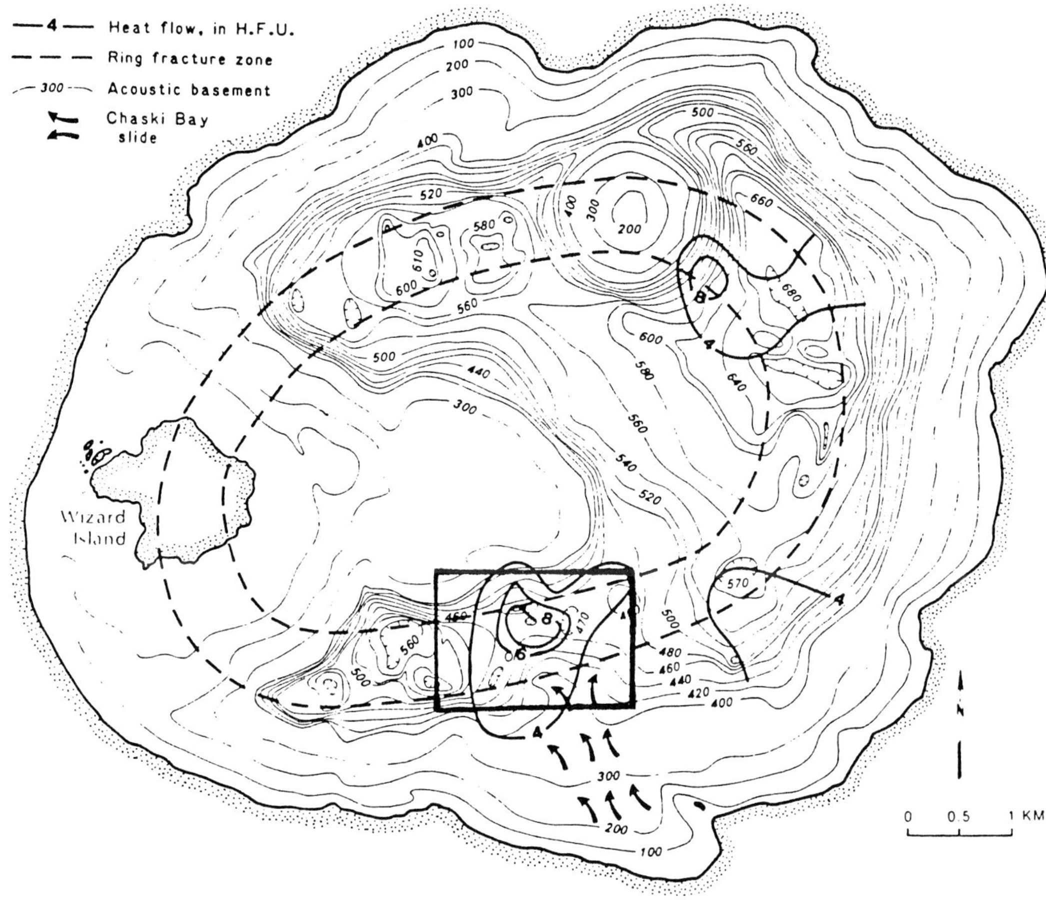

Figure 1. Box encloses detailed study area from Collier, R. W., and J. Dymond. 1989. "Studies of hydrothermal processes in Crater Lake. A report of field studies conducted in 1988 for the National Park Service." Source: Nelson, C. H., P. R. Carlson, and C. R. Bacon. 1988. "The Mount Mazama Climactic Eruption (~6900 yr. B.P.) and resulting convulsive sedimentation on the Crater Lake caldera floor, continent, and ocean basin." Geol. Soc. Amer., Spec. Pap. 229.

a more reasonable explanation than hydrothermal venting.

Collier and Dymond (1989) hypothesize that the bacterial mats may identify where "diffuse flow" or "advective flow" of hydrothermal fluids are ascending slowing through a relatively large area of the lake floor and affect the bulk chemistry of the entire lake. The observations of 13 discrete bacteria colonies ranging in size from 10cm to 2m across do not seem to support the concept of diffuse or advective flow through a larger area. Were diffuse flow taking place over a large area, a relatively uniform blanket of bacterial growth would be anticipated as opposed to a few discrete colonies.

Conceptually, a diffuse flow of hydrothermal fluids into the cold lake would be subject to more rapid secondary sealing processes than would channelized flow.

The Chaski Bay Slide–A Potential Source of Reduced Iron

Ample opportunity exists for waters within the slide to become mineralized. Lithologies within the Chaski Slide were subjected to an earlier stage of hydrothermal alteration that deposited secondary iron pyrite, prior to collapse of the caldera (Keith pers.comm. 1988). Diagenetic alteration of this pyrite is currently taking place within the slide debris. The dissolution of pyrite can release reduced iron in solution by the reaction FeS_2 (Pyrite) + 7/2 O_2 +

$H_2O \rightarrow Fe^{++} + 2SO_4^{=} + 2H^+$ (Krauskopf 1967:112). Data from Thompson *et al.* (1987) show that waters within the Chaski Bay slide are anomalously high in sulfate ($SO_4^{=}$). Two springs issuing subaerially from the slide measured 12 and 26 mg/l $SO_4^{=}$ whereas the other caldera wall springs contained one mg/l of $SO_4^{=}$ or less. Perhaps reduced iron in solution from the dissolution of pyrite is seeping from the nose of the slide and supporting the bacteria.

Alternatively, the bacterial mats may identify where recently exposed concentrations of pyrite (FeS_2) are being oxidized by the Lake water. Krauskopf (1967:275) states, "Locally the reaction is catalyzed by bacteria which make use of the energy released, $2FeS_2 + 15/2\ O_2 + 4H_2O \rightarrow Fe_2O_3 + 4SO_4^{=} + 8H^+\ \Delta F^\circ = -584.2$ kcal." Although this reaction is strongly exothermic, the temperatures reported in the bacterial mats, are seemingly too great to be explained by this reaction alone. If the bacteria are supported directly by the local oxidation of pyrite, no inflow of water need be involved. Albeit possible, this explanation of *in-situ* alteration appears to be less likely than delivery of reduced iron in solution via seasonally active mineral seeps.

Measured Temperatures

As previously mentioned, the bottom of Crater Lake is a notably cold 3.6°C. It should be appreciated that this temperature is significantly lower than would be anticipated in any regional groundwaters which had sufficient residence time to equilibrate thermally with the country rock. Groundwater temperatures measured in wells in the area range from 4.4°C to 11.7°C and are commonly about 10°C (Illian 1970). Data from Thompson *et al.* (1987) show that large, lower elevation springs in the Mazama area commonly have temperatures between 9°C and 12°C. The temperatures of two springs sampled in the subaerial portions of the Chaski Bay slide were 9.5°C and 12°C.

It was reported that the bacterial mats contained waters with temperatures ranging from 4.3°C to 9.5°C (Collier and Dymond 1989). The average of the mat water temperatures reported was 6.2°C. No temperatures in excess of "ordinary" groundwater temperatures have been recorded. As mentioned previously, the bacterial mats identify the specific location where mineralized fluids react with the oxygenated lake water. Were these fluids from a high temperature hydrothermal system, some measurements in excess of normal groundwater temperatures would be expected. Because all of the relatively elevated temperatures measured in the bacterial mats are equivalent to cold groundwater temperatures, these measurements support the cold mineral seep hypothesis.

ABSENCE OF FEATURES CHARACTERISTIC OF HYDROTHERMAL VENTING

Low Seismicity

If hot springs are expected to survive, seismic activity would be required to keep them open. As William and Von Herzen (1983) has stated, "The circulating fluids carry large quantities of dissolved solids that precipitate as the fluids approach the vent area where they undergo rapid cooling and chemical changes. These precipitated minerals will eventually clog the conduit."

This process would be particularly effective on the bottom of a 3.6°C lake, and conditions of diffuse flow of hydrothermal waters would be expected to seal off more rapidly than channelized flow. Recurrent seismic ground rupture is necessary to maintain open vertical conduits for fluids to vent. The Crater Lake area, like most of Oregon, has a low level of recurrent seismicity. Only three macroseismic events have ever been recorded in the Mazama area, registering 1, 2 and 4 on the Richter scale (Jacobson 1986). Even on the highly tectonically active oceanic spreading centers, hot springs apparently persist only for a few weeks to a few decades (Macdonald 1982). Microseismic activity at Crater Lake was briefly monitored during the fall and winter 1969-70 (Blank *et al.* 1971); however, it is the opinion of this author that microseismic events may not provide crustal disturbances sufficient to break secondary seals on the lake floor.

Absence of Sinter

True hot springs entering into a 3.6°C lake would deposit precipitates of minerals dissolved at higher temperatures. Certainly secondary silica and perhaps secondary carbonates and/or sulfides could be anticipated. No such hydrothermal deposits have so far been identified on the floor of Crater Lake. Nelson (1967) examined 130 sediment cores from the floor of Crater Lake and did not report any findings of secondary silica or carbonate precipi-

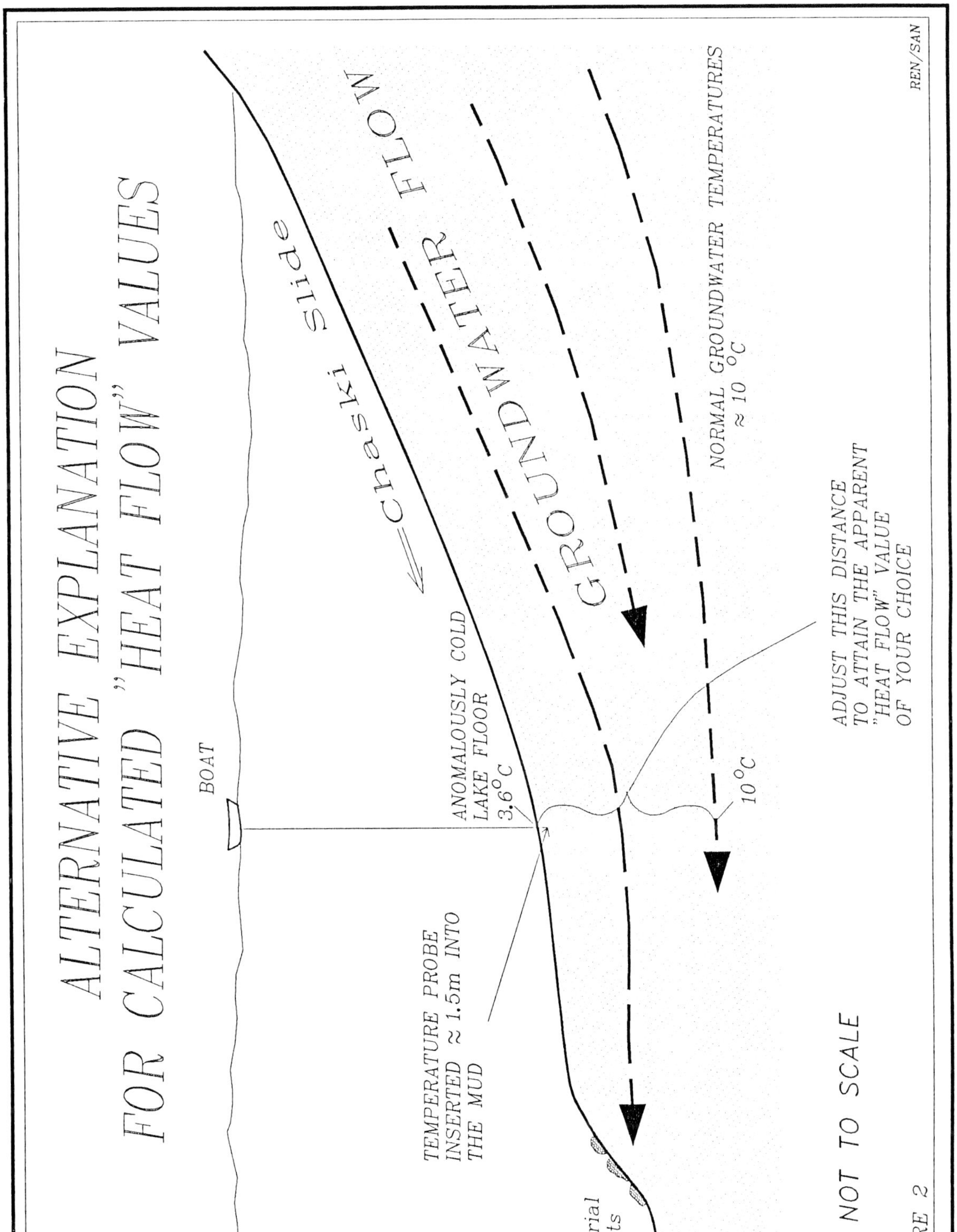

Figure 2. Alternative explanation for calculated "heat flow" values.

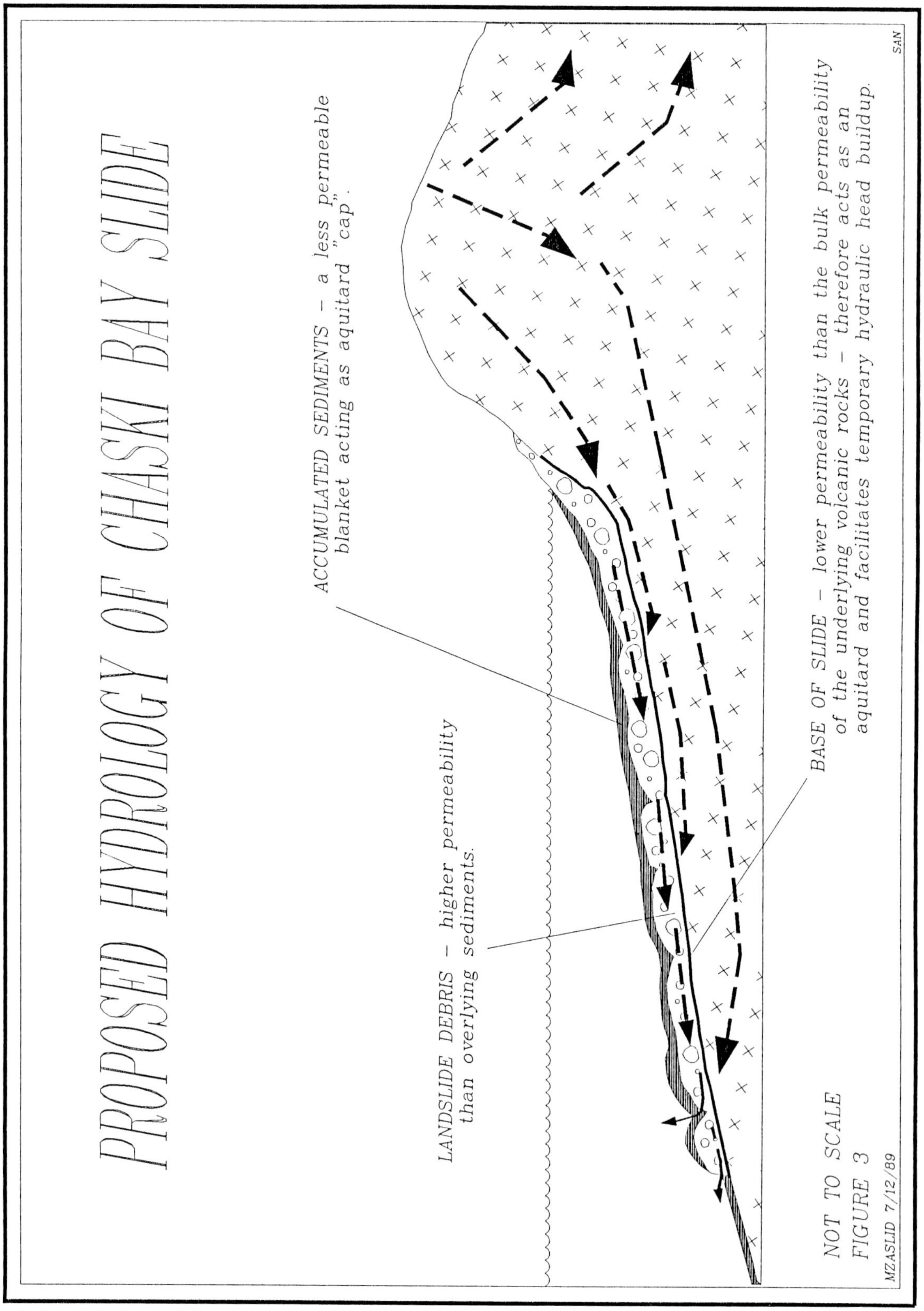

Figure 3. Proposed hydrology of Chaski Bay slide.

tates. Sediment samples were collected by gravity coring by Oregon State University in 1985 and 1986, and by the U.S. Geological Survey in 1979 and 1980 (Collier and Dymond 1988). No secondary silica or carbonate were reported in these samples, despite focused sampling in the area of elevated "heat flow." Since the rusty spots observed on the floor of Crater Lake were not associated with secondary silica or iron silicates, it is reasonable to conclude they were not actual hot springs deposits but rather from cold water reactions.

Absence of Elevated Bicarbonate Concentrations

Carbon dioxide gas is released by hydrothermal alteration and is dissolved in near neutral pH waters as bicarbonate (HCO_3). Carbon dioxide generally accounts for more than 90% of the abundant non-condensable gases present in geothermal reservoirs (Michels 1969). The solubility of CO_2 is proportional to total pressure and inversely proportional to temperature; therefore, the cold depths of Crater Lake should facilitate an identifiable bicarbonate anomaly if a significant hydrothermal input were actually taking place. No bicarbonate anomalies have been demonstrated in Crater Lake. Phillips and Van Denburg (1968) reported that Crater Lake contains 35 ppm HCO_3, the same as nonthermal Davis Lake. They also reported that hydrothermally affected Paulina Lake and East Lake, Oregon, contain 352 ppm HCO_3 and 125 ppm HCO_3 respectively. No measurement of bicarbonate concentrations have been made in the waters in the bacterial mats (Collier and Dymond 1989). These colonies could well be supplied by cold groundwaters.

According to Lupton (1987), samples from Crater Lake show that the deep waters of the lake are enriched in "mantle" helium with elevated $^3He/^4He$ ratios. Because Crater Lake occupies the floor of a relatively young (6900 yr. BP) caldera, it is not surprising that the lake contains enhanced concentrations of ^3He. Expectedly, deep-source helium would continue to emanate from depth for some time after such a volcanic event. The extreme mobility of helium allows it to migrate quite freely through the earth's crust and it does not require aqueous transport. Furthermore, it is not possible to distinguish ^3He released by hydrothermal alteration from ^3He released from magma degassing or directly from the mantle. Elsewhere, it has been repeatedly observed that elevated $^3He/^4He$ ratios in hydrothermal fluids correlate directly with increased bicarbonate and/or dissolved CO_2 concentration (Welhan *et al.* 1988; Kennedy *et al.* 1987). This is not the case in Crater Lake, Oregon, where the absence of significantly enhanced HCO_3 or CO_2 may suggest that ^3He could be seeping into the lake independent of hydrothermal inflow.

Absence of Vertical Hydrothermal Plumes

The deep water temperature profiling demonstrates a gradual increase from 3.5°C at mid-depth in the lake to approximately 3.6°C on bottom. The profiles show uniformly layered subhorizontal isotherms especially when depicted without vertical exaggeration (William and Von Herzen 1983; La Fleur 1987; Collier and Dymond 1988, 1989). If these isotherms do reflect fluid inflows, such inflows would have to be quite cold to remain flat lying on the bottom of a 3.6°C lake. The absence of subvertical isotherms argues in favor of conductive heat transfer or cold groundwater influx.

Absence of Data Indicating True Convective Heat Transfer

The Williams and Von Herzen (1983) study recorded steep temperature gradients in the sediments in the south basin. Elevated thermal gradients are indicative of conductive heat transfer. By contrast, vertically ascending fluids yield nearly isothermal gradients and significantly elevated temperatures would have been recorded. Certainly, small areas of fluid ascension could have gone undetected due to spacing of the sample points; nevertheless, so far no measurements identifying ascending thermal fluids have been recorded (Williams and Von Herzen 1983). The heat flow study, therefore, argues against the concept of ubiquitous "diffuse flow" of hydrothermal fluids to explain the observations made in the south basin. The alternative hypothesis of subhorizontal inflow of groundwaters is not in contradiction with the measured temperature gradients and calculated heat flow values (Fig. 2).

CONCLUSION

Although hydrothermal venting has been proposed as a hypothetical explanation for temperature anomalies, bacterial mats and precipitates of iron and manganese observed in the south basin of Crater Lake, an alternative explanation of subhorizontal through a large landslide deposit is consistent with the data and observations. This alternative explanation is supported by the coincidence of the observed anomalies and the terminus of the landslide and by measured temperatures equivalent to those of normal regional groundwater. The apparent absences of elevated bicarbonate concentrations, sinter deposits, secondary silica, frequently recurrent seismicity, subvertical hydrothermal "plumes" in the deep lake waters and elevated isothermal profiles in the sediments are all seemingly inconsistent with the hydrothermal venting hypothesis but are consistent with the alternative explanation of cold mineral seeps.

LITERATURE CITED

Blank, R. R. Jr., M. M. Brown, T. Matumoto, and J. K. Westhusing. 1971. Seismicity Investigations in the Cascade Mountains and Vicinity, Oregon. Ctr. Volcanol., Dep. Geol., Univ. Oregon, Tech. Rep. NASA/MSC: Contract NAS 9-9690.

Collier, R. W., and J. Dymond. 1988. Studies of hydrothermal processes in Crater Lake: A report of field studies conducted in 1987 for Crater Lake National Park. Oregon State Univ., Coll. Oceanogr., Ref. 88-5.

Collier, R. W. and J. Dymond. 1989. Studies of Hydrothermal Processes in Crater Lake: A report of field studies conducted in 1988 by the National Park Service. Oregon State Univ., Coll. Oceanogr., Ref. 89-2 (March 20, 1989).

Illian, J. R. 1970. Interim report on the ground water in the Klamath Basin, OR. State Engineer, State of Oregon.

Jacobson, R. S. 1986. Map of Oregon seismicity, 1841-1986. Oregon Dep. Geol. Mineral. Ind., GMS-49 (scale 1:1,000,000).

Keith, T. 1988. Assistant Branch Chief, Igneous and Geothermal Processes, U. S. Geol. Surv., Menlo Park, Calif. (oral comm.).

Kennedy, B. M., J. H. Reynolds, S. P. Smith and A. H. Truesdell. 1987. Helium isotopes: Lower Geyser basin, Yellowstone National Park. EOS 68 (25):608. (Abstract).

Krauskopf, K. B. 1967. Introduction to Geochemistry. McGraw-Hill, New York. 721 pp.

La Fleur, J. G. 1987. Legend of the Crater Lake hot springs: Product of model mania. Geotherm. Resources Coun. Trans. 11(Oct. 1987):267-279.

Lupton, J. E., R. Collier and J. Dymond. 1987. Excess ^{3}He in Crater Lake, Oregon: Evidence for geothermal input. EOS 68(50):1722. (Abstract).

Macdonald, Ken C. 1982. Mid-ocean ridges: fine scale tectonic, volcanic and hydrothermal processes within the plate boundary zone. Ann. Rev. Earth Planet. Sci. 10:155-190.

Michels, D. E. 1969. CO_2 and carbonate chemistry applied to geothermal engineering. Lawrence Berkeley Lab., Rep. LBL–11509.

Nelson, C. H. 1967. Sediments of Crater Lake, Oregon. Geol. Soc. Amer. Bull. 78:833-848.

Nelson, C. H., P. R. Carson, and C. R. Bacon. 1988. The Mount Mazama Climactic Eruption (6900 yr. B.P.) and resulting convulsive sedimentation on the Crater Lake caldera floor, continent, and ocean basin. Geol. Soc. Amer. Spec. Pap. 229.

Phillips, K. N. 1968. Hydrology of Crater, East and Davis Lakes, Oregon. U. S. Geol. Surv., Water Supply Pap. 1859-E. 60 pp.

Phillips, Kenneth N., and A. S. Van Denburgh, 1968: See Phillips, K. N., 1968.

Thompson, J. M., L. D. White, and M. Nathenson. 1987. Chemical analyses of waters from Crater Lake, Oregon, and nearby springs. U. S. Geol. Surv., Open File Rep. 87-587.

Welhan, J. A., R. J. Poreda, W. Rison, and H. Craig. 1988. Helium isotopes in geothermal and volcanic gases of the western United States, I. Regional variability and magmatic origin, Jour. Volcanol. Geotherm. Res. 34:185-199.

White, L. D., J. M. Thompson and C. A. Maley, 1985, Evidence for thermal water in Crater Lake, Oregon. EOS 66:1146. (Abstract).

Williams, D. L., and R. P. Von Herzen. 1983. On the terrestrial heat flow and physical limnology of Crater Lake, Oregon. Jour. Geophys. Res. 88(B2):1094-1104.

CHEMICAL AND PHYSICAL PROPERTIES OF THE WATER COLUMN AT CRATER LAKE, OREGON

Robert Collier, Jack Dymond, James McManus
College of Oceanography, Oregon State University
Corvallis, Oregon 97331-5503
and
John Lupton
Marine Sciences Institute, University of California
Santa Barbara, California 93106

The vertical distribution of temperature and density in Crater Lake are primarily controlled by the balance of the strongly-seasonal heat exchange at the lake's surface and a continual conductive and convective input of heat at the lake floor. The net result is a nearly-uniform density and relatively low stability in the deep water. To a first approximation, the concentrations of the major cations and anions in the system are well mixed vertically. The concentrations of most transition metals are very low and the vertical distribution of several trace metals demonstrate the significance of surface (atmospheric) input. Consistent with the well-known clarity of the lake, the concentration of suspended particles, as indicated by a beam-light transmissometer, is generally very low, and during the summer the vertical distribution of particles is primarily controlled by the phytoplankton.

The temperature of the lake water gradually increases below a depth of approximately 300 meters and shows stronger increases ($\Delta T \geq 0.1^{\circ}C$) near the east end of the south basin. These temperature increases are accompanied by an increase in conductivity. The overall form and distribution of these anomalies suggest they are dynamic features that require active input of thermal waters at the SE corner of this basin. There is also an increase in the concentration of the major cations (Na, K, Ca, Mg, Li) which is directly proportional to the conductivity and temperature increase.

The chemical and physical properties of the water column at Crater Lake reflect its unique origin and environment. The lake lies totally within the caldera of Mount Mazama which formed after the Cascade volcano's climactic eruption 6900 years before present. The lake, located at an elevation of 1882 meters, is the deepest lake in the United States at 589m. The lake surface occupies 78.5% of the drainage basin defined by the caldera rim, has a surface area of 53.2 km^2 and a volume of 17.3 km^3 (Phillips 1968). Precipitation input to the lake (171 cm/yr) occurs primarily as snow deposited directly on the lake surface (Phillips 1968). The output of water from the system has been estimated as one-third evaporation and two-thirds seepage (Phillips 1968). Therefore, the residence time of water (with respect to input) is 150 years and the residence time of solutes dissolved in the lake (with respect to seepage) is 225 years. Refinements on the water budget estimates are presented by Redmond (1990).

Despite its remote location, the lake has been studied by numerous limnologists and oceanographers since the turn of the century. A summary of the earlier chemical studies of Crater Lake by the U. S. Geological Survey can be found in a USGS water supply paper by Phillips and Van Denburgh (1968). More recent work by the USGS is included in three papers in this volume (Thompson *et al.*; Nathenson and Thompson; Nathenson). In 1969, a group of investigators studied the fission-produced radioisotope distribution and inventory in the lake in order to study the efficacy of wet vs. dry deposition from the atmosphere (Volchok *et al.* 1970; Simpson

1970). Data collected as part of the on-going 10-year limnological program at the park are reported within the annual reports for this project (Crater Lake National Park). In general, all investigators have reported that the lake is well-mixed with respect to the major ion concentrations. The radioisotope distributions suggest a rapid vertical mixing rate of less than ten years (Volchok *et al.* 1970; Simpson 1970). The concentration of several dissolved ions, notably sodium, chloride, boron, lithium, sulfate and silica (Thompson *et al.* 1990), are anomalously high compared to atmospheric or known surface inputs to the lake. Many of these authors have suggested the possibility of sub-surface thermal springs to account for these concentrations.

The vertical distribution of temperature in the lake was discussed by Neal *et al.* (1972) who presented the first detailed high-precision measurements of the complete water column. The general features they discussed were the strong seasonal thermocline in the upper 100 meters, a decrease in temperature to a minimum value of 3.53°C at mid-depth (295 meters), and an increase in temperature below this point to values as high as 3.69°C. The average temperature of the sediments below these profiles was 3.80°C. These authors postulated that the stability of this hyperadiabatic deep water column was maintained by an increase in dissolved ions or suspended sediments with depth. The history of temperature measurements in the lake was reviewed by Williams and Von Herzen (1983) in a paper presenting detailed water column temperatures and sediment heat-flow analyses. This research demonstrated several areas of the lake bottom with very high heat flow and provided a detailed demonstration of deep lake warming. The plume-like structure and horizontal heterogeneity of several deep profiles were directly attributed to active thermal springs which seemed to provide a warm water layer which was "ponding" in the southwest basin.

This paper will summarize results from on-going research on biogeochemical cycles in the lake which have included studies of the lake water, its suspended particles, and its sediments. We have conducted these studies with the support of the National Park Service's 10-year limnological study (see G. Larson 1990). In particular, we will discuss the distribution of chemical and physical properties within the lake as these are related to vertical stability and to element cycling within the lake. Much of our recent research effort has been focused on the thermal and density structure of the southwest basin area identified by Williams and Von Herzen (1983) as the site of thermal springs. The results of this research will be presented elsewhere (Collier *et al.*, in preparation; Lupton *et al.*, in preparation).

METHODS

Water Temperature, Conductivity and Light Transmission

A profiling instrument package was deployed on a hydrographic wire to measure the conductivity, temperature, and light transmissivity as a function of depth in the water column. This instrument package, which we will refer to as a "CTD," was a SEACAT® model SBE19 (Sea-Bird Electronics, Inc.) coupled to a 25-cm path length beam transmissometer (Sea Tech, Inc.). The CTD records all the data internally (2 scans per second of all sensors) and was also monitored in real-time through a special conducting hydrographic cable attached to the computer on the research boat. The CTD has a temperature resolution of better than 0.001°C; conductivity resolution of 4×10^{-5} Siemens/meter (0.4 µmho/cm); and a pressure resolution of 0.5 decibars. The CTD was recalibrated at the OSU calibration facility after its use during the summer of 1987. The transmissometer has a precision of ±0.2%. Conversion of pressure (decibars) to depths (meters) is carried out by the integration of water densities down through the water column. The depths are effectively equal to 1.02 times the pressure. All physical properties based on the measurement of temperature, conductivity and pressure were calculated using an equation of state for water adapted for application to lakes (Chen and Millero 1986). The temperature of specific water samples collected on the hydrographic wire was estimated from the CTD record collected simultaneously.

Water Samples

Water samples from throughout the water column were collected from the NPS research boat using the standard 1/8" stainless steel hydrographic wire. Various oceanographic water samplers were mounted on the wire depending on the demands of individual chemical analyses. Typically, 5- or 30-liter Niskin®-type samplers (General Oceanics) were used. Some of the sampling for trace-metal analyses used teflon-coated 5- and 20- liter GoFlo®-type (General Oceanics) samplers. In order to avoid contamination and preserve the integrity of each sample, great care is taken in each step of collection, preservation and analysis (Bruland *et al.* 1979). A wide variety of chemical properties have been considered during the course of our research efforts. Table 1 summarizes the dates of the relevant samples we have collected and the investigators involved in their analysis. This paper will discuss the analyses which have been completed to date by the OSU investigators.

TABLE 1. SUMMARY OF WATER SAMPLES COLLECTED FOR CHEMICAL ANALYSES (TO 1987)

Chemical Components	Year Collected	Investigator
Major components		
(NA, K, Mg, CA, $Si(OH)_4$)	1987	Collier, OSU
(SiO_4, Cl, alkalinity, Li)	1987	Thompson, USGS
(high-precision density)	1987	Chen, OSU
Trace components		
(Li, V, Mn, Fe, Ni, Cu, Zn, Mo, Cd, Pb)	1984,5,6,7	Collier, OSU
Trace gases and isotopes		
(He-3, He-4)	1985,1987	Lupton, UCSB
(Rn-222)	1985,1986	Dymond, OSU
(Oxygen and hydrogen isotopes in H_2O)	1987	Thompson & White, USGS
(methane)	1987	Lilley, UW

Chemical Analyses of Water Samples

Major cations (Na, K, Ca, Mg) were analyzed by flame atomic absorption spectrophotometry using a Perkin Elmer model-5000 spectrophotometer. The precision of these instrumental analyses based on replicate samples is approximately 0.5% (ranging from 0.3% for Mg to 0.8% for Na). Lithium was also determined by flame atomic absorption analysis. Silicate concentrations were determined by a colorimetric molybdenum blue method using an Alpkem rapid flow analyzer. Trace metals were determined by atomic absorption using Zeeman-graphite furnace atomization. Some trace metals required substantial chemical preconcentration before analysis, and this was carried out using methods developed for seawater trace metal analysis (Boyle *et al.* 1981; Collier 1985).

RESULTS

During three separate deployment periods in 1987, twenty nine CTD casts (vertical profiles) were carried out (Fig. 1). The data discussed here were recorded on specific CTD casts and on all hydrocasts where water samples were collected. Most sampling focused on the eastern-most portion of the southern basin (see inset to Fig. 1 where previous temperature measurements (Williams and Von Herzen 1983) and our preliminary chemical data (see below) suggested the presence of thermal inputs. Other data were collected in the deep central basin, near Merriam Cone, and near the "saddle" between the two basins.

The new high-precision CTD data collected in 1987 refine the general descriptions of Neal *et al.* (1972) and Williams and Von Herzen (1983) and adds the first detailed measurements of *in situ* conductivity allowing the calculation of densities. The data set focuses on the magnitude and aerial extent of the thermal anomalies in the deep lake and introduces the first significant evidence of high dissolved salts associated with these increases in temperature. Figure 2 shows the full water column distribution of temperature and density at Station 28, located within the South Basin (Fig. 1). These data are based on direct measurements from the CTD package as well as on calculated quantities derived from the equation of state applicable to this lake water (Chen and Millero 1986).

Station 28 was collected late in the summer and demonstrates the fully developed seasonal thermocline in the upper 100 meters (Fig. 2a) and the associated decrease in density within this stable warm surface water layer (Fig. 2b). Below 100 meters depth, the density of the water is very nearly homogeneous with a total change of less than 4 ppm (Fig. 2b, 2c). When the seasonal thermocline erodes and is cooled during the winter, storms will drive significant mixing from above. Preliminary data collected over several months in 1988 demonstrate a number of mixing events which are evident below 150 meters depth which may have been induced by storms. As first noted by Williams and Von Herzen (1983), any significant heating from below by conductive and/or convective inputs may drive convection and mixing of these deep waters. It is still difficult to put time scales on mixing rates based on physical models alone. Rapid mixing is indeed sug-

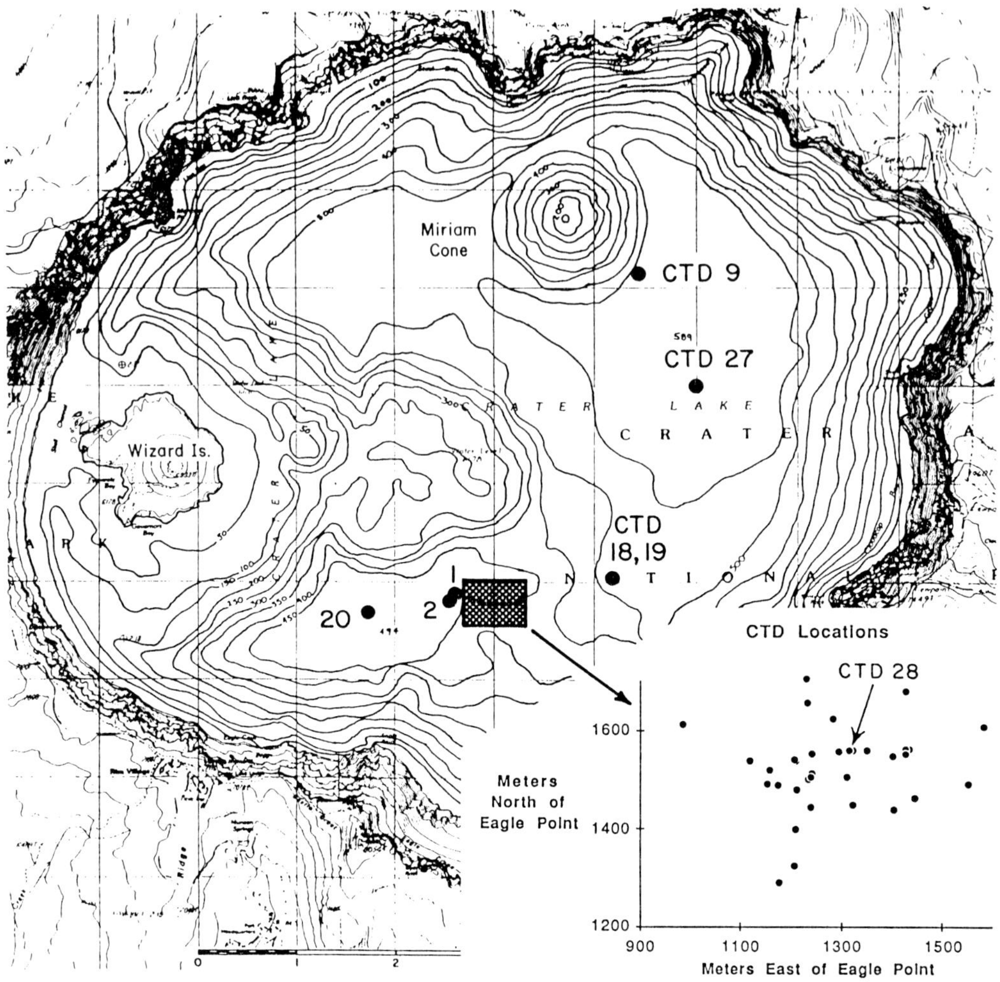

Figure 1. Bathymetric map of Crater Lake showing the locations of 29 CTD profiles collected in 1987. The inset at the lower right shows an expanded view of a special study area (shaded box on map) where most of the data were collected. The topographic map is taken from the USGS 7.5 minute provisional map (1985), and the bathymetric data overlay is from the paper by Byrne (1965).

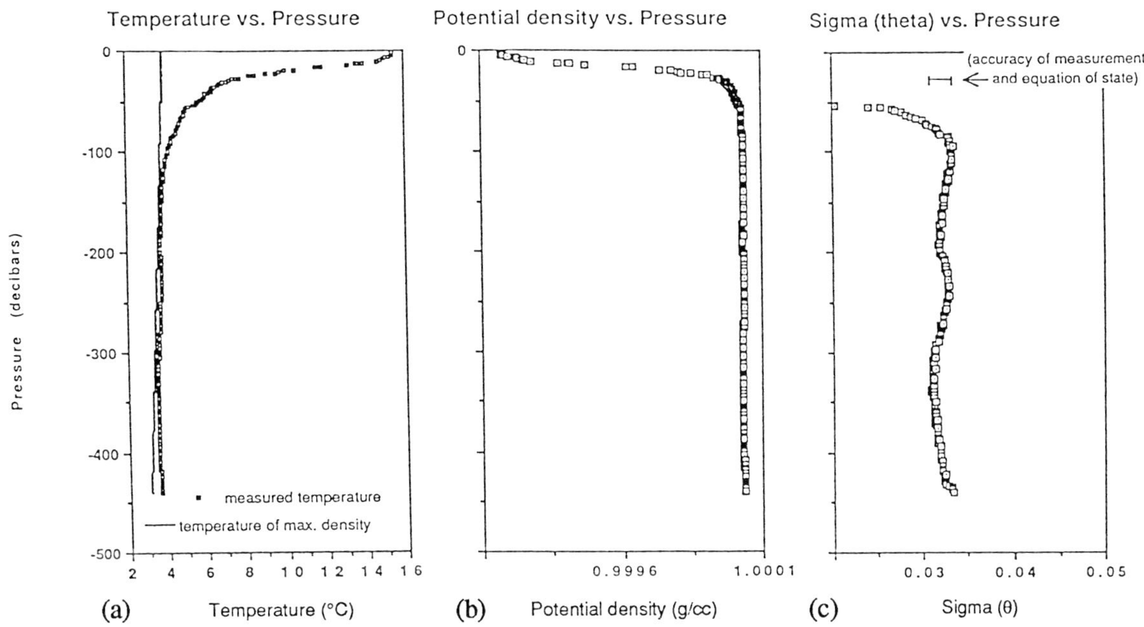

Figure 2. CTD-transmissometer data for the complete water column from the east end of the South Basin (CTD28, Fig. 1). All calculated physical properties are based on the equation of state for limnologic waters developed by Chen and Millero (1986): (A) Vertical distribution of temperature (°C) as a function of pressure (decibars, which is numerically equivalent to [0.98 x depth] in meters). The solid line indicates the temperature of maximum density for Crater Lake water at *in situ* pressure; (B) potential density vs. pressure. The potential density is the density of a water parcel moved adiabatically to a common reference pressure—in this case to the lake surface. This parameter is critical in considering the stability of the water column and the rate of turbulent mixing. It can be seen that the water column below 100 meters has a nearly uniform density; (C) Sigma (θ) vs. pressure. This parameter, derived from b, is defined as Sigma (θ) = (potential density - 1)x 1000.

gested by the vertically-homogeneous tritium data of Simpson (1970), and by dissolved oxygen concentrations which are near-saturation throughout the water column. However, further analysis of these data, new winter data collected in 1988-89, and lake-based meteorological data will help provide constraints on the overall mixing rates. It is very likely that both the intensity of summer heating and the nature and severity of winter storms will cause variations in the mixing rate from year to year. This mixing rate is a critical factor in the interpretation of the overall rates of input for heat, hydrothermal chemicals, and other materials to the deep lake.

Deep Lake Temperature Structure

The upper water column can be contrasted to the expanded view of the lower 100 meters at station 28 shown in Fig. 3. Shown in this figure are: (a) temperature; (b) conductivity; (c) potential density (sigma). All data scales have been significantly expanded over those in Fig. 2. The major feature apparent in Fig. 3a is the increase of temperature with depth. The slow increase, which begins below 300 meters depth, is seen throughout the lake and appears to reflect the net effect of conductive and convective heat inputs. The light dotted line shown in Fig. 3a shows this temperature structure in the central basin. The adiabatic increase in temperature with depth can be calculated from the equation of state (Chen and Millero 1986) and is shown to be an insignificant portion of the observed temperature increase (Fig. 3a). Sharp gradients in temperature, such as that seen at 430 meters in CTD 28 in the South Basin (Fig. 3a), are sometimes followed by a complete reversal in the temperature suggesting a locally unstable system. These results will be discussed in detail elsewhere with respect to their possible hydrothermal origin (Collier *et al.*, in preparation).

The conductivity (Fig. 3b) also increases with depth due to an increase in the concentration of dissolved salts. This is very clearly correlated with

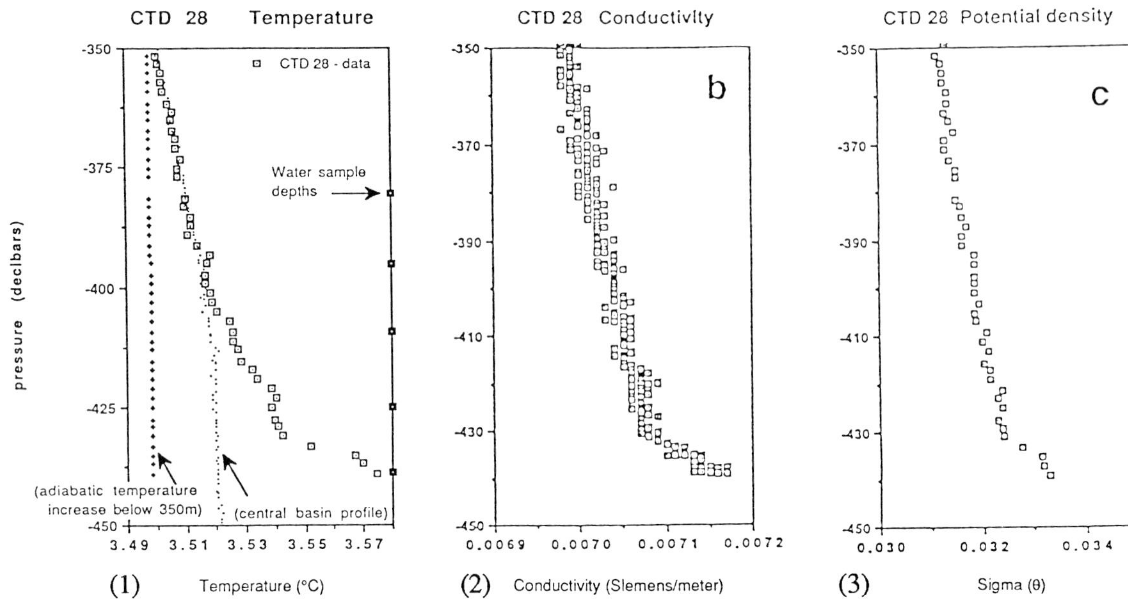

Figure 3. Expanded plots of the CTD data from Fig. 2 which focus on the bottom 100 meters of the water column: (a) Plot of temperature vs. pressure. Three major features are demonstrated: (1) the adiabatic increase in temperature, starting from the minimum temperature in the water column (at 345 decibars), accounts for less than 1 millidegree warming (the adiabatic gradient over this interval averages 3.2 x 10^{-6} degrees/decibar); (2) the general warming below 350 decibars, which is seen throughout the lake, has a gradient which is *two orders of magnitude larger* (3.3 x 10^{-4} degrees/decibar) and is thus *hyperadiabatic;* (3) the gradient seen near the bottom of the thermal study area is approximately 1 x 10^{-2} degrees/decibar - a gradient which is 3400 times larger than the adiabatic increase. The depths of discrete water samples taken along with the CTD cast are noted along side of (a) for reference; (b) *In situ* Conductivity (Siemens/meter =10,000 μmho/cm) vs pressure. This increase is primarily due to the increase in dissolved ions with depth (approximately 5% of the increase is due to the temperature increase); (c) sigma(θ) vs pressure noting the marginal stability of the deep water column (less than 2ppm in density over 100 meters) driven by the combination of heat and dissolved ion increases.

the temperature signal. The net effect of the temperature and conductivity increase is that the potential density (Fig. 3c) increases very slightly with depth such that this water column is stable. Both Neal *et al.* (1972) and Williams and VonHerzen (1983) hypothesized that the lake should have an increase in dissolved solids in order to stabilize the measured temperature increase. However, the total increase in density (less than 2 ppm in 100 meters) produces only a very weak stratification. The static stability, E, is only $2x10^{-8}$ m^{-1} (E is an expression of the density gradient normalized to the absolute density) and the Brunt-Väisälä frequency, N, is $1x10^{-4}$ cycles/sec. N is related to the density gradient and can be thought of as the frequency of oscillation of a water parcel displaced vertically from its equilibrium position in the water column (Ruttner 1974; Pond and Pickard 1978). If the observed temperature increase is due to hydrothermal inputs, the increase in conductivity suggests that the input of warm water may carry enough salts to stabilize the fluid relatively near the lake bottom.

Chemical Distributions

In this research, we have focused on the concentrations of the major cations, silica, and some trace metals (see below). Other analyses focused on questions of mixing and hydrothermal inputs are in progress (Collier *et al.*, in preparation; Lupton *et al.*, in preparation). More data on the composition of the lake and regional springs are presented in Thompson *et al.* (1990). Table 2 presents the mean concentrations of sodium, potassium, magnesium, calcium, lithium and silicon from 46 samples collected in the water column during 1987. The standard deviations around these mean values are all less than 2%, indicating, to a first approximation, that the lake is relatively homogeneous with respect to the distribu-

TABLE 2. MEAN CONCENTRATIONS FOR SODIUM, POTASSIUM, MAGNESIUM, CALCIUM, LITHIUM, AND SILICON IN CRATER LAKE

Element	Na	K	Mg	Ca	Li	Si (SiO_2)
Mean concentration (ppm)	10.51	1.72	2.70	5.06	44.7	17.7
Standard deviation (n=46)	0.18	0.02	0.04	0.06	0.7	0.2

Note: Of 46 samples represented in these means, 39 were collected deeper than 300 meters in the lake. Therefore, these data do not represent a volume-weighted concentration in the lake.

tion of these major ions. A closer look at the vertical profiles (Fig. 4) of these cations in the southern basin reveals a systematic increase in concentration of about 3% below 300 meters which is correlated with the increase in temperature. This increase in ion content accounts for the measured increase in conductance (Fig. 3b). Mass balance calculations for the major ions in the lake (for instance, see M. Nathenson 1990) suggest significant deficits in the known inputs accounting for caldera springs and atmospheric deposition. We will discuss the significance of these near-bottom gradients in resolving this imbalance in terms of hydrothermal inputs from springs (Collier *et al.*, in preparation).

Figure 5a shows the characteristic distribution of water clarity as a function of depth as determined by the transmissometer attached to the CTD package. The total range in light transmission (88-91.3%) demonstrates the extreme clarity of the lake. A value of 91.3%, seen in the deep waters, is essentially equivalent to the theoretical transmission through distilled water. The lower values in the upper 150 meters are primarily related to the phytoplankton populations which have a small shallow maximum near 20 meters depth, possibly related to the species *Nitzschia gracilis*, and the deeper population which is usually coincident with the depth of the productivity maximum (D. Larson 1972; G. Larson 1987, 1990). Transparency changes very little below 150 meters (<0.4%), and it is certain that suspended solids are not a significant contribution to the density structure of the lake during this sampling period.

The lake owes its remarkable clarity primarily to its oligotrophic nutrient status. The upper 200 meters of the water column contain very little nitrate (<0.1μmol/liter) in the summer while phosphate remains relatively high (Fig. 5b). The nutrient status and primary production of the lake are discussed in

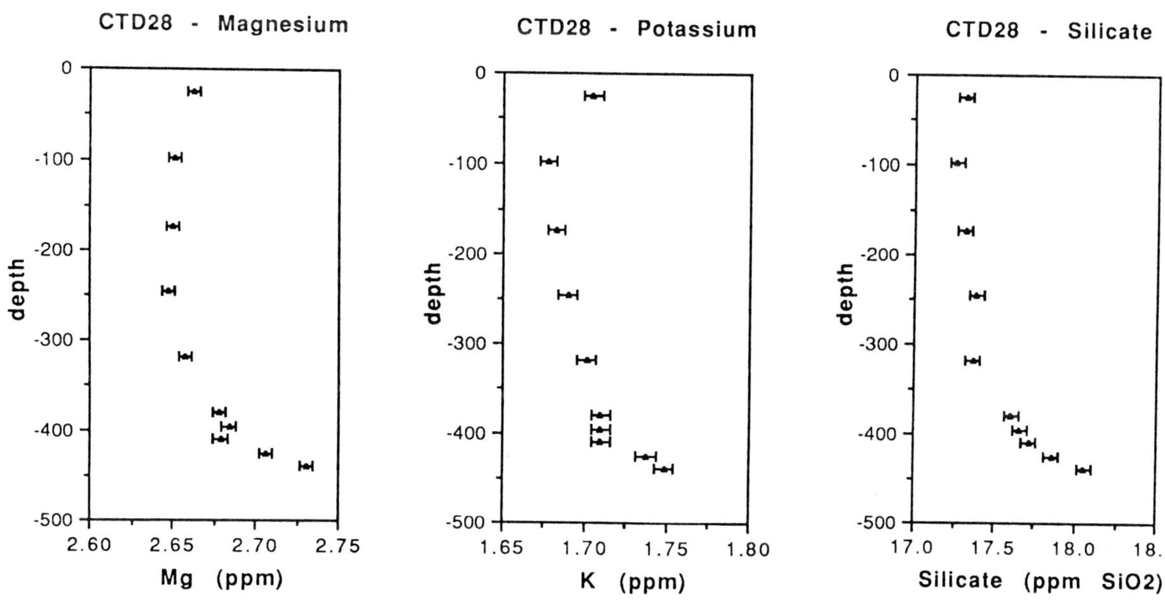

Figure 4. Vertical distribution of magnesium (mg/liter), potassium (mg/liter) and silicate (mg/liter) in the full water column at station CTD28 (Fig. 1).

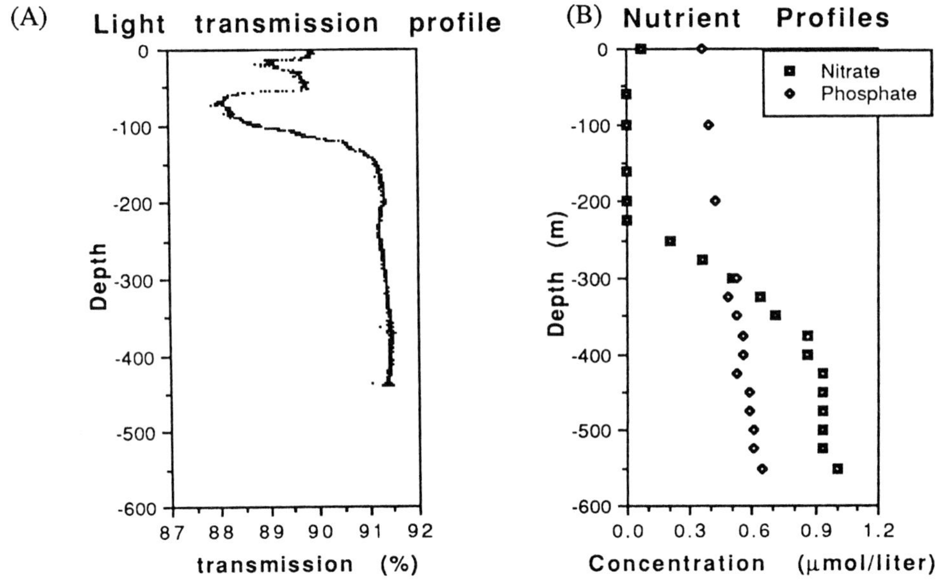

Figure 5. (A) Light transmission (25 cm path, % of air) vs depth. Shallow water minima are due to suspended particles–mostly phytoplankton; while deep water values are very near to distilled water transmissions. (B) The concentration of nitrate and phosphate (µmol/liter) as a function of depth in the central basin. These data were provided by C. Dahm, U. of New Mexico.

detail elsewhere in this volume (see G. Larson; Dahm *et al.*; Geiger and Larson; D. Larson *et al.*). Primary production in this nitrogen-limited system is very sensitive to the introduction of new nitrogenous nutrients from exogenous sources and from exchange with the deep lake reservoir of regenerated nitrate.

Water sample collection for trace metals started in 1983 and has continued through the present field season. These samples included several vertical profiles, numerous near-bottom profiles related to hydrothermal investigations, samples from a transect of surface waters between Phantom Ship and Cleetwood Cove, and a set of samples from several of the caldera springs collected in 1984. The first two years of work related to characterizing bulk lake distributions and the work since then has focused on hydrothermal investigations. Figure 6 shows the vertical profiles for manganese, iron and lead from samples collected during the summer in 1984. These vertical profiles show a clear surface maximum which decreases rapidly below 75 meters (into the thermocline). Significant surface inputs of these metals must exist in order to maintain these maxima. The concentration of both dissolved and particulate aluminum shows a similar surface maximum. Analyses of the major cation concentrations in these profiles suggest that the spring snowmelt run-off event is not the major source of these surface maxima. The compositions of the caldera springs also appear to be insufficient to account for these metal enrichments. Atmospheric deposition of particulate metals remains as a likely source for these surface features. The presence of a significant lead maximum at the surface suggests anthropogenic input from local sources or long-range transport of aerosols.

The concentrations of iron and manganese increase significantly near the bottom of the lake. While this is a common feature in many lakes with deep oxygen depletions and organic-rich anoxic sediments (Sigg 1985), the high oxygen concentrations seen throughout Crater Lake (Larson 1987) should limit this type of redox cycling within the water column. Increases in the concentration of these metals would also be consistent with primary inputs from springs in the deep lake. Elsewhere (Collier *et al.*, in preparation), we will show near-bottom increases in manganese in the S. E. basin that are one to two orders of magnitude larger than the water column profiles shown in Fig. 6 .

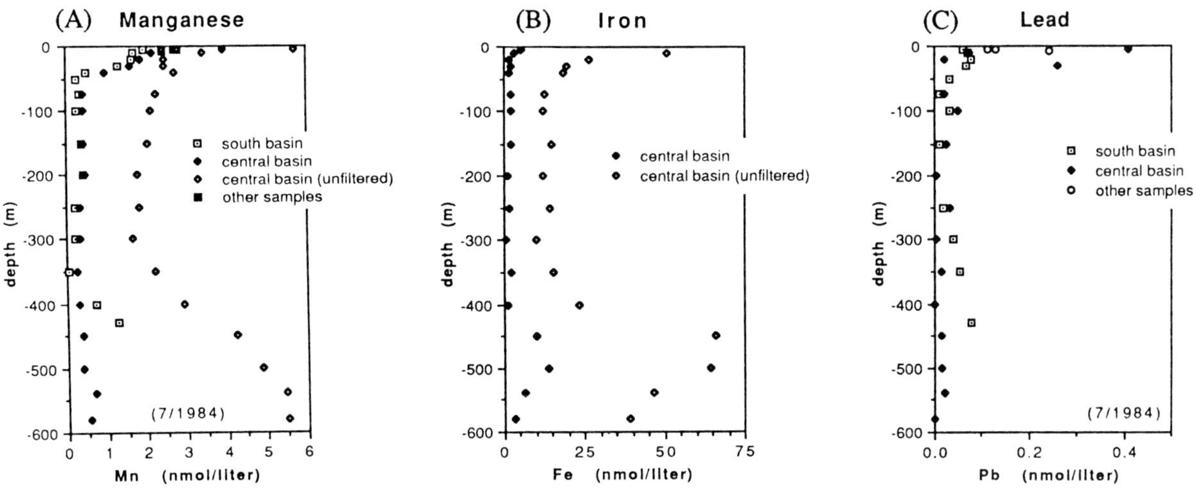

Figure 6. The concentrations (nmol/liter) of (A) manganese, (B) iron, and (C) lead vs depth.

DISCUSSION

A prerequisite to understanding the overall biogeochemical cycles in the lake is a working model of the lake's hydrologic cycle (see Redmond 1990) and the rates and modes of mixing within the water column. Much of our on-going research effort has therefore been focused on the physics of this system. The major features of the system are its strong seasonal heat exchange at the surface and the addition of heat to the deep lake. The system is statically stable as a result of the slight increase in dissolved salts which parallels the temperature increase. The overall system, however, has very low stability and may mix easily in the winter. High dissolved oxygen concentrations (Larson 1987) and homogeneous vertical tritium distributions (Simpson 1970) support this hypothesis. However, the increase in dissolved ions and the associated increase in density with depth may argue for a slower mixing rate.

The seasonal thermocline demonstrated in Fig. 2, represents a barrier to mixing between the surface layers and the deep lake. While this strong pycnocline exists, materials can accumulate in the upper waters of the lake (for instance, atmospheric particles supporting the trace metal distributions) while other nutrients might be depleted due to biological activity. The general depletion of nitrate above 200 meters in the lake is an example of this process. The net vertical flux of particulate organic nitrogen out of the upper water column driven by primary production is balanced by (and limited by) the upward mixing flux of dissolved nitrogen from the deep lake. Nutrient data outside of the summer season are scarce, but relatively little nitrate existed in the upper 200 meters during March 1986 (Larson 1987) when the water column was nearly isothermal. These data would suggest that large amounts of deep lake water had not mixed with the surface during this period of the winter and that estimates of the overall mixing rate may be too high. Future research to answer these questions must include hydrographic data covering the complete annual cycle at the lake. We are currently working on our fifth year of sediment trap deployments in the lake in order to constrain the downward flux of particulate organic matter out of the euphotic zone. These measurements will allow us to apply nutrient cycling models which will better constrain the vertical mixing rates within the basin. As the particulate organic matter decomposes in the deep lake, oxygen is consumed. This consumption rate can be compared to the measured depletion of O_2 (with respect to atmospheric saturation) in order to estimate a mixing rate of oxygenated surface waters into the deep lake. Our research group and others associated with the on-going limnological study of the lake, are currently collecting these data which will allow us to constrain these mixing rates.

Many observations about the lake physics and chemistry made by our research group and earlier investigators, have suggested the existence of a hydrothermal input in the deep lake. The details of these data are presented in Dymond and Collier (1990), Collier *et al.* (in preparation), and Lupton *et al.* (in preparation). We will briefly summarize these results here; especially as they relate to the temperature structure of the lake. Following the lead of Williams and Von Herzen's (1983) heat flow and water column measurements, we have demonstrated both the general warming of the deep lake and have identified specific regions where the water column thermal gradient is more than an order of magnitude larger than this background. These thermal signals are directly correlated with an increase in major ions and are enriched in metals such as manganese and iron. Increases in temperature on the order of 0.3°C are correlated with increases in conductivity of approximately a factor of 2 over spatial scales of meters suggesting the active maintenance of these strong gradients. Enrichments of radon-222 in these waters suggest their recent exit from a spring. Extreme enrichment of He-3 in these waters with respect to atmospheric saturation and the "mantle" isotopic ratios of He-3/He-4 in the added helium suggest an input from the alteration of a fresh magmatic source rock (Lupton *et al.* 1987; Lupton *et al.* in preparation). All of these chemical variations correlate well with the increase in temperature suggesting a common carrier—warm water. Evidence of hydrothermal inputs derived from lake sediments is discussed by Dymond and Collier (1990). Ongoing research will continue to test this and other individual hypotheses which explain these observations. The composition and heat flux of any convective input into the lake could have a significant effect on the stability of the water column and the mixing of deep regenerated nutrients into the euphotic zone.

ACKNOWLEDGMENTS

Many people have made critical (sometimes incredible) contributions to our immediate program and to our general Crater Lake research efforts over the past 5 years. Analytical assistance was provided by R. Conard, P. Collier, S. Holbrook, C. Perhats, D. Pyle and J. Robbins. The staff and management of Crater Lake National Park have provided the critical support and encouragement for this work. Particular thanks to G. Larson, our co-PI on this project, and to J. Milestone at the Park. A host of Park personnel, especially M. Buktenica, have provided long hours of essential field support and critical discussions on Lake processes.

LITERATURE CITED

Boyle, E. A., S. S. Huested, and S. P. Jones. 1981. On the distribution of copper, nickel, and cadmium in the surface waters of the North Atlantic and North Pacific Ocean. Jour. Geophys. Res. 86:8048-8066.

Bruland, K. W., R. P. Franks, G. A. Knauer, and J. H. Martin. 1979. Sampling and analytical methods for the determination of copper, cadmium, zinc, and nickel at the nanogram per liter level in sea water. Analyt. Chim. Acta 105:233-245.

Bruland, K. W. 1983. Trace elements in seawater, *In*: J. P. Riley and G. Skirrow, eds., Chemical Oceanography, 2nd ed., vol. 8. Academic Press, London.

Byrne, J. V. 1965. Morphometry of Crater Lake, Oregon. Limnol. Oceanogr. 10(3):462-465.

Chen, C. T., and F. J. Millero. 1986. Precise thermodynamic properties for natural waters covering only the limnological range. Limnol. Oceanogr. 3(3):657-662.

Collier, R. W. 1985. Molybdenum in the northeast Pacific Ocean. Limnol. Oceanogr., 30(6):1351-1354.

Collier, R., J. Dymond, J. Lupton, A. Chen, M. Lilley, and M. Thompson. 1987. Effects of hydrothermal inputs on the chemistry and physics of Crater Lake, OR. EOS 68(50):1721 (Abstract).

Larson, D. W. 1972. Temperature, transparency, and phytoplankton productivity in Crater Lake, Oregon. Limnol.Oceanogr. 17(3):410-417.

Larson, G. L. 1987. Crater Lake limnological studies. 1986 Ann. Rep., Crater Lake National Park, National Park Service. 81 pp.

Larson, G. L. 1990. Status of the ten-year limnological study of Crater Lake, Crater Lake National Park. Pages 7-18 *in* E. T. Drake *et al.*, eds., Crater Lake: An Ecosystem Study. Pacific Div., Amer. Assn. Advance. Sci., San Francisco, Calif.

Lupton, J. E., R. Collier, J. Dymond. 1987. Excess ^{3}He in Crater Lake, Oregon: Evidence for geothermal input. EOS 68:1722.

Nathenson, M. 1990. Chemical balance for major elements in water in Crater Lake, Oregon. Pages 103-114 *in* E. T. Drake *et al.*, eds., Crater Lake: An Ecosystem Study. Pacific Div., Amer. Assn. Advance. Sci., San Francisco, Calif.

Neal, V. T., S. J. Neshyba, W. W. Denner. 1972.

Vertical temperature structure in Crater Lake, Oregon. Limnol. Oceanogr. 17(3):451-454.

Phillips, K. N. 1968. Hydrology of Crater, East, and Davis Lakes, Oregon. U. S. Geol. Surv. Water-Suply Pap. 1859-E. 60 pp.

Pond, S. and G. L. Pickard. 1978. Introductory Dynamic Oceanography. Pergamon Press, Oxford, UK. 241 pp.

Redmond, K. 1990. Crater Lake climate and lake level variability. Pages 127-141 *in* E. T. Drake *et al.*, eds., Crate Lake: An Ecosystem Study. Pacific Div., Amer. Assn. Advance. Sci., San Francisco, Calif.

Ruttner, F. 1974. Fundamentals of Limnology, 3rd ed. Univ. Toronto Press, Toronto, Can. 307 pp.

Sigg, L. 1985. Metal transfer mechanisms in lakes; the role of settling particles. Pages 283-310 *in* W. Stumm, ed., Chemical Processes in Lakes. J. Wiley-Interscience, New York.

Simpson, H. J. 1970. Tritium in Crater Lake, Oregon. Jour. Geophys. Res. 75(27):5195-5207.

Thompson, J. M., L. D. White, and M. Nathenson. 1987. Chemical analyses of waters from Crater Lake, Oregon, and nearby springs. U. S. Geol. Surv. Open-file Rep. 87-587.

Van Denburgh, A. S. 1968. Chemistry of the Lakes. Pages 58-60 *in* Hydrology of Crater, East, and Davis Lakes, Oregon. U. S. Geol. Surv. Prof. Pap. 1005.

Volchok, H. L., M. Feiner, H. J. Simpson, W. S. Broecker, V. E. Noshkin, V. T. Bowen, and E. Willis. 1970. Ocean fallout the Crater Lake experiment. Jour. Geophys. Res. 75:1084-1091.

Williams, D. L., and R. P. Von Herzen. 1983. On the terrestrial heat flow and physical limnology of Crater Lake, Oregon. Jour. Geophys. Res., 88 (B2):1094-1104.

CRATER LAKE ECOSYSTEM

THE ECOLOGY AND CHEMISTRY OF CALDERA SPRINGS OF CRATER LAKE NATIONAL PARK

Stanley V. Gregory, Randall C. Wildman
Linda R. Ashkenas, Gary A. Lamberti
Department of Fisheries & Wildlife,
Oregon State University, Corvallis, OR 97331

Six springs within the Crater Lake caldera were investigated in August 1986. Spring 42, in particular, exhibited high concentrations of nitrate (287 ppb); the other springs all contained less than 60 ppb. Annual patterns of water chemistry indicate high concentrations of nitrate under snow cover, followed by dilution during snowmelt and a rise in concentration in late summer as discharge decreases. In Spring 48, there was a rapid uptake of nitrate from the source to the outlet into the lake, decreasing from 58 to 18 ppb. Biological activity in the study springs corresponded to the observed pattern of nutrient availability. Abundance of benthic algae and rates of gross primary production were highest in Spring 42. At the relative concentrations of nitrogen and phosphorus in these springs, primary production would be limited by inorganic nitrogen. Primary production was also high in Spring 48. Primary production in the other springs was less than half of that observed in Springs 42 and 48, which may be related to either lower nitrate concentrations or more unstable channels. Aquatic invertebrates, excluding chironomids, were more abundant in Springs 42 and 48, a pattern consistent with the higher primary production in these two springs. Invertebrates were most abundant in Spring 48, which may be related to the greater channel stability in this stream. Hypotheses for both natural and man-caused processes responsible for these patterns of water chemistry and biological activity are discussed.

National parks are faced with an inherent contradiction in land management. They are established to preserve unique natural ecosystems in their pristine state, but they must also attract thousands of visitors to view these resources each year. The very public that visits the National Parks potentially threatens these pristine ecosystems with sewage, automobile emissions, damage to vegetation, disturbance of native wildlife, forest fires, and vandalism. Some of these effects can be minimized by appropriate distribution of visitor facilities within the National Parks, but others cannot be avoided because visitors cannot view the natural attractions without being in close proximity to them. By promoting visitation, land managers may inadvertently accelerate the degradation of the very resources they seek to protect.

Crater Lake National Park, due to the central location of the lake, faces the inherent risks of visitors and visitor support activities altering the natural ecosystem. Crater Lake National Park contains one of the most oligotrophic lakes in the world. Visitor facilities and the park highway lie on the rim of the volcano and are used by more than 600,000 visitors annually (Mohler 1986). Even small numbers of visitors pose risks for such a fragile ecosystem.

In 1986, we examined springs within the caldera of Crater Lake to determine whether human activities in the Rim Village area are altering water chemistry or spring communities. This research was designed to describe water chemistry, aquatic primary production, and invertebrate assemblages in springs within the caldera at Crater Lake. The information provides a foundation for evaluating the current status of the springs around Crater Lake National Park and establishes a benchmark for monitoring future changes in spring chemistry and aquatic biota.

STUDY SITES

In 1986, we studied six springs inside the caldera rim that drain into Crater Lake. Previous studies of their water chemistry had found elevated concentrations of nitrate in several springs in the vicinity of Rim Village and the Lodge (Dr. Cliff Dahm and Dr. Doug Larson, pers. comm.; Dr. Gary Larson, pers. comm.).

National Park staff has numbered the caldera springs within the rim of Crater Lake in a clockwise direction starting at Cleetwood Cove. The study streams were Springs 20, 35, 38, 39, 42, and 48, all located on the southwest wall of the caldera (Fig.1). Spring 20 is in the Chaski Slide area; Spring 35 is in the Eagle Point area; Springs 38, 39, and 42 are in the Rim Village area; and Spring 48 is in the Discovery Point area. Springs 20, 35, and 38 usually exhibit relatively low concentrations of nitrate and have little human activity in their vicinity. Spring 48 occasionally has nitrate concentrations between 50 to 100μg NO_3-N/l, but there is no concentrated human activity in its drainage. Spring 39 is immediately below the Lodge and has exhibited nitrate concentrations in the range of 50 to 150 μg NO_3-N/l. Spring 42 consistently contains higher concentrations of nitrate (generally in the range of 250 to 300 μg NO_3-N/l) than any other stream in the National Park and is located immediately below the Rim Village area.

All springs on the walls of the caldera have extremely high gradients (50%-140%) and are geomorphically unstable. Snow cover persists for approximately 7 months, with an average depth of 3m in winter (Sterns 1963). Avalanches occur frequently (i.e., several times each decade, as evidenced by age of streamside vegetation), and substrates consist of loose accumulations of gravel, cobbles, and boulders. Although water chemistry potentially influences biological activity in these springs, physical instability exerts profound influences on the aquatic organisms. Springs 20, 38, and 39 are relatively open and are lined by small shrubs and herbaceous plants. Sitka alder (*Alnus sinuata*) creates a low, narrow thicket along Spring 35. Springs 42 and 48 flow through mature conifer forests of mountain hemlock (*Tsuga mertensiana*), red fir (*Abies procera*), and white bark pine (*Pinus albicaulis*) with an understory of Sitka alder immediately adjacent to the channels.

METHODS

Water samples were collected in two 1-l polypropylene bottles, held on ice in darkness, and filtered through Whatman GF/F glass fiber filters as soon as possible (less than 8 hr in all cases). All water samples were refrigerated and analyzed within one week. Concentrations of nitrate, ammonium, and soluble reactive phosphorus were determined by automated colorimetric analysis using the cadmium reduction, hypochlorite phenol, and ammonium molybdate methods, respectively. Total reduced nitrogen (Kjeldahl nitrogen) and total phosphorus were measured by digestion and analyzed as ammonium and reactive phosphorus; calcium concentrations were determined by atomic absorption.

Seasonal patterns of nitrate concentration in Spring 42, particularly increases from early summer through fall, have been suggested as indications of possible effects of visitors on spring chemistry. National Park Service data on spring chemistry for 1986 and 1987 were analyzed for seasonal patterns of nitrate and calcium, a cation that would not be expected to exhibit major seasonal patterns related to visitor use or biological processes. Spring 42 was compared to Spring 20, a spring on the southeast caldera wall that is not directly influenced by human activity and is not located in a forest.

Substrates were collected for determination of standing crop of chlorophyll-*a*, an index of the abundance of benthic algae. Three substrate samples consisting of three cobbles each were collected from each site on each sampling date. Substrate samples were submersed in known volumes of 90% acetone for 24 hr at 4°C in the dark. Chlorophyll concentration in the extract was determined by the trichromatic method (Strickland and Parsons 1968). Substrate surface area was measured by wrapping the rocks with aluminum foil, weighing the foil, multiplying by the foil area per unit weight, and dividing by two for surface area exposed to sunlight.

Benthic primary production was measured on substrates collected from each site. Primary production was measured in the laboratory by placing substrates in recirculating chambers with water collected from each site. Chambers were held in a water bath at 13°C; artificial metal arc lamps maintained a light intensity of 400 $\mu E \cdot m^{-2} \cdot s^{-1}$ (photosynthetic saturation). Community respiration, net community primary production, and gross primary production

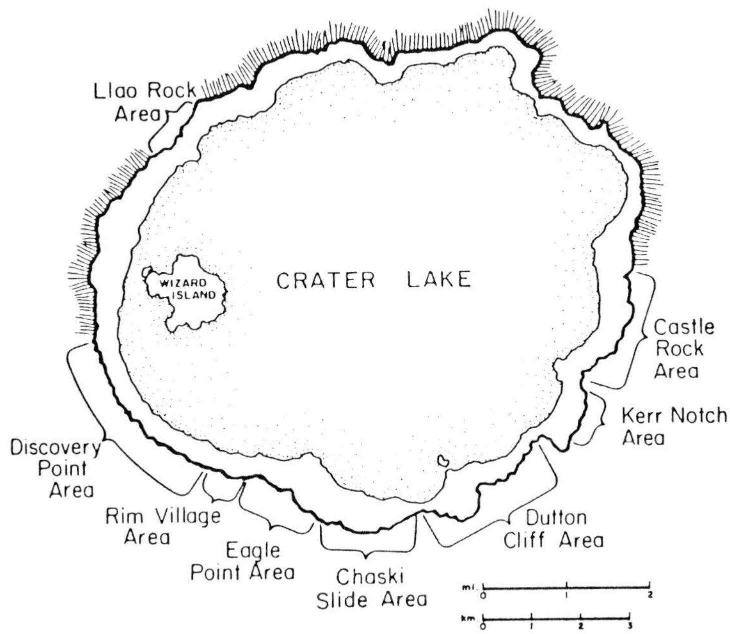

Figure 1. Major study areas within the caldera of Crater Lake (from Larson *et al.* 1986).

were estimated by measuring changes in dissolved oxygen concentrations in the closed chambers for 3-hr incubation periods (Bott *et al.* 1978). Daily P/R ratio was calculated as gross primary production divided by community respiration for a 24-hr period.

Benthic invertebrates were collected from all study sites with a Hess sampler with a 250-µm mesh (a D-frame net with a 500-µm mesh was used in Spring 42). Three samples were collected at each site and preserved in 90% ethanol. Aquatic insects were counted and identified to genus (species if possible), and non-insect invertebrates were identified to class. Invertebrates were assigned to functional feeding groups (i.e., shredders, scrapers, collectors, and predators) according to Cummins and Merritt (1984).

RESULTS

Water Chemistry

In August 1986, the concentration of nitrate in Spring 42 was 287µg NO_3-N·l^{-1} and far exceeded that observed in any other spring within the crater rim (Table 1). No other spring exceeded 60µg NO_3-N·l^{-1} on this sampling date. We climbed to the source of Spring 48 to examine the longitudinal change in water chemistry in a rim spring, and nitrate concentration decreased from 58 µg NO_3-N·l^{-1} to 18 µg NO_3-N·l^{-1} from the source to the outlet. Ammonium concentrations were extremely low (<8 µg N·l^{-1}) or undetectable in all springs. Organic nitrogen concentrations were less than 25 µg N·l^{-1} in the rim springs, and were not detectable in Springs 42 and 48.

Phosphorus concentrations were relatively high in all rim springs, a typical condition in volcanic regions. Most springs contained approximately 40 µg PO_4-P·l^{-1} and 80 µg total P·l^{-1} (Table 1). Spring 39 was somewhat lower in phosphorus than the other streams, and Spring 35 was slightly higher.

TABLE 1. WATER CHEMISTRY OF CALDERA SPRINGS IN AUGUST 1986, EXPRESSED AS µg/l[1]

SPRING	NO_3-N	NH_4-N	TN	SRP	TP	N/P
20	58	1	9	44	65	3.1
35	51	1	16	88	94	1.4
38	7	0	13	57	71	0.3
39	56	0	24	20	29	6.4
42	287	1	0	50	66	13.1
48	18	6	0	43	59	1.3

[1] (NO_3-N=Nitrate nitrogen; NH_4-N= Ammonium nitrogen; TN=Total reduced nitrogen; SRP=Soluble reactive phosphorus; TP=Total phosphorus). N/P ratio is based on molar concentrations of NO_3-N, NH_4-N, and SRP.

In Spring 42 during 1986 and 1987, nitrate exhibited a distinct seasonal pattern, decreasing abruptly in the spring and increasing sharply during summer. Concentrations of nitrate in late summer through late winter generally ranged from 250 to 350 µg NO_3-N·l^{-1} (Fig. 2). Calcium exhibited a similar seasonal pattern, decreasing in spring to approximately 60% of concentrations observed from late summer through winter. In Spring 20, both nitrate and calcium exhibit seasonal patterns that closely resemble those of Spring 42 in timing (Fig. 3). Concentrations of calcium in Spring 20 were higher than those of Spring 42 and nitrate concentrations were lower, reflecting the more xeric watershed of Spring 20.

Benthic Primary Producers

The standing crop of benthic algae was similar in all springs except for Spring 42 (Table 2). Chlorophyll-*a* was more than twice as abundant in Spring 42 than in the other streams, and Spring 35 contained slightly less plant pigment. The abundance of benthic algae was reflected in the rates of benthic metabolism in these streams. Gross primary production was greatest in Spring 48, but was also elevated in Spring 42. Benthic community respiration was also elevated in Spring 42. P/R ratios for most of the streams ranged from 1.0 to 1.75, but the P/R ratio in Spring 48 exceeded 3.5.

Benthic Invertebrates

Benthic invertebrate communities in springs within the rim of the crater were composed primarily of aquatic insects. Benthic invertebrates were most abundant in Springs 20 and 48, but most of the invertebrates in these two springs were small midges (Table 3). Chironomids have short generation times and disperse widely, quickly colonizing disturbed sites. If the longer-lived component of the insect fauna is examined by excluding chironomids from consideration, the pattern changes dramatically. Densities of aquatic insects or total invertebrates (excluding chironomids) were 4 to 10 times higher in Springs 42 and 48 than in the other four springs. Further, invertebrate density in Spring 48 was approximately double that in Spring 42. The densities of aquatic invertebrates in the rim springs were somewhat lower than those observed in our study of streams on the outer slopes of Mount Mazama in 1985, and the aquatic insects other than chironomids were much less abundant in the rim springs. Fewer taxa of benthic invertebrates were observed in the springs within the caldera than in streams on the outer slopes; invertebrate species richness in six sites in Munson, Sun, Dutton, and Goodbye Creeks averaged 28 taxa in contrast to an average of 10 taxa in the caldera springs.

Three major patterns in functional feeding group composition were observed in the six rim springs (Table 4). The most common pattern, an equal dominance of both collectors and scrapers, was observed in Springs 20, 38, and 39. This largely reflects the dominance of chironomids in these streams, because the midges were considered to be 50% collectors, 30% scrapers, and 10% predators (Dr. Ken Cummins, pers. comm.). In Springs 35 and 48, scrapers made up more than 75% of the invertebrate assemblage, resulting from the large number of scraping caddisflies (*Neothrema* and *Imania* in Spring 35; *Neothrema* in Spring 48). Shredders, primarily the stonefly *Zapada columbiana*, comprised half of the invertebrate assemblage in Spring 42.

If chironomids are excluded from the analysis, the proportion of collector-gatherers decreases sharply. Springs 20, 38, 39, and 42 contained high proportions of shredders, reflecting the presence of the stonefly *Zapada columbiana*. The greater relative abundances of scrapers in Springs 35 and 48 are largely comprised of *Neothrema*.

TABLE 2. ABUNDANCE AND METABOLISM OF ALGAL ASSEMBLAGES IN CALDERA SPRINGS IN AUGUST 1986[1]

Spring	Chl-*a*	GPP	CR	P/R
20	14.8	17.3	13.1	0.7
35	4.8	22.9	6.4	1.8
38	10.5	28.4	13.9	1.0
39	10.4	23.6	9.9	1.2
42	33.7	39.9	13.2	1.6
48	14.4	41.5	6.0	3.9

[1](CHL = chlorophyll-*a*, GPP = gross primary production, CR = community respiration, P/R ratio). Abundance of chlorophyll-*a* is expressed in mg/m^2, metabolism as mg O^2/m^2/h, and P/R ratio is based on 14h of primary production and 24h of respiration

DISCUSSION

More than forty perennial springs originate on the crater wall above the surface of Crater Lake, and most of these are located on the south wall of the

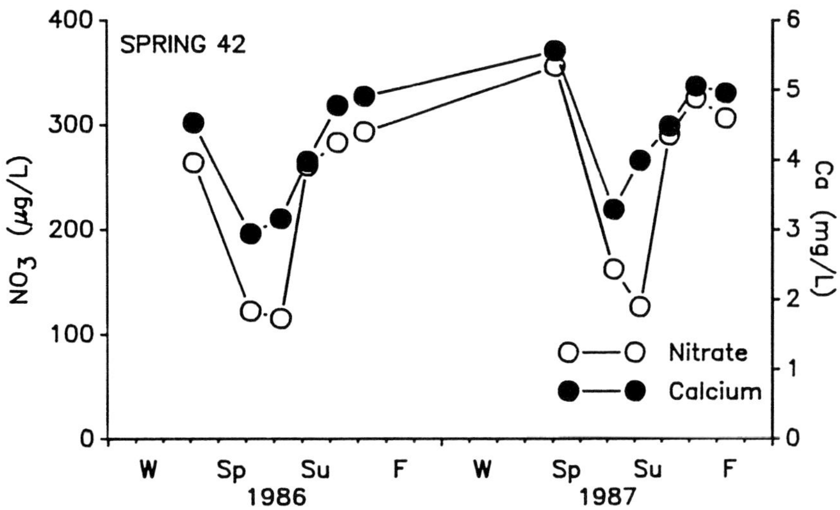

Figure 2. Concentrations of dissolved nitrate nitrogen and calcium in Spring 42 during 1986 and 1987, expressed as µg NO_3-N/L and mg Ca/L

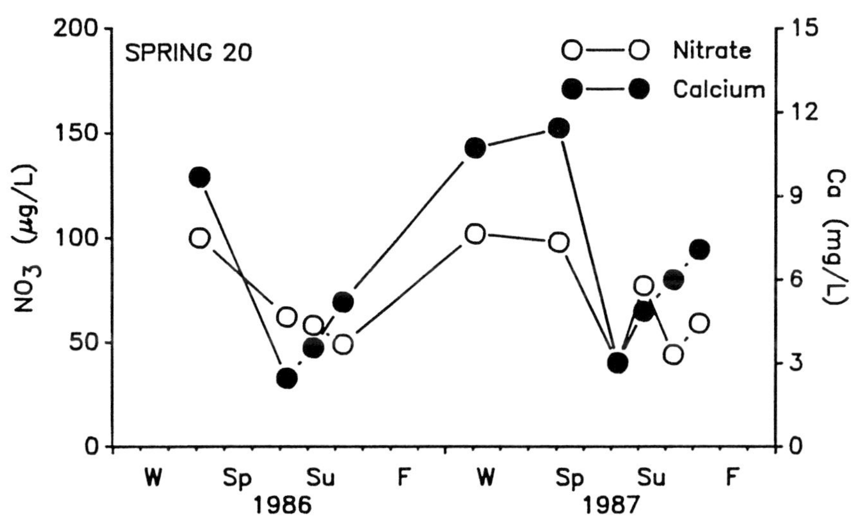

Figure 3. Concentrations of dissolved nitrate nitrogen and calcium in Spring 20 during 1986 and 1987, expressed as µg NO_3-N/L and mg Ca/L.

TABLE 3. DENSITY AND SPECIES RICHNESS OF INVERTEBRATE ASSEMBLAGES IN CALDERA SPRINGS IN AUGUST 1986. DENSITY IS EXPRESSED IN NUMBERS/M^2 (*EXCLUDING CHIRONOMIDS), AND RICHNESS IS EXPRESSED AS NUMBER OF TAXA FOUND IN SAMPLES

Spring	Insect Density*	Invertebrate Density*	Chironomid Density	Total Invertebrate Density	Species Richness
20	149	177	2723	2900	10
35	99	103	2144	2247	10
38	177	197	740	937	10
39	54	68	1548	1616	5
42	754	856	257	1113	9
48	1704	1901	618	2519	15

crater. Phillips and Van Denburgh (1968) did not consider any of these streams to be perennial; but the Park Service staff has monitored these springs since 1983, and many flow year round. Diller and Patton (1902) examined more than 63 springs in mid-July 1901 and estimated their total discharge to be 0.30 m^3/s. All of these streams have extremely high gradients (>50%) and exhibit numerous indications of recent avalanches down the channels.

Previous sampling by the Park Service and other investigators at Oregon State University found that several springs in the vicinity of Rim Village often contained higher concentrations of nitrate than other springs around the lake. Three springs in the Rim Village area (Springs 40, 41, and 42) consistently exhibited nitrate concentrations ranging from 100 to 300µg NO_3-N·l^{-1} in 1983-87. None of the other springs ever exceeded 100 µg NO_3-N·l^{-1}, and most were less than 50 µg NO_3-N·l^{-1}. This study of Springs 20, 35, 38, 39, 42, and 48 in August 1986 found a similar pattern. Spring 42 contained 287 µg NO_3-N·l^{-1}, and the other springs all contained less than 60 µg NO_3-N·l^{-1}. In Spring 48, there was a rapid uptake of nitrate from the source to the outlet into the lake, thus length of stream and relative biological activity in different springs may greatly influence water chemistry observed at stream mouths along the lake.

Comparison of seasonal patterns of nitrate and calcium in Springs 20 and 42 illustrates the pronounced hydrologic effect of snowmelt on spring chemistry. Both biologically active (nitrate) and relatively inactive (calcium) ions exhibit abrupt decreases in concentration during spring and early summer, the period of snowmelt at Crater Lake. After snowmelt, concentrations of both ions increase to levels observed in late winter. Such patterns in spring chemistry would result from dilution during higher surface water discharges associated with snowmelt. Similar seasonal patterns in water chemistry in Spring 20, a spring with negligible human influence, suggest that seasonal changes in

TABLE 4. PROPORTIONS OF INVERTEBRATE FUNCTIONAL FEEDING GROUPS IN CALDERA SPRINGS IN AUGUST 1986. FUNCTIONAL GROUPS EXPRESSED AS PERCENT OF TOTAL NUMBERS IN EACH GROUP (SC = Scrapers, SH = Shredders, C/G = Collector/Gatherers, P = Predators)

	Including Chironomids				Excluding Chironomids			
Spring	SC(%)	SH(%)	C/G(%)	P(%)	SC(%)	SH(%)	C/G(%)	P(%)
20	29	3	57	11	8	46	15	31
35	78	5	11	6	82	5	7	6
38	28	14	48	10	17	66	7	10
39	29	3	58	10	0	73	27	0
42	14	52	15	19	10	67	2	21
48	64	9	20	7	76	11	7	6

spring chemistry reflect, in part, influences of natural hydrologic cycles in springs on the caldera wall.

Biological activity in the study springs corresponded to the observed pattern of nutrient availability. Abundance of benthic algae and rates of gross primary production were high in Spring 42. At the relative concentrations of nitrogen and phosphorus in these springs, primary production would be limited by inorganic nitrogen; therefore, the elevated primary production in Spring 42 may be a result of the higher nutrient supply. Primary production was also high in Spring 48, possibly reflecting greater geomorphic stability relative to the other springs. Such stability may allow development of a more abundant assemblage of primary producers. The lower algal abundance in Spring 48 may be a result of consumption by the more abundant fauna. Primary production in the other springs was less than half of that observed in Springs 42 and 48, which may be related to either lower nitrate concentrations or more unstable channels or possibly to both conditions.

Aquatic invertebrates, excluding chironomids, were more abundant in Springs 42 and 48, possibly a result of the higher primary production in these two springs. Invertebrates were most abundant in Spring 48, which may also reflect greater channel stability in this spring. We hypothesize that Spring 48 was the most geomorphically stable of the six study springs because it exhibited the least evidence of recent avalanches. The faunas of these springs were less abundant and included fewer taxa than those in streams outside the caldera, but the steep, unstable nature of the springs would not be expected to support dense populations and diverse faunas. Chironomids were a major portion of the invertebrate assemblage in most of these springs, and the midges would be well suited to the harsh environments of the rim springs because of their short life histories and wide array of feeding habits.

Springs within the crater in the vicinity of Rim Village are more productive than similar adjacent streams. These streams contained higher concentrations of nitrate, supported greater amounts of benthic algae and higher rates of primary production, and had more abundant and diverse invertebrate faunas than their counterparts. Causes for these patterns of production cannot be proven by the research, but potential mechanisms can be identified and evaluated.

If it is assumed that these patterns in aquatic production are natural and not caused by man's activities in the Park, the characteristics of these areas of the Crater Lake caldera must be unique in some respect. Springs that have exhibited the highest concentrations of nitrate are immediately below the Rim Village area and are located on the southwest wall of the crater. Because of the northerly aspect of these slopes on the south wall, they are cooler and more moist than those on the north side of the crater. The slopes on the southwest wall are vegetated by a mature forest of hemlock, fir, and pine, but the slopes on the southeast wall are sparsely vegetated. Although there were trees and shrubs along Springs 20, 35, 38, and 39, their basins were less vegetated than those of Springs 42 and 48. Forests may have been present on the southwest wall over the last several thousand years and possibly have built up a nitrogen pool in the soil that is reflected in the spring chemistry.

Nitrate concentrations in all caldera springs are far greater than those in streams outside the caldera. This may reflect the shallower soils and sparser vegetation within the caldera, which would account for lower demand for nutrients in the terrestrial ecosystem. The three springs that always have nitrate concentrations in excess of 100 $\mu g\ NO_3\text{-}N \cdot l^{-1}$ are located immediately below Rim Village, and other springs in the forested southwest wall of the crater only occasionally have nitrate concentrations that approach the lower concentrations found in the springs below Rim Village. Nitrogen-fixing alder occur on all springs studied, and it is unlikely that this natural source of nitrogen would account for the differences between springs.

We have recently conducted nutrient uptake studies in streams in the McKenzie River drainage to the north of Crater Lake, and nitrate was released in habitats where there was rapid depletion of ammonium (particularly in lateral depositional areas or depositional areas associated with debris dams), indicating a high potential for microbial nitrification. We also measured nitrate concentrations in the range of 100-200 $\mu g\ NO_3\text{-}N \cdot l^{-1}$ in old-growth, headwater streams in the Bull Run watershed on the north flanks of Mt. Hood (unpublished data, Bruce McCammon, U.S. Forest Service). After clearcutting, nitrate output from watersheds is often elevated, starting one to three years after harvest and continuing for approximately five years (Fredriksen 1975; Likens *et al.* 1977). This response has been attributed to increased nitrification. When a watershed is disturbed, the supply of ammonium for autotrophic nitrifiers increases and vegetative uptake of nitrate decreases because of the lowered demand by terrestrial plants (Vitousek *et al.* 1982). Patterns of nitrate concentrations found in the springs of the caldera of Crater Lake might be related to similar phenomena. On the more mesic forests on the southwest wall of the caldera, greater soil moisture and availability of organic matter would enhance decomposition, providing a more abundant source of ammonium to nitrifiers. The sparse vegetation associated with the drier sites on the north wall of the caldera produce little organic matter for soil decomposers, limiting the supply of ammonium for nitrifiers; therefore low concentrations of nitrate might be observed in the springs. The extended duration of the snowpack and north aspect contribute to higher soil moisture through the summer in the watersheds on the south rim of the caldera. The higher soil moisture and greater production of organic matter in the forests within the south rim may create more favorable conditions for nitrification.

If it is assumed that the differences in the chemistry and biota of the caldera springs are not explained by natural factors, human activity in the Park might account for observed responses. Park Headquarters and residences have been located in the Munson Valley for more than 55 years, and addition of nutrients to the basin is inevitable. Sewage facilities are required for the public and park staff, but wastes from these facilities potentially can enter groundwater and contribute to the gradual eutrophication of Crater Lake. In 1975, a sewer line was obstructed, and sewage flowed out a manhole and into Munson Springs, the water supply for the visitor facilities and Park Headquarters, resulting in an outbreak of more than 1000 cases of diarrhea (Craun 1981). In 1986, sewer pipes below the Park Headquarters froze, ruptured, and delivered raw sewage into Munson Creek. Recent examinations of the sewage pipes with remote cameras indicated several breaks and leaks in the pipes, therefore release of sewage to the groundwater is certain, though the amount of sewage delivery is unknown. Such events and conditions present obvious human health hazards, but they also demonstrate the potential for chronic contamination of the ecosystem by man's use of the National Park.

If contamination of groundwater by sewage was substantial, elevated concentrations of nitrate would be expected. If groundwater within the caldera rim remains aerobic, nitrate from a sewage source would be the dominant form of inorganic nitrogen in transport because of the greater mobility of this species of nitrogen ions. If microbial activity makes groundwaters anaerobic, ammonium would be the dominant form of nitrogen in transport, being converted to nitrate in shallow, aerobic subsurface soils near the spring sources. No other spring around the lake exhibits nitrate concentrations as high as those of Spring 42. None of the other forms of nitrogen, either ammonium or organic nitrogen, are elevated in Spring 42, and organic nitrogen is usually undetectable, a pattern that would be consistent with microbial nitrification. It is important to note that concentrations of both organic and inorganic phosphorus in Spring 42 resemble those of other springs. If human wastes were contaminating groundwater sources for Spring 42, concentrations of other nutrients would be expected to change as well.

Seasonal changes in nitrate concentrations have been discussed as possible indications of the influence of human sewage because visitor activity also increases from early summer through fall, and the cumulative effect of visitor use would be greatest in late summer. This mechanism can be examined by comparison of nitrate concentration patterns in different springs and comparison with chemical species that would not be expected to change seasonally due to biological activity. Nitrate concentrations are lowest during spring and early summer regardless of the degree of human activity above the spring source. In addition, the dominant cation, calcium, exhibits seasonal patterns that are almost identical to those of nitrate. These observations indicate that hydrologic dilution during snowmelt is the most likely mechanism responsible for these seasonal trends in spring chemistry.

This study does not prove that human activity in Crater Lake National Park is responsible for differences in water chemistry and aquatic biota of springs below Rim Village, but it does identify a critical resources issue for the management of Crater Lake National Park. The water chemistry and aquatic productivity of streams are elevated in the area of greatest human activity, but it is not the productivity of these streams that is of primary concern. If these ecological patterns are linked to human activity in these areas, nutrient loading into Crater Lake could lead to the gradual eutrophication of this unique ecosystem.

Several opportunities for research and management have been identified in this study of water chemistry and biological communities of the caldera springs. The most immediate, short-term need is better understanding of the dynamics of nitrogen and groundwater associated with the springs of the caldera. In the longer term, a more thorough baseline study of water chemistry in the surface waters of Crater Lake National Park will contribute to the evaluation of their current status and provide an invaluable reference for future management of the Park. Rather than solely responding to immediate environmental concerns, resource managers must establish a broad base of ecological information and research for effective management of Crater Lake's unique ecosystems in the future.

LITERATURE CITED

Bott, T. L., J. T. Brock, C. E. Cushing, S. V. Gregory, D. King, and R. C. Petersen. 1978. A comparison of methods for measuring primary productivity and community respiration in streams. Hydrobiologia 60:3-12.

Craun, G. F. 1981. Outbreaks of waterborne disease in the United States 1971-1978. Jour. AWWA 73(7):360-369.

Cummins, K. W., and R. W. Merritt. 1984. Ecology and distribution of aquatic insects. Pages 59-75 *in* W. R. Merritt and K.W. Cummins, eds., An Introduction to the Aquatic Insects of North America. Kendall/Hunt Publishing Co., Dubuque, Iowa.

Diller, J. S., and H. B. Patton. 1902. The geology and petrography of Crater Lake National Park. U.S. Geol. Surv. Prof. Pap. 3. 164 pp.

Fredriksen, R. L. 1975. Nitrogen, phosphorus, and particulate matter budgets of five coniferous forest ecosystems in the Western Cascades Range, Oregon. Ph.D. Thesis. Oregon State Univ., Corvallis, Ore.

Likens, G. E., F. H. Bormann, R. S. Pierce, J. S. Eaton, and N.M. Johnson. 1977. Biogeochemistry of a forested ecosystem. Springer-Verlag, New York. 146 pp.

Mohler, W. 1986. Water supply and waste disposal facilities, Crater Lake National Park, February 1986. (internal rep.)

Petersen, R. C., and K. W. Cummins. 1974. Leaf processing in a woodland stream. Freshwater Biol. 4:343-368.

Phillips, K. N., and A. S. Van Denburgh. 1968. Hydrology of Crater, East and Davis Lakes, OR. U. S. Geol. Surv. Water Supply Pap. 1859-E. 68 pp.

Sterns, G. L. 1963. The climate of Crater Lake National Park. Crater Lake Nat. Hist. Assoc. 12 pp.

Strickland, J. D. H., and T. R. Parsons. 1968. A Practical Handbook of Seawater Analysis. Fish. Res. Bd. Can., 311 pp.

Vitousek, P. M., J. R. Gosz, C. C. Grier, J. M. Melillo, and W. A. Reiners. 1982. A comparative analysis of potential nitrification and nitrate mobility in forest ecosystems. Ecol. Monogr. 52:155-177.

CHEMICAL AND ISOTOPIC COMPOSITIONS OF WATERS FROM CRATER LAKE, OREGON, AND NEARBY VICINITY

J. Michael Thompson, Manuel Nathenson, and L. Douglass White
U. S. Geological Survey
345 Middlefield Road
Menlo Park, California 94025

Crater Lake, Oregon, has no surface outlet and loses its inflow by evaporation and leakage. In order to understand the hydrology of the lake and the leakage of the lake in relation to nearby cold springs, water samples were collected for chemical and isotopic analyses. No spring analyzed had evidence of more than 10 percent Crater Lake water. Chemical and isotopic analyses show that Crater Lake is well mixed. Crater Lake also has anomalously high chloride, boron, lithium, sulfate, and silica concentrations compared to nearby Diamond Lake and to cold springs discharging on the flanks of Mount Mazama. This elevated chloride may be caused by input of thermal water. Weight ratios of Cl/Li are within the range of western United States hot springs and significantly below those for surrounding cold spring waters. Estimates of total heat flow out of the lake bottom range from 670 to 1380mW/m^2, also suggesting addition of thermal water to the lake bottom.

Crater Lake, Oregon, is located in the 6800 year-old caldera of Mount Mazama (Bacon 1983). The lake receives 85% of its inflow by direct precipitation with the remainder coming as inflow from the surrounding drainage area. The lake covers 78% of its drainage area. The lake has no surface outlet, but loses 72% of its inflow by leakage and 28% by evaporation (Phillips 1968). Van Denburgh (1968) recognized that chloride and sulfate and perhaps silica and sodium were anomalously high in the lake and suggested that these constituents may be contributed by thermal springs at depth in the lake. Based on unpublished analyses, Van Denburgh also suggested that the lake is quite uniform in chemical composition both areally and vertically.

Our purpose here is to present chemical and isotopic data for Crater Lake and cold springs emanating on the flanks of Mount Mazama in order to understand the lake hydrology and the relationship of the lake water to nearby cold springs. The pertinent questions are: (1) Are the lake chemistry and isotopic compositions anomalous compared to nearby cold springs? (2) How well mixed both chemically and isotopically is the water in Crater Lake? and (3) Do the dissolved chemical constituents in Crater Lake water indicate an input of thermal water? In a companion paper these data are also used to study how weathering produces the observed cold-spring water chemistry (Nathenson and Thompson 1990).

Table 1 contains the complete chemical and isotopic data previously discussed in abstracts by Thompson and White (1983), Salinas *et al.* (1984), and White *et al.* (1985). These abstracts contained only preliminary answers to the questions posed above. Additionally, inconsistencies in some of the previously reported data sets have been identified, and the values have been redetermined and are given in Table 1.

SAMPLING METHODS

Cold-spring waters and lake waters were collected using methods similar to those described in Thompson (1985). Temperatures of springs were determined using a conventional, total immersion, mercury-in-glass thermometer. In 1981 and 1982 field measurements of the spring water pH were made

TABLE 1a. CHEMICAL ANALYSES OF SPRINGS IN THE VICINITY OF MOUNT MAZAMA

Sample Numbers	Name or Locality	Date	l pH	SiO_2	Ca	Mg	Na	K	Li	HCO_3	SO_4	Cl	F	B	Cond. (μS/cm)	Temp. °C	$\delta^{18}O$ ‰	δD ‰
				<------					in mg/L					------>				
	1981 Samples																	
JCL-81-1	Annie Spring	8 Aug 81	7.2	38	2.9	1.0	2.6	0.8	<0.01	15	4	0.4	0.2	0.1	144	4	-13.89	-99.4
JCL-81-2	Diamond Lake, S End	9 Aug 81	7.3	3.6	1.7	1.0	3.2	.8	<.01	30	1	.2	.1	.1	121	22.5	-10.85	-83.2
JCL-81-3	Boundary Springs	9 Aug 81	7.6	34	4.3	2.4	3.3	.5	<.01	25	3	.2	.2	<.1	120	5	-13.76	-98.1
JCL-81-4	Lightning Springs	9 Aug 81	7.1	26	1.7	.3	1.5	.8	<.01	12	1	.2	.1	.1	84	4	-14.23	-96.6
JCL-81-5	Lodgepole Picnic area	9 Aug 81	7.2	36	1.9	.5	2.8	1.5	<.01	17	<0.5	.3	.1	<.1	102	5	--	--
JCL-81-6	Maklaks Spring	10 Aug 81	6.9	24	1.8	.4	1.8	.8	<.01	12	1	.2	.1	<.1	105	10.5	--	--
JCL-81-7	Headwater of Lost Creek	10 Aug 81	7.2	32	1.8	.6	1.9	.4	<.01	22	1	.3	.2	<.1	109	7.5	--	--
JCL-81-8	Vidae Falls	10 Aug 81	7.1	34	2.1	.7	2.0	.8	<.01	16	<0.5	.2	.2	.2	117	9	--	--
JCL-81-9	Thousand Springs	11 Aug 81	7.3	34	4.8	2.5	2.5	1.1	<.01	27	2	.2	.2	<.1	129	5	-13.70	-99.2
JCL-81-10	Source of Wood River	11 Aug 81	7.3	40	5.6	2.7	6.1	1.0	.01	34	5	3.2	.2	.2	132	9.5	-14.87	-107.6
JCL-81-11	Steel Bay, C.L.	13 Aug 81	7.0	26	.4	.3	1.9	.4	<.01	14	2	.3	.2	<.1	105	18	-13.75	-101.7
JCL-81-12	N. 'Pumice Castle' C.L.	13 Aug 81	8.6	36	1.6	.8	2.5	.6	.01	18	<0.5	.3	.2	<.1	102	9	--	--
JCL-81-13	S. 'Pumice Castle' C.L.	13 Aug 81	8.2	40	1.6	1.1	2.7	1.0	<.01	27	1	.4	.2	<.1	110	6.5	-15.45	-110.5
JCL-81-14	'Chaski Slide-E', C.L.	13 Aug 81	7.1	26	4.9	1.7	2.0	.9	<.01	19	12	.1	.2	<.1	125	12	--	--
JCL-81-15	'Chaski Slide-W', C.L.	13 Aug 81	6.2	22	10.1	3.2	3.3	.4	<.01	20	26	.2	.2	<.1	145	9.5	-13.88	-105.2
JCL-81-16	'The Watchman Spring'	13 Aug 81	6.6	34	1.6	.4	2.1	1.0	<.01	16	<0.5	.2	.2	<.1	110	5	--	--
JCL-81-17	Dutton Cliff	13 Aug 81	7.8	36	1.1	.9	4.2	.6	<.01	21	1	.2	.2	<.1	115	14	--	--
JCL-81-24	Spring near C. L. Lodge	17 Aug 81	6.3	30	2.0	.3	1.9	.7	<.01	24	<0.5	.3	.2	.2	107	6	--	--
	1982 Samples																	
JCL-82-1	Cascade Spring	31 Aug 82	7.06	40	2.5	.9	2.9	1.4	<.01	37	<.2	.2	<.1	.4	--	3.5	-15.11	-108.4
JCL-82-2	Cattle Crossing Cafe	1 Sep 82	7.15	40	2.7	2.9	10.8	.7	<.01	63	<.2	.2	.1	1.1	--	COLD	-14.04	-101.1
	1983 Sample																	
JCL-83-1	Crater Spring	7 Aug 83	6.34	35	3.0	1.1	3.0	1.6	<.01	32	<2	.8	.1	<.1	--	3.0	-13.56	-97.4
	1984 Samples																	
JCL-84-1	Annie Spring	3 Aug 84	5.39	40	2.0	1.4	3.0	2.2	<.01	30	<1	1.2	.7	.2	44	3	-13.9	-99.5
JCL-84-2	Tecumseh Spring	3 Aug 84	7.88	34	7.5	1.8	12.5	1.4	<.01	58	3.4	4.9	.2	.2	95	11	-14.7	-106.8
JCL-84-3	Source of Crooked Crk	3 Aug 84	7.90	36	8.0	2.4	15.6	1.9	<.01	53	6.2	8.4	.2	.4	126	11	-14.7	-108.0
JCL-84-4	Source of Wood River	3 Aug 84	6.74	46	2.1	2.4	6.6	1.9	<.01	47	1.8	2.8	.1	.2	50	12	-15.1	-105.5
JCL-84-5	Reservation Spring	3 Aug 84	7.58	40	15.7	1.9	10.8	2.1	.01	50	4.6	5.8	.1	.1	103	8	-14.6	-106
JCL-84-6	Source of Spring Crk	3 Aug 84	7.51	41	3.2	1.8	8.5	1.3	<.01	46	2.4	3.3	.1	.1	60	6	-14.3	-105
JCL-84-7	Annie Creek at boundary	3 Aug 84	N.R.	40	6.4	1.1	3.4	1.6	<.01	32	1.8	.5	<.1	.2	50	10	-14.2	-98
JCL-84-8	Pothole Spring	4 Aug 84	6.68	43	2.7	0.90	2.9	1.7	.01	29	0.2	.5	<.1	.1	30	3	-15.1	-110
JCL-84-9	Unnamed spring nr road	4 Aug 84	6.79	45	3.0	1.3	3.6	2.1	<.01	34	0.1	.5	<.1	.1	47	4	-15.2	-108
JCL-84-10	Unnamed spring, source of Crk 1/4 mi S of Scott Crk	4 Aug 84	6.94	31	17.8	0.53	2.4	1.4	<.01	21	0.4	.5	<.1	.2	29	6	-15.0	-103
JCL-84-11	Unnamed spring on Minnehaha Creek nr Soda spring	4 Aug 84	N.R.	99	23.7	5.1	89	9.7	.03	417	0.8	4.2	0.1	.2	320	10	-13.2	-90
JCL-84-12	Mare's Egg Spring	5 Aug 84	7.70	35	10.8	9.0	4.2	1.5	<.01	55	0.3	.5	.1	.2	77	4	-14.4	-101
JCL-84-13	Four-mile Spring	5 Aug 84	7.96	32	6.0	2.4	4.4	1.5	<.01	54	0.5	1.4	<.1	.2	74	5	-14.1	-98
JCL-84-14	Ranger Spring	5 Aug 84	N.R.	39	14.2	1.0	3.0	2.0	<.01	34	<0.1	.9	<.1	.2	47	2	-13.6	-95
JCL-84-15	Cedar Springs	5 Aug 84	6.37	39	11.6	1.2	3.4	1.6	<.01	44	<0.1	.5	<.1	<.1	60	7	-13.4	-101
JCL-84-16	Geyser Spring	6 Aug 84	N.R.	31	7.5	2.7	3.3	1.2	<.01	57	0.1	.5	<.1	.2	75	5	-12.9	-91
	1985 Sample																	
JCL-85-7	Soda Spg on Minneh. Cr	6 Aug 85	5.31	71	271	243	106	31.5	0.06	2280	16.	17.7	<.1	0.4	3620	10	-14.3	-102

TABLE 1b. CHEMICAL ANALYSES OF CRATER LAKE WATERS

Sample Numbers	Name or Locality	Date	l pH	SiO_2	Ca	Mg	Na	K	Li	HCO_3	SO_4	Cl	F	B	Cond. (µS/cm)	Temp. °C	$\delta^{18}O$ ‰	δD ‰
				<------ in mg/L										------>				
1981 Samples																		
JCL-81-18	E Basin, surface	14 Aug 81	6.0	18.2	6.4	2.8	9.1	1.2	0.04	30	10	9.6	0.2	0.2	155	18	-9.40	-79.4
JCL-81-19	E Basin, 579 m	14 Aug 81	7.2	17.6	7.4	2.9	9.2	1.3	.04	30	8	9.6	.2	.3	154	--	-9.59	-79.6
JCL-81-20	SW Basin, 448 m	14 Aug 81	8.6	17.8	7.8	2.8	9.5	1.1	.04	34	8	9.9	.2	.3	156	--	-9.55	-78.2
JCL-81-21	SW Basin, 489 m	15 Aug 81	7.9	18.2	7.4	3.0	9.7	1.2	.04	30	7	9.6	.2	.2	157	--	-9.53	-79.9
JCL-81-22	SW Basin, 448 m	15 Aug 81	7.4	17.4	7.5	2.7	9.6	1.2	.04	24	8	9.4	.2	.2	140	--	-9.67	-79.9
JCL-81-23	SW Basin, 468 m	16 Aug 81	7.0	19.6	7.6	2.8	9.9	1.7	.04	33	5	9.4	.2	.2	155	--	-9.49	-79.6
1983 Samples																		
JCL-83-2	SW Basin, surface	8 Aug 83	7.24	20.5	6.5	2.4	9.4	1.6	.03	39	10	10.0	.1	.4	--	14.5	-9.84	-78.4
JCL-83-3	SW Basin, 50 m	8 Aug 83	7.77	18.8	6.7	2.4	9.2	1.6	.03	42	10	10.1	.1	.4	--	11	-9.68	-79.1
JCL-83-4	SW Basin, 100 m	8 Aug 83	7.73	21.5	7.1	2.3	9.3	1.6	.03	45	10	10.1	.1	.5	--	9	-9.74	-79.3
JCL-83-5	SW Basin, 150 m	8 Aug 83	7.65	19.1	6.8	2.3	9.2	1.6	.03	42	10	10.1	.1	.4	--	10	-9.62	-78.9
JCL-83-6	SW Basin, 200 m	8 Aug 83	7.60	19.8	7.2	2.6	9.2	1.6	.03	45	10	9.8	.1	.4	--	10	-9.86	-78.7
JCL-83-7	SW Basin, 250 m	8 Aug 83	7.57	24.4?	9.4	2.6	9.2	1.5	.03	45	10	10.1	.1	.6	--	7	-9.67	-77.3
JCL-83-8	SW Basin, 300 m	8 Aug 83	7.66	20.0	7.5	2.6	9.0	1.6	.03	31	10	9.8	.1	.6	--	7	-9.76	-78.6
JCL-83-15	E Basin, surface	8 Aug 83	7.55	19.3	5.5	3.7	9.3	1.7	.04	47	10	10.4	.1	.6	--	16	-9.74	-78.2
JCL-83-11	E Basin, 50 m	8 Aug 83	7.82	17.8	5.6	3.7	9.3	1.8	.03	37	10	10.3	.1	.4	--	9	-9.69	-78.3
JCL-83-12	E Basin, 100 m	8 Aug 83	7.77	18.5	6.0	3.7	9.2	1.7	.03	32	10	10.1	.1	.4	--	8	-9.65	-78.3
JCL-83-9	E Basin, 150 m	8 Aug 83	7.64	18.3	7.0	2.6	9.2	1.8	.03	39	10	9.9	.1	.5	--	8	-9.71	-77.6
JCL-83-10	E Basin, 200 m	8 Aug 83	7.82	17.9	7.1	2.5	9.1	1.6	.04	47	10	10.2	.1	.5	--	8	-9.76	-78.0
JCL-83-13	E Basin, 250 m	8 Aug 83	7.68	18.5	5.9	3.8	9.3	1.7	.03	39	10	10.2	.1	.4	--	8	-9.73	-77.9
JCL-83-14	E Basin, 300 m	8 Aug 83	7.57	20.3	6.4	3.8	9.3	1.6	.04	42	10	10.0	.1	.3	--	8	-9.73	-77.9
1984 Samples																		
JCL-84-17	SW Basin, surface	7 Aug 84	7.12	18.8	5.8	2.3	10.4	1.7	.05	41	8	10.0	.1	.4	--	--	-9.6	-79
JCL-84-22	SW Basin, 50 m	7 Aug 84	N.R.	19.6	6.5	2.4	10.2	1.5	.05	38	8	10.0	.1	.5	--	--	-9.8	-79
JCL-84-21	SW Basin, 200 m	7 Aug 84	7.01	18.5	6.6	2.2	10.9	1.7	.05	41	8	9.8	.1	.4	--	--	-9.8	-79
JCL-84-20	SW Basin, 300 m	7 Aug 84	6.66	18.9	5.5	2.2	10.2	1.5	.05	42	8	10.0	.1	.5	--	--	-9.7	-79
JCL-84-19	SW Basin, 400 m	7 Aug 84	6.65	19.2	10.1	2.4	10.4	1.7	.05	42	8	10.1	.1	.6	--	--	-10.0	-80
JCL-84-18	SW Basin, 500 m	7 Aug 84	6.72	19.7	13.8	2.6	10.6	1.8	.05	42	8	10.6	.1	.5	--	--	-9.8	-79

with non-bleeding, low-ionic-strength, pH-indicating dyes (E. M. Colorphast pH strips[1]). Beginning in 1983 all field pH measurements were made with a gel-filled pH electrode and a portable pH meter. Temperature and pH were generally determined at each spring site. The alkalinity of the 1984 and the few 1985 samples was also determined in the field. At each spring site a filtered, unacidified water sample for anion analysis was collected by passing the water through a 0.45μm membrane filter. Additionally, a filtered, acidified sample for cation analysis was collected by adding concentrated, Baker[1] trace-metal quality HCl to the filtered water. An untreated sample for deuterium and oxygen-18 analysis was also collected in a glass bottle at each site. The methods of chemical and isotopic analyses employed are described in Thompson, White, and Nathenson (1987).

Samples of lake water were collected in 2-liter Van Dorn sample bottles attached to a metal cable and retrieved either by hand (1981 samples) or mechanically (all others). The depth of sampling in 1983 was limited to 300 m by the available cable; in 1984 the hand winch and cable were replaced with a mechanically driven one, permitting retrieval of samples from the lake bottom. Samples were collected and treated as described above: one bottle for anion analysis, another for cation analysis and a third for isotopic analysis. In 1985 a one-liter raw water-sample was collected from each spring site and lake point and evaporated at about 90°C to approximately 50 mL for B and Li analysis (Thompson *et al.* 1987).

COLD-SPRING WATERS

Water samples of cold springs were collected in the Crater Lake area from 1981 through 1985 at the locations shown in Fig. 1. An effort was made to sample all large discharging springs. Major-ion concentrations and water isotopes for cold-spring waters are reported in Table 1a and for lake waters in Table 1b.

The cold spring waters discharging on the flanks on Mount Mazama have low total dissolved solids and are essentially a sodium-calcium-magnesium bicarbonate water (Table 1a). Generally, the waters are neutral to slightly alkaline. The waters contain much less than 10 mg/L of dissolved sulfate and chloride and substantially less than 1 mg/L of dissolved boron and lithium. Fluoride is slightly above the detection limit of 0.1 mg/L. Dissolved silica in the waters is higher than would be expected for quartz solubility control. Aluminosilicates and glassy volcanic rocks appear to be the source of the SiO_2 in the cold-spring waters (Nathenson and Thompson 1990).

In general, spring waters discharging above the surface elevation of Crater Lake are remarkably similar to those discharging below it on the outer flanks of the volcano. Few chemical differences exist between intracaldera spring water and extracaldera spring water. However, in the vicinity of Chaski Slide, a large piece of hydrothermally altered volcanic rock that slid sometime after the climactic eruption, water passing over the slide material is relatively enriched in calcium and sulfate (samples JCL 81-14 and 81-15 on Fig. 1 and Table 1a). Not all waters discharging from the caldera walls were analyzed for deuterium and oxygen-18 because samples were not collected at the spring orifice and the extent of evaporation was unknown.

Springs above the lake all have chloride concentrations less than 0.5 mg/L (Fig. 2) whereas the lake has chloride concentration around 10 mg/L. The chloride concentration in the cold springs is similar to that measured in precipitation in western Oregon (Junge and Werby 1958), so that the chloride in the cold springs appears to be that which was contained in the precipitation. Assuming 28 to 33 percent evaporation (Phillips 1968; Simpson 1970a), in a steady-state lake, evaporation can increase the chloride content of the inflowing water by no more than 50 percent. Thus the chloride concentration in the lake is anomalously high compared to that in the available water supply.

Both Crater Lake and Diamond Lake, a lake about 20 km north of Crater Lake and 300 m lower in elevation, show the effect of evaporation on their isotopic ratios, and they have distinctly different values from cold-spring samples both above and below the surface elevation of Crater Lake (Fig. 3). The evaporation trend, as determined by Craig (1961), has an empirical slope of five on a δD-$\delta^{18}O$ plot. It is possible to use values for the isotopes to calculate this evaporation trend (e.g., Gonfiantini 1986); however, there are a number of other param-

[1]Brand names are for information purposes only and do not constitute a recommendation by the U.S. Geological Survey.

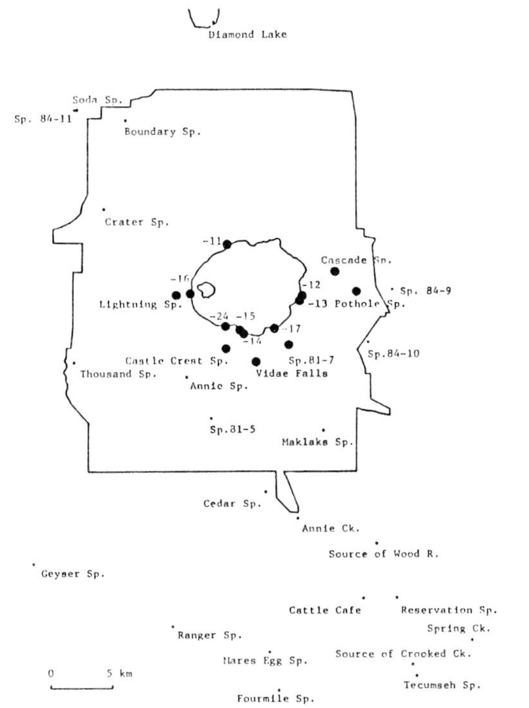

Figure 1. Locations of spring samples in the Crater Lake area. Large dots show spring locations that are higher in elevation than the surface of Crater Lake. Outline is the boundary of Crater Lake National Park. Numbers around Crater Lake are last two digits of sample number in the JCL 81- series in Table 1a.

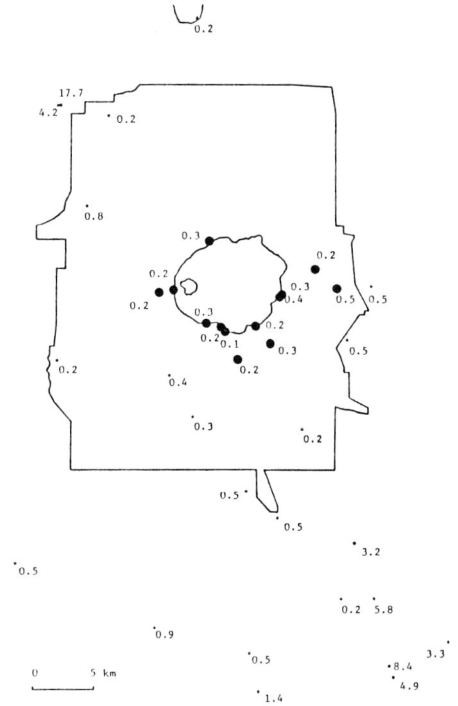

Figure 2. Chloride concentrations (mg/L) in spring waters from the Crater Lake area.

eters that have not been measured that are required to perform such a calculation (see for instance, IAEA 1979). Mixing between Crater Lake water and meteoric waters would be along a straight line in this diagram. The intersection of the meteoric water line (MWL) and the evaporation line is near the value for Annie Spring.

When deuterium is compared to chloride for Crater Lake and for the cold-spring samples (Fig. 4), a few samples show elevated chloride, but the combination of possible mixing ratios from Figs. 3 and 4 permits some of these to be ruled out as containing a significant fraction of Crater Lake water. Based solely on Cl and δD, the unnamed spring on Minnehaha Creek (Spring 84-11 on Fig. 1, with Cl=4.2 mg/L, δD=–90‰) could be a mixture of 40% normal spring water and 60% Crater Lake water (Fig. 4). However, its plotted value in Fig. 3 is near the MWL, indicating that it has no more than a small percentage of Crater Lake water. Also, the high Cl and HCO_3 in this spring is probably related to that in the nearby Soda Spring. The six data points below the lines on Fig. 4 have similar deuterium values with varying amounts of chloride; they cluster together on Fig. 3 with similar values of deuterium and oxygen-18 isotopes along the MWL. This additional Cl may be derived from sediments in the Upper Klamath Lake Basin. Thus, based on the isotopes and chloride, there is no single cold spring that contains more than 10% Crater Lake water.

CRATER LAKE WATER

Crater Lake can be characterized as being a low total-dissolved-solids, sodium-calcium-magnesium bicarbonate-chloride-sulfate water containing less than 1.0 mg/L boron and less than 0.1 mg/L lithium. Using the 1981 east basin samples (Table 1b) as an example, the concentrations of SiO_2, Mg, Na, K, HCO_3, and Cl and conductivity are almost identical between surface and bottom waters, and the concentrations of Ca and SO_4 are approximately the same. The concentrations of any of these constituents usually do not vary by more than the error of the

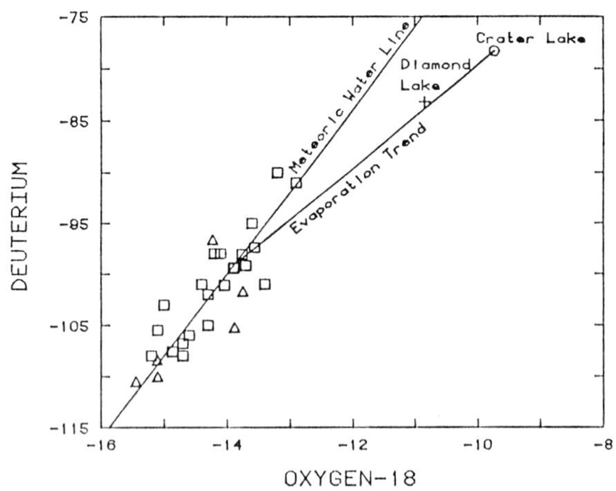

Figure 3. Values for deuterium versus oxygen-18 isotopes (‰) for Crater Lake (average of 1983 values), Diamond Lake and nearby cold springs. Springs above the elevation of the surface of Crater Lake are shown as triangles; springs below the surface elevation are shown as squares. Meteoric water line is $\delta D = 8\delta^{18}O + 12$ ‰ which has a slightly different intercept than the meteoric water line of Craig (1961). The evaporation trend line has a slope of 5, based on the results for other lakes (Craig 1961).

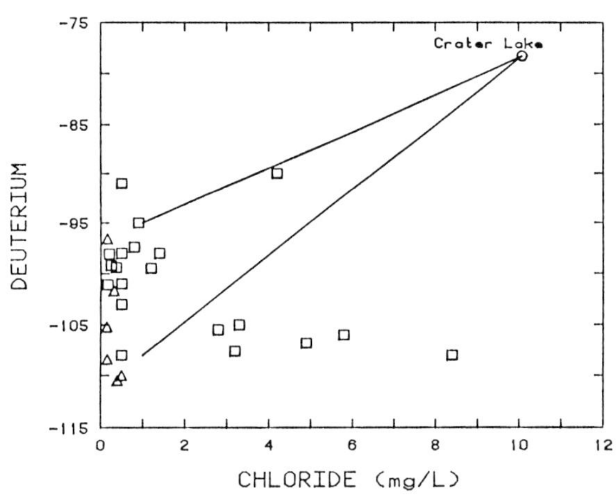

Figure 4. Deuterium (‰) versus chloride concentration (mg/L) for Crater Lake (average of 1983 values) and nearby cold springs. Springs above the elevation of the surface of Crater Lake shown as triangles; springs below the surface elevation shown as squares. Lines shown bound the mixing zone of Crater Lake water with the available range of deuterium isotopes in cold-spring waters.

determination, which generally does not exceed 10%.

Some of the constituents in Table 1b appear to show significant differences either between years or as a function of depth in a single year's sampling. However, these differences are not confirmed by the other constituents. For example, the 1984 data show that Ca concentrations at 300 m and above are about 6 mg/L, but values at 400 and 500 m are 10.1 and 13.8 mg/L, respectively. The other cations and bicarbonate do not change significantly with depth, and this lack of variation indicates the high values for Ca must be an artifact of the sampling, preservations, and/or analytical procedures.

Variations in earlier results for Cl concentrations between years were shown by Thompson, White, and Nathenson (1987) to be caused by the use of different analytical procedures. A study of Cl methods showed that the automated $AgNO_3$ procedure gave the best results. Considering the period of time over which the analytical work has been done and the low levels of constituents in Table 1b, no real differences exist.

Figure 5 shows isotopic and chemical data as a function of depth. The reported isotopic values do not vary by more than 2 standard deviations on replicate samples. These data indicate that Crater Lake is well mixed. Some variation from year to year, most likely analytical error, is seen in Fig. 5.

Silica analyses of Crater Lake bottom and surface water samples in 1981, point samples collected from the east basin and the southwest basin at 50 m intervals to 300 m depth in 1983, and point samples collected from the southwest basin at about 100 m intervals to the bottom in 1984 all indicate Crater Lake is well mixed with respect to silica (Fig. 5). Published SiO_2 values of Salinas *et al.* (1984) for 1983 samples contained lower SiO_2 concentrations than reported by Larson (1984, Table 3) for his 1983 samples and our 1984 samples. For that reason the 1983 lake water samples were reanalyzed for SiO_2. The corrected values are reported in Table 1b; however, they may be questionable because the redeterminations were made in 1987 after 4 years of storage. The source of this error in the original values is currently unknown and tentatively is attributed to analyst error. However, the chlorophyll and dissolved oxygen values reported by Salinas *et al.* (1984) are correct. The corrected SiO_2 values for the 1983 samples are similar to those reported by Larson (1984) and the 1984 samples of the lake water (Table 1b).

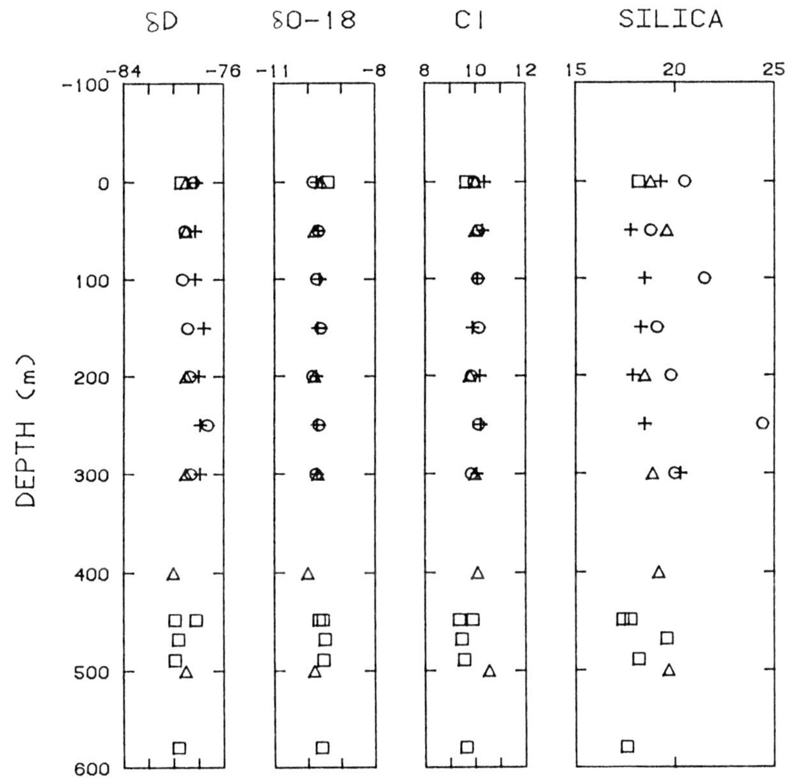

Figure 5. Isotope and chemical data for Crater Lake. Squares (1981), plus symbols (1983 east basin), circles (1983 southwest basin), and triangles (1984 southwest basin). Isotopic data are in ‰ and chemical data are in mg/L.

The high degree of mixing of Crater Lake, as shown by major-ion chemistry and light stable isotopes, is supported by the tritium data of Simpson (1970a and b) (Fig. 6). Except for the near-surface samples, the tritium content of the deeper lake water is constant at 24 TU from 50 m to total depth. In the seven months previous to the date of sampling the lake, the tritium concentration of atmospheric precipitation averaged 171 TU (Simpson 1970a). It seems likely that recent precipitation was the source of the peak concentration of 31 TU (Fig. 6). Assuming that precipitation having 171 TU was added to the lake (already at a tritium concentration of 24 TU) to produce the peak concentration of 31 TU, the recent precipitation would have been diluted by 20 volumes of low tritium-containing Crater Lake water. This amount of dilution cannot be detected by chemical or stable isotopic procedures that we report because the techniques are not sufficiently sensitive to detect it. For example, one part precipitation having a chloride concentration of 0.2 mg/L added

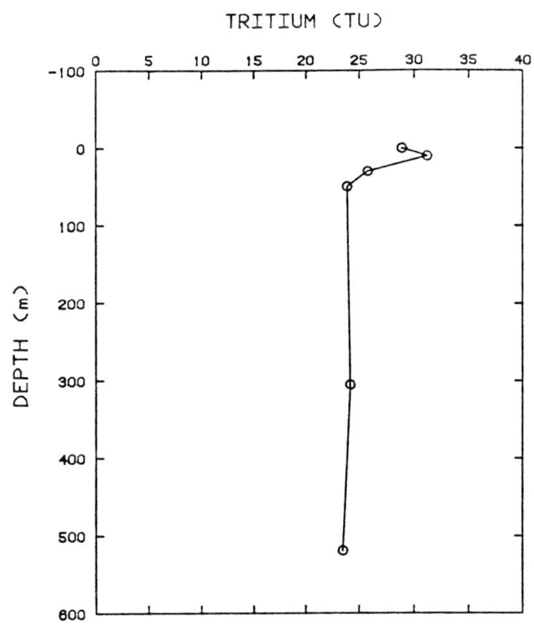

Figure 6. Tritium profile for Crater Lake obtained in 1967 by Simpson (1970a).

to 20 parts of Crater Lake water containing 10 mg/L chloride would have a resulting concentration of 9.5 mg/L. This small difference is well within the analytical uncertainty of the chloride determinations reported here (Thompson *et al.* 1987).

Comparing Crater Lake water to nearby cold-spring waters and to Diamond Lake water, a lake about 20 km north of Crater Lake at an elevation of 1580 m (USGS 1956), Crater Lake has higher Cl, SO_4, Ca, Mg, Na, Li, and B concentrations (Tables 1a and 1b). Crater Lake, at a higher elevation of 1882 m (USGS 1956), should contain either less chloride than a typical lower elevation lake and nearby cold springs or a similar chloride, but not more. Both lakes are significantly lower in dissolved SiO_2 than surrounding cold-spring waters. For Diamond Lake, which is quite a biologically productive lake, this is probably a result of diatom metabolism. For Crater Lake, diatoms also consume silica, but its silica concentration is actually anomalously high. The inflow of spring and ground water measured by Phillips (1968) is 15% of the total inflow to Crater Lake. Precipitation carries negligible silica whereas the cold springs above Crater Lake contain about 35 mg/L. Using the assumptions in Phillips (1968), the inflow from the cold springs would yield a silica concentration of only 7 mg/L in Crater Lake whereas the measured concentration is 18 mg/L. This extra silica must be provided by the same inflow that supplies the added chloride, sodium, and other constituents calculated by Nathenson (1990).

The anomalous constituents in Crater Lake led Van Denburgh (1968) to conclude that Cl and SO_4, and perhaps SiO_2 and Na, "may have been contributed to the lake by thermal springs or fumaroles . . . ." This suggestion, which is supported by the heat flow data of Williams and Von Herzen (1983), has caused much controversy. However, if the interpretation of Williams and Von Herzen's heat flow data is correct, it provides a mechanism for the relatively uniform chemical and isotopic composition of the lake, namely Rayleigh convection. They calculated a Rayleigh number of 6.3×10^{14}, whereas 1.0×10^{3} is sufficient to initiate convection.

The chemical composition of Crater Lake water appears to have stayed relatively constant since it was first analyzed by N. M. Finkbiner in 1912 (Van Winkle and Finkbiner 1913). Nathenson (1990), using a statistical analyses of the historical Crater Lake chemical data set, concludes that the lake water has been relatively constant between 1912 and 1986.

CHEMICAL EVIDENCE OF THERMAL COMPONENTS IN CRATER LAKE WATER

Elevated concentrations of boron and lithium are typically found in thermal waters of volcanic origin (e.g., White *et al.* 1976; Ellis and Mahon 1977, p. 58-116). Because Crater Lake water is enriched in boron and lithium compared to local meteoric water and because the concentrations of boron and lithium are either at or below the detection limit for these dissolved constituents in the cold-spring waters, we evaporated water from 8 cold-spring and 2 lake water profiles, collected at 100 m intervals, from 1 liter to approximately 50 mL. This reduced volume was then analyzed for boron and lithium. The concentration of lithium and boron were then significantly above the detection limits and are reported in Table 2.

The concentrations of boron in Crater Lake water (Tables 1 and 2) is at least twice that of the water from nearby cold springs and the lithium concentration is at least 10 times that of the cold springs. If the chloride, boron, and lithium are derived from a thermal source, then the Cl/B and Cl/Li weight ratios would be expected to be similar to ratios from known hot springs in volcanic areas (Table 2). Unfortunately, the cold water Cl/B ratios range from 6 to 33 and the lake water ratios range from 17 to 31. This overlap invalidates the use of the Cl/B ratio for identifying thermal components in the lake waters.

The Cl/Li ratio appears to be more diagnostic. In cold-spring waters the Cl/Li ranges from 540 to 4600, and in Crater Lake the ratio ranges from 220 to 280 (mean=242, std. dev.=20)(Table 2). The Cl/Li weight ratio is substantially lower in lake water than in cold-spring water. Typical Cl/Li weight ratios for thermal waters from other volcanic areas range from 80 to 410 (mean=246, std. dev.=115)(Table 2). The Crater Lake Cl/Li weight ratios are near the mean Cl/Li ratios for a variety of hot-spring waters from volcanic settings in the western United States. This also suggests that the additional chloride may be contributed by a thermal water.

The other anionic indicators of thermal waters are SO_4 and HCO_3. Sulfate can arise from biogenic oxidation of sulfur and sulfides (Schoen 1969;

TABLE 2. ANALYSES OF B AND Li IN PARTIALLY EVAPORATED SAMPLES OF LAKE WATER COLLECTED IN 1985, RECALCUALTED TO ORIGINAL COMPOSITIONS, AND COMPARED TO VALUES FOR OTHER WESTERN U. S. HOT SPRINGS

Source		Temp °C	Cl	B (mg/L)	Li	Cl/B weight ratio	Cl/Li
		Cold Spring Waters					
Mare's Egg Spring		5	1.1	0.18	0.0016	6.1	690
Fourmile Spring		12	0.9	.10	.0006	9.0	560
Tecumseh Spring		9.5	4.4	.21	.0029	21	1500
Crooked Creek		10	11.1	.34	.0044	33	4600
Wood River		7	8.4	.37	.0130	23	650
Reservation Spring		8	6.4	.37	.0079	17	810
Castle Crest Spring		3	0.8	.14	.0008	6	1000
Annie Spring		2.5	1.2	.04	.0022	30	550
		Crater Lake Waters					
Crater Lake, E Basin,	surface	—	9.5	.46	.041	21	230
	100 m	—	10.0	.49	.042	20	230
	200 m	—	11.4	.46	.041	25	280
	300 m	—	9.9	.58	.043	17	230
	400 m	—	10.0	.54	.043	18	230
	500 m	—	9.7	.51	.043	19	230
	590 m	—	10.3	.45	.043	23	240
Crater Lake, SW Basin,	surface	—	9.5	.42	.037	23	260
	100 m	—	9.8	.42	.043	23	230
	200 m	—	9.8	.42	.043	23	230
	300 m	—	9.9	.59	.045	17	220
	400 m	—	11.1	.36	.040	31	280
	500 m	—	11.6	.53	.046	22	260
		Typical Thermal Waters					
Growler Hot Springs, Lassen N.F.[1]		95	2430	71	7.7	34	320
Loowit Hot Springs, Mt. St. Helens[2]		84	395	2.0	.97	197	410
Geyser Spring, Seigler Hot Springs[3]		43	294	15	1.6	20	180
Long Valley, unnamed[4]		60	250	13	2.5	19	100
Ear Spring, Yellowstrone[5]		93	414	4.2	5.1	99	81
Gamma Hot Springs, Mt. Baker[6]		65	755	9.0	2.8	84	270
Ohanapecosh Hot Spring, Mt. Rainier[6]		48	880	12	2.9	73	300
Baker Hot Springs, Mt. Baker[6]		44	110	2.7	.36	41	310

[1] Thompson, 1985
[2] unpublished, data of Thompson
[3] Thompson, Goff and Donnelly-Nolan, 1981
[4] Mariner and Willey, 1976
[5] Thompson and Yadav, 1979
[6] Mariner, Presser and Evans, 1982

Schoen and Rye 1970; Brock and Mosser 1975). Atmospheric CO_2 can also dissolve in the lake. We do not have the requisite isotopic data to determine what fraction of HCO_3 and SO_4 could be contributed by this deep thermal fluid.

Crater Lake is a near neutral (pH~7.5) sodium chloride-sulfate lake. This observation negates the possibility that acidic fumarolic gases such as HCl and H_2S are being discharged into the lake bottom. If HCl were being added to the lake, then the ionization of the HCl would make the lake acidic (pH<7) and elevated in Cl (Simpson 1970b). The oxidation of H_2S, which generates sulfuric acid, also would tend to make the lake acidic and elevated in SO_4. Thus, Na^+ and Cl^- appear to enter the lake together, probably dissolved in water. NaCl is not transported in a low temperature (t<150°C), low pressure (P<15 bars) gas. Additionally, the excess SiO_2 discussed earlier suggests transport of SiO_2 in water because little SiO_2 is transported in a vapor phase.

The application of chemical geothermometry to the composition of Crater Lake water is inappropriate to estimate thermal water temperatures because the effects of dilution, addition of constituents from springs on the crater wall and consumption of silica by diatoms is not considered. Nathenson (1990) uses a chemical balance for the lake to calculate what the additional load of each major constituent must be to obtain the current composition of the lake. The composition of this inflow is calculated for various rates of flow. If the flow is low and total dissolved solids are high, the calculated temperature is near 240°C. If the flow is high and the total dissolved solids are low, the temperature is near 60°C. His assumptions are bounded by Williams and Von Herzen's' heat flow data and the bulk lake water chemistry.

CONCLUSIONS

Compared to nearby cold springs and Diamond Lake to the north, Crater Lake has anomalously high concentrations of dissolved Na, Li, Cl, SO_4, and B. Additionally, the δD and $\delta^{18}O$ values for the lake water are significantly higher (heavier) than for cold-spring waters. The isotopic difference between lake water and cold-spring water is caused by evaporation. Because the intersection of the MWL and the evaporation line is near -99 deuterium and -13.9 for oxygen-18, the source of water for Crater Lake is similar to for Annie Spring. Diamond Lake water also plots along the evaporation line. The chemical enrichments in Crater Lake, however, cannot be explained solely by evaporation.

Crater Lake appears to be well mixed based on chemical and isotopic analyses. The concentrations of SiO_2, Cl, Na, Li, SO_4, and B do not vary significantly as a function of depth. The δD and $\delta^{18}O$ values are remarkably uniform throughout the lake water. Tritium data indicate that recent precipitation rapidly mixes with lake water in the near surface. The heat flow values reported by Williams and Von Herzen (1983) are sufficient to cause small density gradients that allow the lake to convect to the deepest levels. This Rayleigh convection appears to mix the lake water thoroughly over a 1-year period because no major-ion chemical gradients are found in Crater Lake (Volchok *et al.* 1970).

Thermal water generally contains moderate to high concentrations of dissolved boron, chloride, and lithium. Crater Lake also appears to have an anomalously high Li concentration compared to other waters in this area. As observed in Table 1a and Table 2, some nearby cold springs have somewhat elevated chloride concentrations and similar Cl/B weight ratios thus negating any meaningful comparison. The mean Cl/Li weight ratio for Crater Lake is 242, which is comparable to thermal waters from volcanic environments, 81-410, and is substantially lower than the lowest cold-spring ratio (550) at Annie Spring. The dissolved sulfate may originate from a thermal source or from dissolution of sulfate-containing minerals.

ACKNOWLEDGMENTS

We thank R. W. Decker, R. O. Fournier, R. H. Mariner and D. K. Nordstrom for helpful and thoughtful reviews. We also thank the U. S. National Park Service for use of their pontoon boat and to John Salinas and Jerry McCrea, the boat operators. Cindy Maley, Rob O'Connor and Seta Simonian assisted with the laboratory analyses.

LITERATURE CITED

Bacon, C. R. 1983. Eruptive history of Mount Mazama and Crater Lake caldera, Cascade Range, U.S.A. Jour. Volcanol. Geotherm. Res.18:57-115.

Brock, T. D., and J. L. Mosser. 1975. Rate of sulfuric-acid production in Yellowstone National Park: Geol. Soc. Amer. Bull. 86:194-198.

Craig, Harmon. 1961. Isotopic variations in meteoric waters. Science 133:1702-1703.

Ellis, A. J., and W. A. J. Mahon. 1977. Chemistry and Geothermal Systems: Academic Press, New York. 392 pp.

Gonfiantini, Roberto. 1986. Environmental isotopes in lake studies. Pages 113-168 *in* P. Fritz and J. Ch. Fontes, eds., Handbook of Environmental Isotope Geochemistry, vol. 2. Elsevier Scientific Publ. Co., Amsterdam, Netherlands.

International Atomic Energy Agency (IAEA). 1979. Isotopes in Lake Studies. Proceedings of an advisory group meeting on the application of nuclear techniques to the study of lake dynamics. Internat'l Atomic Energy Agency, Vienna, 29 Aug.-2 Sept., 1977. 285 pp.

Junge, C. E., and Werby, R. T. 1958. The concentration of chloride, sodium, potassium, calcium, and sulfate in rain water over the United States: Jour. Meteorol. 15:417-425.

Larson, D. W. 1984. Second annual report on the limnology and water quality monitoring program at Crater Lake National Park, Oregon. National Park Service Pacific Northwest Region, Seattle, Wash. 108 pp.

Mariner, R. H., T. S. Presser, and W. C. Evans. 1982. Chemical and isotopic composition of water from thermal and mineral springs of Washington. U. S. Geol. Surv. Open-File Rep. 82-98. 18 pp.

Mariner, R. H., and L. M. Willey. 1976. Geochemistry of thermal waters in Long Valley, Mono County, California. Jour. Geophys. Res. 81:792-800.

Nathenson, M. 1990. Chemical balance for major elements in water in Crater Lake, Oregon Pages 103-114 *in* E. T. Drake *et al.*, Crater Lake: An Ecosystem Study. Pacific Div., Amer. Assn. Advance. Sci., San Francisco, Calif.

Nathenson, M., and J. M. Thompson. 1990. Chemistry of Crater Lake, Oregon, and nearby springs in relation to weathering. Pages 115-126 *in* E. T. Drake *et al.*, Crater Lake: An Ecosystem Study. Pacific Div., Amer. Assn. Advance. Sci., San Francisco, Calif.

Phillips, K. N. 1968. Hydrology of Crater, East, and Davis Lakes, Oregon. U. S. Geol. Surv. Water-Supply Pap. 1859-E, 60 pp.

Salinas, John, L. D. White, and J. M. Thompson. 1984. Studies on chemical, isotopic, and limnological parameters of Crater Lake, Oregon. EOS 65:885. (Abstract)

Schoen, Robert. 1969. Rate of sulfuric acid formation in Yellowstone National Park: Geol. Soc. Amer. Bull., 80:643-650.

Schoen, Robert, and R. O. Rye. 1970. Sulphur isotope distribution in solfataras, Yellowstone National Park. Science 170:1082-1084.

Simpson, H. J. 1970a, Tritium in Crater Lake, Oregon. Jour. Geophys. Res. 75:5195-5207.

Simpson, H. J., Jr. 1970b. Closed basin lakes as a tool in geochemistry. Ph.D. Thesis. Columbia Univ., New York. Pages 44-70.

Thompson, J. M. 1985. Chemistry of thermal and nonthermal springs in the vicinity of Lassen Volcanic National Park. Jour. Volcanol. Geotherm. Res., 25:81-104.

Thompson, J. M., F. E. Goff, and J. M. Donnelly-Nolan. 1981. Chemical analyses of waters from springs and wells in the Clear Lake volcanic area. Pages 183-191 *in* R. J. McLaughlin and J. M. Donnelly-Nolan, eds., Research in the Geysers - Clear Lake Geothermal Field, Northern California: U. S. Geol. Surv. Prof. Pap. 1141.

Thompson, J. M., and Sandhya Yadav. 1979. Chemical analyses of waters from geysers, hot springs, and pools in Yellowstone National Park, Wyoming, from 1974 to 1978. U. S. Geol. Surv. Open-File Rep. 79-704. 49 pp.

Thompson, J. M., and L. D. White. 1983. Does Crater Lake, Oregon, have a discernible outlet? EOS 64:895. (Abstract)

Thompson, J. M., L. D. White, and M. Nathenson. 1987. Chemical analyses of waters from Crater Lake, Oregon, and nearby springs. U. S. Geol. Surv. Open-File Rep. 87-587. 26 pp.

U. S. Geological Survey. 1956. (Map) Crater Lake National Park and Vicinity, (scale 1:62500).

Van Denburgh, A. S. 1968. Chemistry of the Lakes. Pages 41-45 *in* Hydrology of Crater, East, and Davis Lakes, Oregon. U. S. Geol. Surv. Water-Supply Pap. 1859-E.

Van Winkle, Walton, and N. M. Finkbiner. 1913. Composition of the water of Crater Lake, Oregon: Jour. Indus. & Engineer. Chem. 5:198-199.

Volchok, H. L., M. Feiner, H. J. Simpson, W. S. Broecker, V. E. Noshkin, V. T. Bowen, and E. Willis. 1970. Ocean Fallout–The Crater Lake experiment. Jour. Geophys. Res. 75:1084-1091.

White, D. E., J. M. Thompson, and R. O. Fournier. 1976. Lithium contents of thermal and mineral springs. Pages 58-60 *in* J. D. Vine, ed., Lithium Resources and Requirements by the Year 2000. U. S. Geol. Surv. Prof. Pap. 1005.

White, L. D., J. M. Thompson, and C. A. Maley. 1985. Evidence for thermal water in Crater Lake, Oregon. EOS 66:1146. (Abstract)

Williams, D. L., and R. P. Von Herzen. 1983. On the terrestrial heat flow and Physical limnology of

Crater Lake, Oregon. Jour. Geophys. Res. 88: 1094-1104.

CHEMICAL BALANCE FOR MAJOR ELEMENTS IN WATER IN CRATER LAKE, OREGON

Manuel Nathenson
U.S. Geological Survey
345 Middlefield Road
Menlo Park, CA 94025

Crater Lake, Oregon, contains anomalous amounts of dissolved constituents compared to what would be expected from the evaporative concentration of the available water supply. Using published data for dissolved constituents collected since 1912, it is shown that the lake is presently in approximately steady state including a source term for each constituent. In addition, a significant amount of silica is consumed by diatoms that fall to the lake floor. The resulting values for the inflow of chemicals can be converted to an equivalent flow of water with dissolved constituents. Geothermometer temperatures calculated for this inflow can be used to calculate thermal power. The thermal power of the inflow is easily able to supply the thermal energy needed to supply the estimated convective heat flow into the lake. The input of dissolved constituents could be from a small flow of water with a high temperature and a high concentration of dissolved solids, reflecting the existence of a current hydrothermal system; or from a high flow of water with a low temperature and a low concentration of dissolved solids.

The climactic eruption of Mount Mazama that formed the Crater Lake caldera took place 6800 years ago (Bacon 1983). The time at which Crater Lake filled to its present level is unknown. Lake levels since 1878 vary within a range of 4.6 m but show no systematic evolution to shallower or deeper levels (Phillips 1968). The lake is quite fresh (around 80 mg/L total dissolved solids); the concentrations of dissolved chloride and sulfate, however, are anomalous for surface water of humid mountain regions and have been interpreted to indicate an input of warm water at depth (Van Denburgh 1968). Crater Lake has no surface outlet and loses about 72% of its water supply by leakage and 28% by evaporation (Phillips, 1968). Although evaporation concentrates the input of dissolved constituents to Crater Lake by about 40%, this amount is insufficient to explain the concentrations of dissolved constituents found (Simpson 1970a). The purpose of this study is to calculate a chemical balance for the major elements in Crater Lake in order to obtain the amounts of each constituent that must be added to the lake beyond the amounts found in precipitation and runoff. Based on these values, the inflow of fluid can be constrained within a broad range of values. Calculations for the total heat flow into the lake can also be used to indicate how large an inflow is necessary to provide the anomalous heat flow found by Williams and Von Herzen (1983).

CRATER LAKE MIXING

In using chemical data obtained from surface samples to calculate a chemical balance for Crater Lake, the variability of constituents as a function of depth in the lake (maximum depth 589 m) must first be assessed. Thompson *et al.* (1987, 1990) present several profiles of chloride, silica, and deuterium and oxygen isotope data indicating that the lake is well mixed to total depth. A more sensitive indicator of the degree of mixing are the six values for tritium obtained by Simpson (1970b) in August, 1967, from the surface to a depth of 519 m. Between the surface and a depth of less than 50 m, he found a spike of tritium with a maximum value of 31 TU, but the values at 50, 305, and 519 m were all 24 TU. The

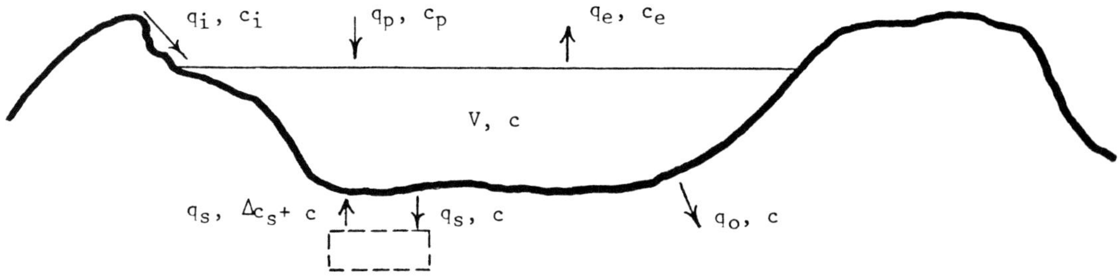

Figure 1. Diagram representing the water and chemical balance of Crater Lake. Flows q and concentrations c for the various sources denoted by subscripts p precipitation, i runoff and underflow, e evaporation, o outflow, and s for inflow. The inflow is assumed to be lake water that circulates within the rock to add dissolved constituents. The volume of the lake V is assumed to be at a uniform concentration c.

total tritium in the upper 50 m (average concentration times thickness) above this background value of 24 TU is about 169 TU•m—a value similar to the precipitation input of 154 TU•m for the first half of 1967. Thus the spike of tritium appears to have been a result of tritium input over the first half of 1967. Comparing the peak concentration in the lake of 31 TU to the input over the first half of 1967 (171 TU average concentration in 0.9 m of precipitation) indicates that the recent precipitation is already diluted by about 20 parts of low-tritium lake water.

H. J. Simpson (written comm., 1989) has provided a somewhat different interpretation. He proposes that the layer of higher than average tritium is only 10 m thick, because the value at 30 m is not significantly different from the background value. Based on temperature data that he obtained (Simpson 1970a), the mixed layer is about 7 m thick at the time of the tritium measurements. The tritium above background would then be about 60 TU•m, an amount corresponding to the input from rainfall since sometime in the spring. In either case, the high tritium water is rapidly diluted by lake water at the background value.

The tritium data can also be used to study the degree of whole-lake circulation. Simpson's accounting for the input of tritium to Crater Lake shows that of the 24 TU found in 1967 in the entire lake, about 10 TU was contributed during the peak-fallout years 1963 and 1964. This large slug of tritium was well mixed to total depth by the time of his measurements 3 years later. He estimates the time scale of mixing is on the order of one to two years. Leventhal and Libby (1970) estimated a somewhat longer time scale of 5 years for whole-lake circulation. For a one-year time period, they estimated that the lake mixes only to a depth of 18% of the mean depth, based on a calculation for how surface samples should evolve as high-tritium precipitation is added to low-tritium lake water. This result was based on two samples of tritium obtained at the surface in 1965. Their surface samples taken at a later date, however, indicated mixing to nearly the total depth of the lake in a one-year time period. It is possible that the two earlier measurements made by Leventhal and Libby are anomalous in some way and are unrelated to the depth of lake-water mixing. If they obtained a little-mixed patch of recent precipitation, they could have obtained the high values that they measured. In the same month that they made their earlier measurements, there were 7 cm of precipitation with a tritium content of 450 TU (Simpson 1970b).

Measurements of temperature as a function of depth in Crater Lake are equivocal concerning whole-lake circulation. The annual wave of temperature change from solar heating and cooling reaches to a depth of approximately 300 m, and it is clear that the lake water circulates to that depth (Neal *et al.* 1972). Below 300 m to approximately 500 m, temperatures increase slightly with depth and do not change with the seasons. Near the lake floor, there are anomalies of the order of 0.1°C that come and go. Neal *et al.* (1972) believed that the steadiness of the temperature profile below 300 m indicates that the bottom part of the lake does not mix. However, Williams and Von Herzen (1983) suggest that the increasing temperatures at depths below 300 m indicate that the lake is undergoing Rayleigh convection caused by the input of thermal energy at depth. For high Rayleigh-number convection, the lake need not turn over at a certain time of the year but could be in continuous, small scale turbulent convection. The temperature data are thus equivocal concerning the state of the bottom half of the lake, but the chemical data show that the lake water is well mixed.

HYDROLOGIC AND CHEMICAL BALANCE

Phillips (1968) carried out a hydrologic balance of the lake that was confirmed with minor differences by Simpson (1970b). The hydrologic balance can be used to develop a chemical balance for the lake. The chemical balance will be developed both for a steady-state equilibrium model and for a model showing how the concentration of dissolved constituents would evolve through time if the input of dissolved solids is not in steady-state equilibrium with the current concentration of the lake.

The lake covers 78% of its drainage area, so that most of the input to the lake is from direct precipitation. Precipitation at a rate q_p with a dissolved chemical concentration of c_p falls on the lake (Fig. 1). The surrounding drainage area adds water both as runoff and as percolation at a rate q_i with a concentration c_i. The water supply is balanced by evaporation q_e, which is assumed to carry no chemicals, and by leakage q_o which has the same concentration as that in the entire lake c. Lake water is assumed to circulate below the floor at a concentration c, equal to the average lake concentration to provide the inflow at a rate q_s with a concentration $c+\Delta c_s$. The hydrologic balance of the lake is:

$$q_i + q_p - q_o - q_e = 0 \qquad (1)$$

For a lake volume V, the rate of change of some dissolved constituent c is:

$$V\frac{dc}{dt} = q_i c_i + q_p c_p + q_s \Delta c_s - q_o c \qquad (2)$$

The steady-state solution to equation (2) is obtained by setting the right-hand side equal to zero:

$$(q_s \Delta c_s)_{ss} = q_o c - q_i c_i - q_p c_p \qquad (3)$$

The transient solution to equation (2) for the concentration c as a function of time for an initial condition of concentration $c = C_o$ at time $t = 0$ (where C_o is not equal to the steady state concentration) is:

$$\frac{q_o c - q_i c_i - q_p c_p - q_s \Delta c_s}{q_o C_o - q_i c_i - q_p c_p - q_s \Delta c_s} = e^{-q_o t/V} \qquad (4)$$

Since we do not know when the problem began, it is useful to define $t = t' - t_o$ where t' is the time in years from t_o, the calendar time when the inflow $q_s \Delta c_s$ changed. Since data are available back to 1912, I will set t_o arbitrarily at 1900. Because equation (4) is an exponential, different combinations of C_o and t_o can result in the same values of c as a function of t'. Thus the initial concentration and the time that the problem starts are somewhat arbitrary. This result makes good physical sense in that we should not be able to tell today if the lake has evolved from a high concentration C_o a long time ago or from a lower (but still high) concentration C_o a shorter time ago.

All the flows q and the volume of the lake are given in Table 1 (Phillips 1968). The time constant V/q_o in equation (4) is 219 years, indicating that any change will take quite some time to become apparent.

TABLE 1. PHYSICAL CHARACTERISTICS OF CRATER LAKE (PHILLIPS 1968)

Surface elevation		1882 m
Greatest measured depth		589 m
Average water depth		325 m
Volume		17.3 km^3
Surface area		53 km^2
Area of watershed (including lake)		68 km^2
Water balance		($10^7 m^3/y$)
Runoff	q_i	1.7
Direct precipitation	q_p	9.3
Evaporation	q_e	3.1
Leakage	q_o	7.9

Chemical data for Crater Lake have been taken by a number of investigators since 1912 and in a systematic sampling program by the U. S. Geological Survey since 1967. The data are presented in Table 2. The concentrations of the runoff c_i given in Table 3 are an average of the 8 springs directly above the lake reported in Thompson *et al.* (1987). Concentrations for a number of constituents in precipitation collected at Medford, Oregon, (90 km southwest of Crater Lake) are given in Junge and Werby (1958) and are reproduced in Table 3. Junge and Werby (1958) did not measure all the major constituents, and I have estimated values for silica, magnesium, and bicarbonate based on a comparison with the values in the Sierra Nevada given in Feth *et al.* (1964).

TABLE 2. CHEMICAL DATA FOR MAJOR CONSTITUENTS (IN mg/L)
FOR SAMPLES FROM CRATER LAKE, OREGON

Data from Van Winkle and Finkbiner (1913), Pettit (1936), Thornton (1965), Van Denburgh (1968), U. S. Geological Survey (1969, 1970, 1972a, b, c, 1974a, b, c, d, 1976, 1977, 1978, 1979, 1981a, b, and 1983), Hubbard *et al.* (1983, 1984, 1986), Alexander *et al.* (1987), and Thompson *et al.* (1987).

Date	SiO_2	Ca	Mg	Na	K	HCO_3	SO_4	Cl
Aug 27, 1912	18	7.1	2.8	11	2.2	34	11	11
1934	–	–	–	–	–	–	–	10
Jul 18, 1940	18.2	–	–	–	–	–	–	–
Sep 6, 1961	18	7.0	2.6	11	1.7	37	10	10
Aug 5, 1964	16	7.0	2.5	11	1.6	35	10	9.5
Aug 5, 1964	16	–	–	–	–	–	–	9.8
Aug 5, 1964	16	7.0	2.5	11	1.6	35	10	9.5
Sep 20, 1965	16	6.8	2.8	11	2.0	36	10	10
Jun 30, 1967	17	6.8	2.7	12	2.3	37	10	11
Aug 15, 1967	18	6.9	2.7	11	1.9	36	10	10
Sep 16, 1967	17	8.3	2.2	11	1.6	37	10	11
Oct 5, 1967	18	7.0	2.9	11	2.0	36	9.6	10
Oct 24, 1967	18	6.8	2.9	11	2.2	38	10	10
Nov 24, 1967	–	6.8	2.8	11	–	40	–	11
May 14, 1968	16	7.0	2.8	11	1.8	36	10	11
Jul 1, 1968	18	6.5	2.7	11	1.8	38	10	10
Aug 9, 1968	18	6.6	2.8	12	1.7	37	10	12
Sep 14, 1968	19	6.7	2.7	11	1.7	36	10	10
Jun 2, 1969	18	6.6	2.5	11	1.7	36	11	9.0
Jul 1, 1969	18	6.6	2.8	11	2.1	36	10	10
Jul 22, 1969	17	6.6	2.8	11	1.7	36	10	9.0
Aug 20, 1969	18	6.6	2.8	11	1.7	38	10	10
Jun 13, 1970	18	6.5	2.7	9.6	2.0	37	6.0	9.5
Jul 1, 1970	19	6.8	2.8	11	1.9	34	10	9.5
Aug 26, 1970	19	6.7	2.8	10	1.9	37	10	8.0
Oct 1, 1970	20	6.8	2.8	12	2.2	37	11	10
Aug 12, 1971	18	7.1	2.4	11	1.8	45	9.8	9.0
Sep 7, 1971	19	6.9	2.6	10	2.2	45	11	6.7
Oct 16, 1971	19	6.8	2.3	11	1.6	40	11	10
Jun 30, 1972	19	6.8	2.6	9.5	1.7	33	11	9.0
Sep 11, 1972	18	7.3	2.7	10	1.8	36	10	10
Oct 30, 1973	18	8.0	2.6	17	2.3	37	11	13
Jul 2, 1974	19	7.7	2.9	19	2.2	39	14	11
Aug 27, 1974	18	7.6	2.1	11	1.8	36	12	12
Oct 21, 1974	18	7.6	2.7	15	1.8	37	11	9.9
Aug 22, 1975	19	6.4	2.7	10	1.7	36	11	11
Oct 9, 1975	16	6.5	2.8	16	2.1	36	12	10
Jun 29, 1976	18	9.6	2.7	10	1.8	42	7.8	9.3
Aug 30, 1976	16	6.9	2.4	11	1.9	37	10	11
Oct 15, 1976	19	7.2	2.7	11	1.8	32	11	11

TABLE 2 (continued). CHEMICAL DATA FOR MAJOR CONSTITUENTS (IN mg/L) FOR SAMPLES FROM CRATER LAKE, OREGON

Date	SiO_2	Ca	Mg	Na	K	HCO_3	SO_4	Cl
Jun 10, 1977	17	9.2	2.6	11	1.8	35	9.8	11
Jul 13, 1977	18	7.7	2.6	11	2.0	34	9.7	10
Sep 8, 1977	18	9.9	2.8	11	2.0	34	9.4	12
Oct 11, 1977	18	7.9	3.1	11	1.8	35	10	9.9
Jun 13, 1978	14	6.7	2.6	11	1.8	37	13	11
Aug 9, 1978	18	6.9	2.6	10	1.8	33	10	9.9
Oct 11, 1978	17	6.6	2.6	11	1.6	33	12	10
Jun 12, 1979	16	7.5	2.7	10	2.2	33	10	10
Aug 15, 1979	17	6.4	2.5	10	1.9	24	14	15
Oct 10, 1979	18	6.8	2.2	10	2.0	28	8.8	11
Jul 1, 1980	17	6.7	2.7	10	1.6	34	11	9.4
Oct 10, 1980	17	6.9	2.6	9.9	1.9	35	10	11
Jun 2, 1981	18	9.3	2.7	11	1.8	49	12	9.0
Jul 29, 1981	18	7.4	2.6	10	1.5	37	9.0	9.7
Aug 15, 1981	18.1	7.4	2.8	9.5	1.3	30	8	9.6
Oct 14, 1981	18	6.9	2.5	9.7	1.9	32	6.0	16
Jul 7, 1982	18	7.0	2.5	10	1.8	38	11	9.7
Aug 25, 1982	17	6.7	2.6	11	1.8	32	11	9.9
Oct 14, 1982	17	7.3	2.5	10	1.7	–	12	10
Jun 30, 1983	18	6.7	2.7	10	2.0	–	9.7	9.7
Aug 8, 1983	19.6	6.8	2.9	9.2	1.6	41	10	10.1
Aug 19, 1983	17	7.7	2.7	11	1.8	34	17	9.7
Oct 12, 1983	17	6.6	2.6	10	2.0	34	12	9.7
Jul 17, 1984	17	7.0	2.6	10	1.6	35	12	9.7
Jul 17, 1984	18	7.1	2.7	10	1.7	34	9.9	9.6
Aug 7, 1984	19.1	8.0	2.4	10.5	1.7	41	8	10.1
Oct 10, 1984	17	7.2	2.7	10	1.6	37	11	9.8
Jun 18, 1985	18	7.3	2.8	10	1.8	37	9.7	9.8
Aug 15, 1985	17	8.0	2.7	10	1.9	38	10	9.6
Oct 16, 1985	18	7.1	2.7	11	1.8	37	10	9.6
Mar 5, 1986	17	7.0	2.5	10	1.8	48	8.8	9.4
Jul 1, 1986	17	6.5	2.5	10	1.7	37	11	8.6
Aug 25, 1986	18	7.0	2.0	11	1.8		20	9.9

Before using the measured concentration of constituents in Crater Lake to calculate the amount of added inflow $q_s \Delta c_s$, it is worthwhile to calculate the concentrations in the lake assuming the only inputs were precipitation and runoff from the surrounding drainage basin, in order to evaluate varying rates of evaporation. Equation (3) can be rearranged to:

$$c = \frac{(q_i c_i + q_p c_p)}{q_o}, \qquad (5)$$

assuming that $q_s \Delta c_s = 0$. Table 3 shows the calculated concentration for two values of the ratio $q_e/(q_i + q_p)$. The calculated concentrations are all much lower than the measured concentrations, indicating that the additional inflow to Crater Lake carries the full range of major element constituents. Simpson (1970a) showed previously that evaporation alone could not produce the measured concentrations. The calculation for 50% of the available water supply being lost to evaporation shows that even a large

TABLE 3. CONCENTRATION OF DISSOLVED CONSTITUENTS (mg/L) IN THE RUNOFF c_i, PRECIPITATION c_p, AND CALCULATED AND MEASURED FOR CRATER LAKE

Values for concentration in the runoff c_i are average of data for 8 springs directly above the lake from Thompson *et al.* (1987). Values for concentration in precipitation c_p are from Junge and Werby (1958) or estimated from Feth *et al.* (1964) by comparison to measured values of other constituents. Calculations for Crater Lake concentrations assume that the only sources of constituents are precipitation and runoff; and that the fraction of the water supply lost to evaporation is either 28% as determined by Phillips (1968) or 50%, to illustrate the sensitivity of the result.

			Crater Lake		
	c_i	c_p	Calculated for Evaporation of		
			28%	50%	Measured
SiO_2	31	0.1[1]	6.8	9.8	17.7
Ca	2.9	0.52	1.2	1.8	7.2
Mg	1.1	0.1[1]	0.4	0.5	2.6
Na	2.6	0.15	0.7	1.1	10.9
K	0.7	0.10	0.3	0.4	1.8
HCO_3	20	1[1]	5.5	7.9	36.4
SO_4	5.4	0.80	2.1	3.0	10.5
Cl	0.25	0.21	0.3	0.4	10.2

[1] Estimated concentration

error in the water balance of Phillips (1968) does not make it possible to explain the measured concentrations by evaporation alone. For comparison, Nathenson (1989) has shown that evaporative concentration of precipitation and runoff can explain the measured concentrations in Lake Tahoe if allowance is made for the loss of silica to consumption by diatoms.

In order to use the data to determine values for the added inflow $q_s \Delta c_s$ and initial concentration C_o, a linear least-squares fit was done of c versus exp $(-t'/219)$; the results of these fits, however, will be presented in nonlinear c versus t' plots. Figures 2 through 9 show the concentration of each constituent versus time. Three lines are shown on each figure: (1) the average value for the constituent (horizontal line), (2) a curve based on the best fit to the data, and (3) a curve assuming no current input beyond that in precipitation and the contribution from the surrounding drainage area ($q_s = 0$). The initial concentration in 1900 for this latter model is calculated by assuming that the average value of the constituent occurred in 1976, about the mid-point of when most of the data were taken. The value for the steady state inflow $(q_s \Delta c_s)_{ss}$ is calculated based on the average concentration and is given in Table 4. The parameters for the best fit curves are also given in Table 4. An unfortunate characteristic of the USGS data is that they are reported to only two significant figures for some constituents, so that concentrations 10 mg/L and above are reported to 10% while concentrations below this value are reported to 1%. This characteristic of the data is reflected in the plots when most of the data line up at only a few values (silica, sodium, sulfate, and chloride).

Most of the variation in all the constituents is clearly caused by the level of precision in the analyses. The means and standard deviations for the data are given in the first two columns of Table 4, and the third column gives the standard deviation normalized by the mean. The normalized standard deviations are all around 10%, which is about what one would expect for the precision of water chemistry data. Thus, one could explain the data as reflecting a constant concentration with no time variation. In fitting the best line, values for the inflow $q_s \Delta c_s$ and the initial concentration C_o in 1900 are obtained, and these values are given in Table 4 along with their associated standard errors. A useful parameter for assessing the overall pattern of inflow values is the ratio of the inflow for the best fit to the steady state value $q_s \Delta c_s/(q_s \Delta c_s)_{ss}$. In general, this ratio is close to one, indicating the inflow is approximately the steady state value. The value of the ratio is greater than one for calcium, bicarbonate, and sulfate, and the best-fit result is that the concentration of these ions increases with time (Figs. 3, 7, and 8). Looking at all the plots, the ability of the data set to distinguish between a steady state value for the inflow and

TABLE 4. CALCULATED VALUES FOR MAJOR CONSTITUENTS IN WATER FROM CRATER LAKE

Values for mean, standard deviation (SD), and their ratio are given for each chemical. Values for concentration in the runoff c_i are average of data for 8 springs directly above the lake from Thompson *et al.* (1987). Values for concentration in precipitation c_p are from Junge and Werby (1958) or estimated from Feth *et al.* (1964) by comparison to measured values of other constituents. Calculated value for steady state inflow $(q_s\Delta c_s)_{ss}$ and mean and standard error for best-fit inflow $q_s\Delta c_s$ are given along with mean and standard error for initial concentration C_o in 1900.

Chemical	mg/L Mean	mg/L SD	SD/Mean	c_i mg/L	c_p mg/L	$(q_s\Delta c_s)_{ss}$ mg/s	$q_s\Delta c_s$ mg/s Mean	$q_s\Delta c_s$ mg/s SE	C_o mg/L Mean	C_o mg/L SE	$\frac{q_s\Delta c_s}{(q_s\Delta c_s)_{ss}}$
SiO_2	17.7	1.0	0.06	31	0.1[1]	27,200	24,700	5,800	18.1	2.3	0.91
Ca	7.2	0.7	0.10	2.9	0.52	14,800	19,100	4,400	6.5	1.7	1.29
Mg	2.6	0.2	0.07	1.1	0.1[1]	5,700	4,000	1,140	2.9	0.5	0.70
Na	10.9	1.6	0.15	2.6	0.15	25,500	13,300	9,600	12.9	3.8	0.52
K	1.8	0.2	0.11	0.7	0.10	3,900	1,140	1,200	2.3	0.5	0.29
HCO_3	36.4	3.9	0.11	20	1[1]	77,200	82,400	24,000	35.5	9.4	1.07
SO_4	10.5	2.0	0.19	5.4	0.80	21,100	30,900	12,000	8.9	4.7	1.46
Cl	10.2	1.3	0.13	0.25	0.21	24,700	23,100	7,000	10.5	2.8	0.94

[1] Estimated concentration

an inflow that is in the range of 0.7 to 1.5 times steady state is limited. This is confirmed by the large standard errors for $q_s\Delta c_s$ and C_o. The best fits for sodium and potassium both indicate inflows $q_s\Delta c_s$ near zero. In the case of sodium (Fig. 5), this is not consistent with the measurement in 1912, while the 1912 value for potassium is consistent with inflow $q_s\Delta c_s$ near zero. The average of all the determinations of the ratio of inflows is 0.9 ± 0.1 standard error. Thus, the mean value is within one standard error of the steady state value. Visually, the data do not generally support the model involving no current input.

Some of the data in Figs. 2-9 appear to be outliers, and it is worth trying to remove some to obtain improved fits. Because increased chloride concentrations tend to increase sodium concentrations, one technique to check for outliers is to compare the sodium and chloride concentrations. A plot of sodium versus chloride showed that 6 of 64 pairs have either an anomalously high value of chloride or of sodium, but not of both. The remaining 58 pairs are tightly grouped, and it seems likely that the outliers are bad values. Using these two constituents to check each other is virtually a replicate determination, and the 6 pairs with one or the other constituent as outliers can be deleted from the data set and the correlations recalculated. Table 5 shows the results. Deletion of these outliers has lowered the standard deviations of the concentrations from over 10% to less than 10%. The standard errors for $q_s\Delta c_s$ and C_o are also lower. The ratio of the best fit flow to the steady state flow is nearly the same for sodium but lower for chloride after removing the outliers. That

TABLE 5. VALUES OF MEAN CONCENTRATIONS AND FLOWS FOR SODIUM AND CHLORIDE DATA WITH OUTLIERS REMOVED.

Chemical	mg/L Mean	mg/L SD	SD/Mean	c_i mg/L	c_p mg/L	$(q_s\Delta c_s)_{ss}$ mg/s	$q_s\Delta c_s$ mg/s Mean	$q_s\Delta c_s$ mg/s SE	C_o mg/L Mean	C_o mg/L SE	$\frac{q_s\Delta c_s}{(q_s\Delta c_s)_{ss}}$
Na	10.6	0.6	0.06	2.6	0.15	24,700	13,300	3,500	12.4	1.4	0.54
Cl	10.0	0.9	0.09	0.25	0.21	24,200	18,600	4,700	10.9	1.8	0.77

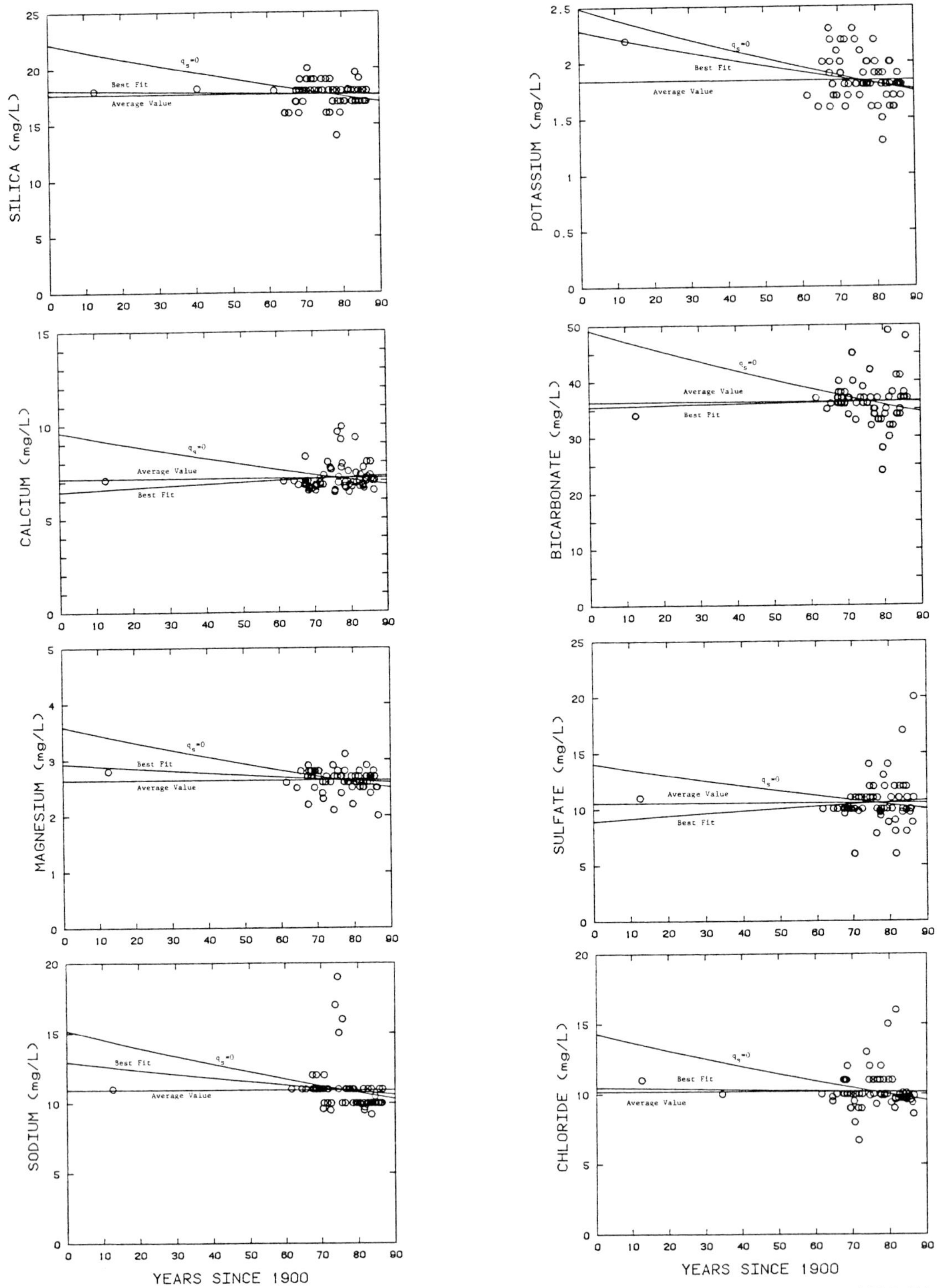

Figures 2–9 (Upper left to lower right). Concentration of dissolved constituent versus years since 1900. Lines are shown for steady state (average value), best fit to the data, and no current inflow ($q_s = 0$). Note that the best-fit curve is not just determined by a few points at early time with high leverage because of the large number of points in recent years. The best-fit curve does not pass through the early data for calcium, sodium, bicarbonate, sulfate, and chloride.

removing the outliers tends to bring the values for the ratio for sodium and chloride in Table 5 into closer agreement suggests that the flow may actually be less than steady state. Without some further data, the simplest hypothesis is to assume that the lake is in approximately steady state for the input of added constituents, recognizing that the flow may be a bit less than the steady state value.

DISCUSSION

The chemical balance of Crater Lake discussed above neglects any losses of constituents. Nathenson (1989) has shown that the consumption of silica by diatoms is a significant term in the chemical balance of Lake Tahoe, and the occurrence of diatoms in the sediments of Crater Lake (Nelson 1967) indicates that it might be a significant term here also and should be calculated. In a core collected on the platform near Wizard Island, Steve Robinson (written comm., 1988) measured a recent sedimentation rate of 17 cm per 1000 years, and P. Bradbury (written comm., 1988) has estimated that the diatoms comprise 10 to 30% by volume in the top of the core. The bulk density of silica in diatoms is about 0.25 g/cm^3 over a broad range of sizes of diatoms based on the equation in Conley *et al.* (1987). Assuming that 20% of the sediments are diatoms and that diatoms are deposited over the entire area of the lake, the amount of silica consumed each year by diatoms is 14,300 mg/s or 53% of the steady state value of 27,200 mg/s from Table 4. Thus the consumption of silica by diatoms is a significant term in the silica balance and cannot be ignored. The value calculated here is quite uncertain because the estimate is based on data from only one core. As more core data become available, it should be possible to refine this estimate.

Based on the total silica flow of 41,500 mg/s (steady-state inflow plus that consumed by diatoms), it is possible to put limits on the range of inflow that will carry this silica into the lake. The maximum silica concentration would occur if there were very high temperatures just beneath the surface of the sediments in Crater Lake. At a water depth of 450 m, the pressure is 47 bars and the boiling temperature is 259°C. The silica concentration for quartz solubility would be 537mg/L (Fournier 1981), and the associated flow would be 80 L/s. The highest flow value is that which would be obtained for the lowest concentration of silica in the inflow. Based on the similar chemistry of Crater Lake water to that found in the Wood River area springs (Nathenson and Thompson 1990), the minimum silica concentration should be similar, about 40 mg/L; and the flow would be 1900 L/s. The range of flow calculated for these silica concentrations is 80 to 1900 L/s. At the low end, this flow is a small term in the hydrologic balance whereas at the high end, it is a substantial fraction of the water supply of 3500 L/s.

The chemical characteristics and the thermal power of the inflow can be calculated for various flow rates. The concentration of each constituent is calculated from $c + (q_s \Delta c_s)_{ss}/q_s$, using the values for $(q_s \Delta c_s)_{ss}$ and c from Table 4, except a value of 41,500 mg/s is used for silica. Table 6 shows the resulting fluid compositions for several values of flow along with geothermometer temperatures calculated from formulas in Fournier (1981) and Giggenbach (1986). The silica geothermometer depends on the silica concentration, whereas the other geothermometers are calculated using ratios of concentrations. Thus, the silica geothermometer is sensitive to dilution, but the other geothermometers are relatively insensitive (Na-K-Ca and Na-K-Ca-Mg) or completely insensitive (Na-K, K-Mg) to actual concentrations. Until the actual amount of dilution is established, the silica geothermometer temperatures are not very meaningful. For higher concentrations of silica, equilibrium with quartz is assumed, whereas for lower concentrations, equilibrium with chalcedony is assumed (Fournier 1981). The Na-K geothermometer gives uniformly high temperatures (>250°C). At the lower inflows, the Na-K-Ca geothermometer temperature is near 200°C; the magnesium correction, however, lowers this to temperatures of less than 50°C. The Mg-corrected Na-K-Ca geothermometer temperatures range from 41° to 67°C, and the K-Mg geothermometer temperatures range from 53° to 82°C. At high inflows (600-1900 L/s), four of the geothermometers are in reasonable agreement indicating a temperature of 60° to 70°C, with only the Na-K geothermometer giving a discordant temperature. At low inflows (80-120 L/s), the silica, Na-K, and Na-K-Ca geothermometers give high temperatures (≈200°C) whereas the Na-K-Ca-Mg and K-Mg geothermometers give low temperatures. One interpretation of the conflicting

TABLE 6. CALCULATED COMPOSITION (IN mg/L) OF FLUID ENTERING CRATER LAKE FOR SEVERAL VALUES OF INFLOW

Geothermometer temperatures (°C) are quartz, chalcedony, Na-K, Na-K-Ca, Na-K-Ca with the magnesium correction (Fournier 1981), and K-Mg (Giggenbach 1986). Thermal power calculated for silica (quartz or chalcedony) and Na-K-Ca-Mg geothermometer temperatures.

Inflow (L/s)	80	120	200	300	400	600	1000	1900
			Calculated inflow composition (mg/L)					
SiO_2	540	360	230	156	122	87	59	40
Ca	192	131	81	57	44	32	22	15
Mg	74	50	31	22	17	12	8	6
Na	330	220	138	96	75	53	36	24
K	51	34	21	15	12	8	6	4
HCO_3	1000	680	420	290	230	165	114	77
SO_4	270	186	116	81	63	46	32	22
Cl	320	220	133	93	72	51	35	23
			Geothermometer temperatures (°C)					
t (quartz)	260	230	188	164	148			
t(chal.)						102	81	61
t(Na-K)	257	257	257	257	257	258	259	260
t(Na-K-Ca)	196	192	97	88	83	75	67	59
t(Na-K-Ca-Mg)	42	41	50	53	55	58	67	59
t(K-Mg)	82	77	72	68	65	61	57	53
			Power (MW_t) calculated based on geothermometer temperatures					
Silica t	86	111	155	202	240	250	320	455
Na-K-Ca-Mg t	13	19	39	62	86	136	265	439

geothermometer temperatures at low flows is that the fluid is low-temperature (60°-70°C) with a high flow rate. Another interpretation is that the fluid has a high temperature source, with the high-temperature fluid mixing with lake water while still in contact with rock and reequilibrating some of the geothermometers before flowing into the lake.

The thermal power of the inflow (Table 6) is calculated using temperatures for both silica and Na-K-Ca-Mg geothermometers to give representative values. The measurements of Williams and Von Herzen (1983) can also be used to calculate a thermal power. They measured heat flows in Crater Lake at 62 locations and found 15 values above 210 mW/m^2. Excluding the values above $540mW/m^2$, the average heat flow measured was 138 mW/m^2. Assuming that this average represents the conductive heat flow, they estimated that the total heat flow, both convective and conductive, might be 5 to 10 times this average or 670 to 1380 mW/m^2. Assuming that the convective heat flow is between 500 and $1200mW/m^2$ and that it occurs over only half the area of the caldera (because substantial areas have low heat flow), the calculated thermal power is 13 to 32MW. These values are at the low end of the values shown in Table 6, and the comparison indicates that the inflow is easily capable of supplying the convective heat flow estimated by Williams and Von Herzen (1983).

CONCLUSIONS

The concentration of all major-element constituents in Crater Lake are too high to be explained by evaporative concentration of supply from precipitation and runoff. The available data can be most easily explained by a steady-state chemical balance of the lake, though there is a suggestion that the inflow of constituents may be somewhat less than steady state. The calculated inflow ranges from 80 to 1900 L/s, based on the silica inflow of 41,500 mg/s. Although the rate of silica consumption by diatoms is fairly uncertain, changes in this value would change the details of the calculation but not the overall pattern. Calculations indicate that the inflow is easily capable of supplying the thermal power inferred from the heat flow measurements of Williams and Von Herzen (1983). The available data are unable to constrain whether the inflow is large with a low temperature or small with a high temperature. The low-flow model involves an inflow that is much less than any term in the hydrologic balance, while the high-flow model could occur only by circulating lake water to dissolve constituents in the subsurface. Resolution of which model is correct awaits further data on the thermal and chemical characteristics of the inflow system underlying Crater Lake.

ACKNOWLEDGMENTS

I would like to thank Steve Ingebritsen, Robert Fournier, and Patrick Muffler for helpful reviews of the manuscript. Jack Dymond's review of an early version helped to clarify the analysis.

LITERATURE CITED

Alexander, C. W., R. L. Moffatt, P. R. Boucher, and M. L. Smith. 1987. Water resources data for Oregon, water year 1985. U. S. Geol. Surv. Water-Data Rep. OR-85-1, Vol. 1. Eastern Oregon. 218 pp.

Bacon, C. R. 1983. Eruptive history of Mount Mazama and Crater Lake caldera, Cascade Range, U.S.A. Jour. Volcanol. Geotherm. Res. 18:57-115.

Conley, D. J., S. S. Kilham, and E. Theriot. 1987. Potential influence of diatom silica cell quota on dissolved silica limitation along salinity gradients. EOS 68:1685. (Abstract)

Feth, J. H., C. E. Roberson, and W. L. Polzer. 1964. Sources of mineral constituents in water from granitic rocks, Sierra Nevada, California and Nevada. U. S. Geol. Surv. Water-Supply Pap. 1535-I. 70 pp.

Fournier, R. O. 1981. Application of water geochemistry to geothermal exploration and reservoir engineering. Pages 109-143 *in* L. Rybach and L. J. P. Muffler, eds., Geothermal Systems: Principles and Case Histories. John Wiley, New York.

Giggenbach, W. F. 1986. Graphical techniques for the evaluation of water/rock equilibration conditions by use of Na, K, Mg, and Ca-contents of discharge waters. Proc. 8th New Zealand Geotherm. Workshop, Univ. Auckland Geotherm. Inst., pp. 37-43.

Hubbard, L. L., T. D. Parks, D. L. Weiss, and L. E. Hubbard. 1983. Water resources data for Oregon, water year 1982. U. S. Geol. Surv. Water-Data Rep. OR-82-1, vol.1. Eastern Oregon. 206 pp.

Hubbard, L. L., T. D. Parks, D. L. Weiss, and L. E. Hubbard. 1984. Water resources data for Oregon, water year 1983. U. S. Geol. Surv. Water-Data Rep. OR-83-1, vol. 1. Eastern Oregon. 202 pp.

Hubbard, L. L., M. L. Smith, and L. E. Hubbard. 1986. Water resources data for Oregon, water year 1984. U. S. Geol. Surv. Water-Data Rep. OR-84-1, vol. 1. Eastern Oregon. 224 pp.

Junge, C. E., and R. T. Werby. 1958. The concentration of chloride, sodium, potassium, calcium, and sulfate in rain water over the United States. Jour. Meteorol. 15:417-425.

Leventhal, J. S., and W. F. Libby. 1970. Tritium fallout in the Pacific United States. Jour. Geophys. Res. 75:7628-7633.

Nathenson, M. 1989. Chemistry of Lake Tahoe, California-Nevada, and nearby springs. U. S. Geol. Surv. Open-File Rep. 88-641. 27 pp.

Nathenson, M., and J. M. Thompson 1990. Chemistry of Crater Lake, Oregon, and nearby springs in relation to weathering. Pages 115-126 *in* E. T. Drake *et al.*, eds., Crater Lake: An Ecosystem Study. Pacific Div., Amer. Assn. Advance. Sci., San Francisco, Calif.

Neal, V. T., S. J. Neshyba, and W. W. Denner. 1972. Vertical temperature structure in Crater Lake, Oregon. Limnol. Oceanogr. 17:451-453.

Nelson, C. H. 1967. Sediments of Crater Lake, Oregon. Geol. Soc. Amer. Bull. 78:833-848.

Pettit, Edison. 1936. On the color of Crater Lake water. Proc. Nat'l Acad. Sci. 22:139-146.

Phillips, K. N. 1968. Hydrology of Crater, East, and Davis Lakes, Oregon. U. S. Geol. Surv. Water-Supply Pap. 1859-E. 60 pp.

Simpson, H. J., Jr. 1970a. Closed basin lakes as a tool in geochemistry. Ph.D. Thesis. Columbia Univ., New York. 325 pp.

Simpson, H. J. 1970b. Tritium in Crater Lake, Oregon. Jour. Geophys. Res. 75:5195-5207.

Thornton, E. B. 1965. Investigations at Crater Lake, hydrologic benchmark. U. S. Geol. Surv., Water Res. Div. Bull., pp. 23-26.

Thompson, J. M., L. D. White, and M. Nathenson. 1987. Chemical analyses of waters from Crater Lake, Oregon, and nearby springs. U. S. Geol. Surv. Open-File Rep. 87-587. 26 pp.

Thompson, J. M., M. Nathenson, and L. D. White. 1990. Chemical and isotopic compositions of waters from Crater Lake, Oregon, and nearby vicinity. Pages 91-102 *in* E. T. Drake *et al.*, eds., Crater Lake: An Ecosystem Study. Pacific Div., Amer. Assn. Advance. Sci., San Francisco, Calif.

U. S. Geological Survey. 1969. Quality of surface waters of the United States, 1964, Parts 9-11. Colorado River Basin to Pacific Slope Basins in California. U. S. Geol. Surv. Water-Supply Pap. 1958. 615 pp.

U. S. Geological Survey. 1970. Quality of surface waters of the United States, 1965, Parts 9-11. Colorado River Basin to Pacific Slope Basins in California. U. S. Geol. Surv. Water-Supply Pap. 1965. 678 pp.

U. S. Geological Survey. 1972a. Quality of surface waters of the United States, 1967, Parts 9-11. Colorado River Basin to Pacific Slope Basins in California. U. S. Geol. Surv. Water-Supply Pap. 2015. 702 pp.

U. S. Geological Survey. 1972b. Quality of surface waters of the United States, 1968, Part 11. Pacific Slope Basins in California. U. S. Geol. Surv. Water-Supply Pap. 2099. 359 pp.

U. S. Geological Survey. 1972c. 1971 Water resources data for Oregon Part 2. Water quality records, 131 pp.

U. S. Geological Survey. 1974a. Quality of surface waters of the United States, 1969, Part 11. Pacific Slope Basins in California. U. S. Geol. Surv. Water-Supply Pap. 2149. 349 pp.

U. S. Geological Survey. 1974b. Quality of surface waters of the United States, 1970. Part 11. Pacific Slope Basins in California. U. S. Geol. Surv. Water-Supply Pap. 2159. 397 pp.

U. S. Geological Survey. 1974c. 1972 Water resources data for Oregon Part 2. Water quality records. 137 pp.

U. S. Geological Survey. 1974d. 1974 Water resources data for Oregon Part 2. Water quality records. 146 pp.

U. S. Geological Survey. 1976. Water resources data for Oregon, water year 1975. U. S Geol. Surv. Water-Data Rep. OR-75-1. 586 pp.

U. S. Geological Survey. 1977. Water resources data for Oregon, water year 1976. U. S. Geological Survey Water-Data Rep. OR-76-1. 592 pp.

U. S. Geological Survey. 1978. Water resources data for Oregon, water year 1977. U. S. Geol. Surv. Water-Data Rep. OR-77-1. 607 pp.

U. S. Geological Survey. 1979. Water resources data for Oregon, water year 1978. U. S. Geol. Surv. Water-Data Rep. OR-78-1. 650 pp.

U. S. Geological Survey. 1981a. Water resources data for Oregon, water year 1979. U. S. Geol. Surv. Water-Data Rep. OR-79-1. 743 pp.

U. S. Geological Survey. 1981b. Water resources data for Oregon, water year 1980. U. S. Geol. Surv. Water-Data Rep. OR-80-1, vol. 1. Eastern Oregon. 258 pp.

U. S. Geological Survey. 1983. Water resources data for Oregon, water year 1981. U. S. Geol. Surv. Water-Data Rep. OR-81-1, vol. 1. Eastern Oregon. 242 pp.

Van Denburgh, A. S. 1968. Chemistry of the lakes, Pages 41-44 *in* K. N. Phillips, Hydrology of Crater, East, and Davis Lakes, Oregon. U. S. Geol. Surv. Water-Supply Pap. 1859-E.

Van Winkle, W., and N. M. Finkbiner. 1913. Composition of the water of Crater Lake, Oregon. Jour. Indus. Engineer. Chem. 5:198-199.

Williams, D. L., and R. P. Von Herzen. 1983. On the terrestrial heat flow and physical limnology of Crater Lake, Oregon. Jour. Geophys. Res. 88:1094-1104.

CHEMISTRY OF CRATER LAKE, OREGON, AND NEARBY SPRINGS IN RELATION TO WEATHERING

Manuel Nathenson and J. Michael Thompson
U.S. Geological Survey
345 Middlefield Road
Menlo Park, CA 94025

The chemistry of most cold springs in the Crater Lake area is explained by the process of weathering of volcanic glass and clinopyroxene to kaolinite clay by water containing dissolved carbon dioxide. At higher concentrations of bicarbonate, the simple weathering process is modified by saturation to limit the amount of silica that can be retained in solution. Springs that do not satisfy this model are the soda spring on Minnehaha Creek and a group of springs in the Wood River area. The soda spring is an unlikely source of the dissolved constituents in Crater Lake, but its chemistry does demonstrate that, locally, a substantial amount of carbon dioxide is available in the subsurface to drive chemical reactions. The springs in the Wood River area are chemically quite similar to Crater Lake. The process that generates the chemistry of the Wood River area springs could be the same as the process that generates the chemistry of the inflow to Crater Lake, or the similarity of the chemistry could be a coincidence. Until the source water of both is characterized, it is not possible to settle this question.

In a recent report, Thompson *et al.* (1987, 1990) presented data and some interpretations for the chemistry of Crater Lake, Oregon, and nearby cold springs. The major focus of that study was to use chemical and isotopic data for water samples as tracers to study how well-mixed Crater Lake is and if the water from Crater Lake could be discharging in nearby cold springs. In addition, these data were used to show that Crater Lake is anomalously high in some constituents compared to nearby springs and Diamond Lake to the north. The purpose of this paper is to study the chemistry of cold springs in the Crater Lake area in relation to processes that control their chemistry. In a companion paper (Nathenson 1989), the same techniques have been applied to the Lake Tahoe area, where the notion of weathering as a process for generating the composition of cold springs in an area of relatively uniform geology was reported in a series of classic studies by Feth *et al.* (1964), Garrels (1967), and Garrels and MacKenzie (1967). Simpson (1970) studied the same processes in Crater Lake and Lake Tahoe but started by dividing the springs into the ephemeral and perennial classification of Feth *et al.* (1964). The comparison with the Lake Tahoe area is useful because both lakes are deep, cold, at high altitude, and are surrounded by rocks of relatively uniform (but different) geology. The major difference is that the composition of Lake Tahoe is reasonably explained by the conservative processes of addition of dissolved constituents by stream flow and precipitation and their removal by stream outflow, evaporation, and consumption of silica by diatoms, whereas Crater Lake requires an additional input of dissolved constituents (Nathenson 1990).

To study the processes that form the chemistry of cold-spring waters at Crater Lake, the first step is to divide the samples into similar chemical groups as was done in Nathenson (1989). Based on modified Schoeller plots (e.g., Hem 1985) and locations, the samples have been divided into five groups. Two groups are springs above and below the surface elevation of Crater Lake (Table 1). Chemically, these two groups are quite similar, but it is useful to separate them to be able to show this similarity. Within these two geographic groups, there are a number of anomalous samples that are presented separately. Within the springs above the lake, the two Chaski Slide samples have high sulfate concentrations. Within the springs below the lake, Boundary and Thousand Springs are similar to each other but chemically distinct from the other springs. The sample of Annie Creek at the park boundary is not a spring sample but was collected to study the modification of water composition as it changes downstream. The group of springs below the lake includes only those north of Cedar Springs (Fig. 1). The springs south of the park boundary are divided into two groups based on chemical similarity and location (Table 1, Fig. 1). The group southwest of Cedar Springs shows chemical patterns similar to that for weathering but modified by solubility considerations. The Wood River area group has chemical patterns that are similar to Crater Lake and different from that produced by weathering. The two springs on Minnehaha Creek northwest of the park boundary are considered separately because of their high bicarbonate concentration relative to all the other springs.

ANALYTICAL METHODS

The data of Thompson *et al.* (1987) were collected and analyzed over several years with the objectives of searching for lake water in the springs and for any signal of thermal water. The analytical methods used are discussed in Thompson *et al.* (1987). Because of the small concentrations of dissolved constituents in these waters, using these analyses for studying the process of weathering requires a degree of precision that was not designed into the analytical procedures. Some of the variability in plots of data that will be shown is real but some is an analytical artifact. It is not always possible to distinguish between the two. It is clear that some values are outliers and should not be considered for their implications for processes. Table 1 presents the analytical data of Thompson *et al.* (1987) rearranged into groups based on geographic and chemical similarity. A number of values are shown in parentheses because they are outliers and will not be included in trying to establish patterns. A plot (not shown) of cations and anions in equivalents versus specific conductance established that all the 1981 conductance values are in error. Specific conductance increases systematically with concentrations of anions and cations; however, the intercept is substantially

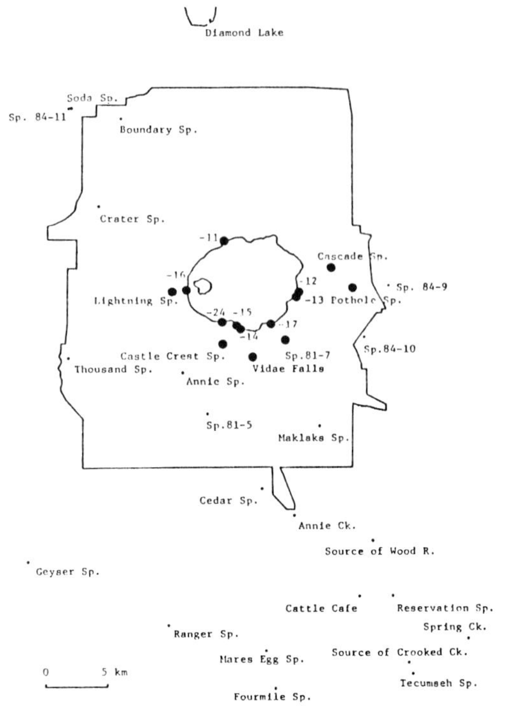

Figure 1. Locations of springs sampled in the Crater Lake area (Thompson *et al.* 1990). Large dots show spring locations that are higher in elevation than the surface of Crater Lake. Numbers near the lake are the last two digits of sample number series JCL 81-. Outline is boundary of Crater Lake National Park.

above zero. The calcium value for sample 84-10 is rejected based on a 97% error in charge balance and because it is anomalous relative to other data in the group. The magnesium value for sample 84-12 is rejected because it is much higher than any cold spring whereas other constituents are in the normal range. The calcium value for sample 84-5 is rejected because it is double the values in its group whereas other constituents are in a similar range.

In addition to these random outliers, there is some systematic bias in the analyses. Figure 2 shows a plot of cations versus anions for the samples from the groups listed as above and below the surface elevation of Crater Lake in Table 1 without the values for the Chaski Slide springs (81-14, 81-15) and samples 84-10 and 84-7. Because the anions are dominated by bicarbonate whereas the cations are made up of several constituents, this plot is close to a plot of cations versus bicarbonate. About half of the analyses are close to the line, but the other half are systematically below the line. Either the bicarbonate

TABLE 1. CHEMICAL ANALYSES OF WATERS IN THE VICINITY OF MOUNT MAZAMA

Sample Numbers	Name or Locality	Date	l pH	SiO_2	Ca	Mg	Na	K	Li	HCO_3	SO_4	Cl	F	B	Cond.	Temp.	$\delta^{18}O$	δD
				←------	------	------	------	------	in mg/L	------	------	------	------	------→	(μS/cm)	°C		
Springs above surface elevation of Crater Lake																		
JCL-81-4	Lightning Springs	9 Aug 81	7.1	26	1.7	0.25	1.5	0.8	<0.01	12	1	0.2	0.13	0.1	(84)	4	-14.23	-96.6
JCL-81-7	Headwater of Lost Creek	10 Aug 81	7.2	32	1.8	.61	1.9	.4	<.01	22	1	.3	.15	<.1	(109)	7.5	--	--
JCL-81-8	Vidae Falls	10 Aug 81	7.1	34	2.1	.70	2.0	.8	<.01	16	<.5	.2	.15	.2	(117)	9	--	--
JCL-81-11	Steel Bay, C.L.	13 Aug 81	7.0	26	.4	.30	1.9	.4	<.01	14	2	.3	.18	<.1	(105)	18	-13.75	-101.7
JCL-81-12	N. 'Pumice Castle' C.L.	13 Aug 81	8.6	36	1.6	.83	2.5	.6	.01	18	<.5	.3	.24	<.1	(102)	9	--	--
JCL-81-13	S. 'Pumice Castle' C.L.	13 Aug 81	8.2	40	1.6	1.1	2.7	1.0	<.01	27	1	.4	.19	<.1	(110)	6.5	-15.45	-110.5
JCL-81-16	'The Watchman Spring	13 Aug 81	6.6	34	1.6	.42	2.1	1.0	<.01	16	<.5	.2	.15	<.1	(110)	5	--	--
JCL-81-17	Dutton Cliff	13 Aug 81	7.8	36	1.1	.92	4.2	.6	<.01	21	1	.2	.16	<.1	(115)	14	--	--
JCL-81-24	Spring near C. L. Lodge	17 Aug 81	6.3	30	2.0	.29	1.9	.7	<.01	24	<.5	.3	.17	.2	(107)	6	--	--
JCL-84-8	Pothole Spring	4 Aug 84	6.68	42.7	2.7	.90	2.9	1.7	<.01	29	.2	.5	<.1	.1	30	3	-15.1	-110
JCL-82-1	Cascade Spring	31 Aug 82	7.06	40.5	2.5	.90	2.9	1.4	<.01	37	<.2	.2	.03	.4	--	3.5	-15.11	-108.4
	Mean composition			34	1.7	0.7	2.4	0.9		21	0.8	0.3						
JCL-81-14	'Chaski Slide-E', C.L.	13 Aug 81	7.06	26	4.9	1.7	2.0	.9	<.01	19	12	.1	.19	<.1	(125)	12	--	--
JCL-81-15	'Chaski Slide-W', C.L.	13 Aug 81	6.2	22	10.1	3.2	3.3	.4	<.01	20	26	.2	.20	<.1	(145)	9.5	-13.88	-105.2
Springs below lake elevation and north of Cedar Spring																		
JCL-81-1	Annie Spring	8 Aug 81	7.2	38	2.9	1.0	2.6	0.8	<0.01	15	4	0.4	0.17	0.1	(144)	4	-13.89	-99.4
JCL-84-1	Annie Spring	3 Aug 84	5.39	40.5	2.0	1.4	3.0	2.2	<.01	30	<1	1.2	.7	.2	44	3	-13.9	-99.5
JCL-81-5	Lodgepole Picnic area	9 Aug 81	7.2	36	1.9	.48	2.8	1.5	<.01	17	<.5	.3	.14	<.1	(102)	5	--	--
JCL-81-6	Maklaks Spring	10 Aug 81	6.9	24	1.8	.41	1.8	.8	<.01	12	1	.2	.14	<.1	(105)	10.5	--	--
JCL-83-1	Crater Spring	7 Aug 83	6.34	35.1	3.0	1.1	3.0	1.6	<.01	32	<2	.8	.1	<.1	--	3.0	-13.56	-97.4
JCL-84-9	Unnamed spring nr road	4 Aug 84	6.79	45.4	3.0	1.3	3.6	2.1	<.01	34	.1	.5	<.1	.1	47	4	-15.2	-108
JCL-84-10	Unnamed spring, source of Cr 1/4 mi S of Scott Cr	4 Aug 84	6.94	31.4	(17.8)	.53	2.4	1.4	<.01	21	.4	.5	<.1	.2	29	6	-15.0	-103
	Mean composition			36	2.4	0.9	2.7	1.5		23	1.3	0.6						
JCL-81-3	Boundary Springs	9 Aug 81	7.6	34	4.3	2.4	3.3	.5	<.01	25	3	.2	.15	<.1	(120)	5	-13.76	-98.1
JCL-81-9	Thousand Springs	11 Aug 81	7.3	34	4.8	2.5	2.5	1.1	<.01	27	2	.2	.16	<.1	(129)	5	-13.70	-99.2
	Mean composition			34	4.6	2.4	2.9	0.8		26	2	0.2						
JCL-84-7	Annie Creek at boundary	3 Aug 84	N.R.	39.8	6.43	1.1	3.4	1.6	<.01	32	1.8	.5	<.1	.2	50.3	10	-14.2	-98
Cedar Springs area springs																		
JCL-84-12	Mare's Egg Spring	5 Aug 84	7.70	34.8	10.8	(9.0)	4.2	1.5	<0.01	55	0.3	0.5	0.05	0.2	77	4	-14.4	-101.
JCL-84-13	Four-mile Spring	5 Aug 84	7.96	31.7	6.0	2.4	4.4	1.5	<.01	54	.5	1.4	<.1	.2	74	5	-14.1	-98.
JCL-84-14	Ranger Spring	5 Aug 84	N.R.	38.9	14.2	1.0	3.0	2.0	<.01	34	<.1	.9	<.1	.2	47	2	-13.6	-95.
JCL-84-15	Cedar Springs	5 Aug 84	6.37	39.2	11.6	1.2	3.4	1.6	<.01	44	<.1	.5	<.1	<.1	60	7	-13.4	-101.
JCL-84-16	Geyser Spring	6 Aug 84	N.R.	30.7	7.5	2.7	3.3	1.2	<.01	57	.1	.5	<.1	.2	75	5	-12.9	-91
	Mean composition			35	10.0	1.8	3.7	1.6		49	0.2	0.8						
Wood River area springs																		
JCL-81-10	Source of Wood River	11 Aug 81	7.3	40	5.6	2.7	6.1	1.0	0.01	34	5	3.2	0.18	0.2	(132)	9.5	-14.87	-107.6
JCL-84-4	Source of Wood River	3 Aug 84	6.74	45.8	2.1	2.4	6.6	1.9	<.01	47	1.8	2.8	.10	.2	50	12	-15.1	-105.5
JCL-82-2	Cattle Crossing Cafe	1 Sep 82	7.15	40.0	2.7	2.9	10.8	.7	<.01	63	<.2	.2	.13	1.1	--	COLD	-14.04	-101.1
JCL-84-2	Tecumseh Spring	3 Aug 84	7.88	34.2	7.5	1.8	12.5	1.4	<.01	58	3.4	4.9	.17	.2	95.5	11	-14.7	-106.8
JCL-84-3	Source of Crooked Cr	3 Aug 84	7.90	36.3	8.0	2.4	15.6	1.9	<.01	53	6.2	8.4	.16	.4	126	11	-14.7	-108.0
JCL-84-5	Reservation Spring	3 Aug 84	7.58	39.5	(15.7)	1.9	10.8	2.1	.01	50	4.6	5.8	.14	.1	103	8	-14.6	-106.
JCL-84-6	Source of Spring Cr	3 Aug 84	7.51	40.8	3.2	1.8	8.5	1.3	<.01	46	2.4	3.3	.12	.1	60	6	-14.3	-105
	Mean composition			40	4.9	2.3	10.1	1.5		50	3.4	4.1						
Springs on Minnehaha Creek																		
JCL-85-7	Soda Spg on Minnehaha Cr	6 Aug 85	5.31	71	271.	243	106.	31.5	0.06	2280	16.	17.7	0.04	0.43	3620	10	-14.3	-102
JCL-84-11	Unnamed spring on Minnehaha Creek nr Soda spring	4 Aug 84	N.R.	98.6	23.7	5.1	89	9.7	.03	417	.8	4.2	.12	.2	320	10	-13.2	-90.
Lakes																		
JCL-84-av	Crater Lake	7 Aug 84	6.8	19	7	2.3	10.4	1.7	0.05	41	8.	10.0	0.1	0.5	--	--	-9.8	-79.
JCL-81-2	Diamond Lake, S End	9 Aug 81	7.3	3.6	1.7	1.0	3.2	.8	<.01	30	1	.2	.13	.1	(121)	22.5	-10.85	-83.2

( ) Based on balance errors and chemical patterns, these values are considered to be outliers. Outliers are not included in calculated mean compositions.

concentrations are systematically high, the cations are systematically low, or there is some combination of the two errors. Without additional studies, it is not possible to establish which is the case. Because of this bias, small differences in chemistry should not be interpreted.

SPRING CHEMISTRY AND ROCK WEATHERING

The springs above and below the surface elevation of Crater Lake are chemically similar. Figure 3 shows modified Schoeller plots for these two groups. The ionic species are plotted in milliequivalents per liter (meq/L), so that relative proportions of constituents will correspond to those in a Piper diagram (e.g., Hem 1985, p. 179). Silica is shown in millimoles per liter (mmol/L), because it is not an ionic species. For singly charged species, the value of the concentration in meq/L is the same as the value in mmol/L; but for doubly charged species, the concentration in meq/L is twice that in mmol/L. In general, the samples have nearly equal concentrations of Ca and Na with less Mg and even lower K. Cl and SO_4 are quite low, and HCO_3 and SiO_2 tend to increase together systematically. Although constituents show systematic trends, it is

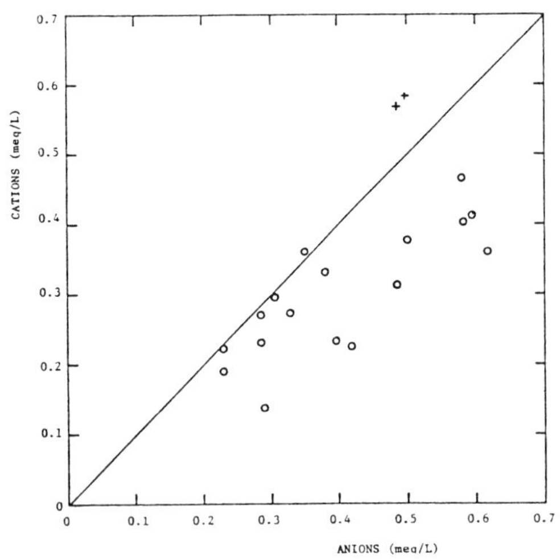

Figure 2. Cations versus anions for springs above and below lake elevation as given in Table 1. Plus symbols are values for Boundary and Thousand Springs. Samples 84-7, 84-10, and Chaski Slide springs not plotted.

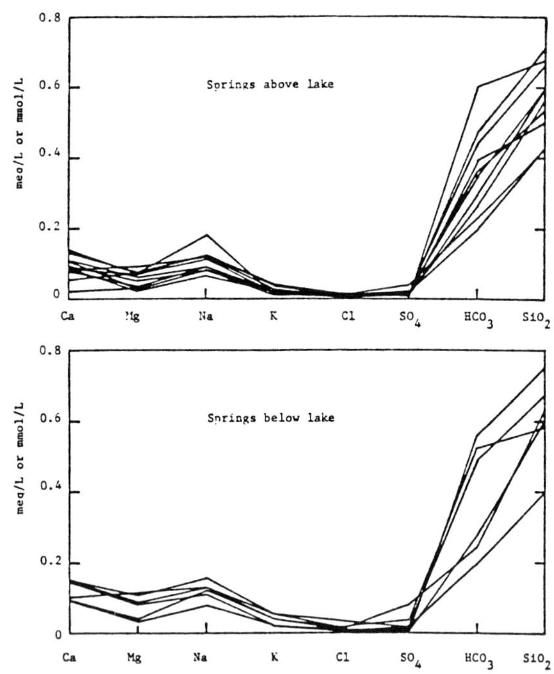

Figure 3. Modified Schoeller plots for springs above and below lake. Ionic species are in meq/L and silica is in mmol/L. Samples 84-7 and 84-10 not plotted. Other springs in these groups shown in Figure 4.

useful to calculate mean compositions to use for comparison and modeling. The mean compositions given in Table 1 for these two groups are similar. The charge-balance errors for the means of these two groups are 31 % and 19 %, so the systematic bias shown in Fig. 2 is in both groups.

Figure 4 shows plots for the Chaski Slide springs from the group of springs above the lake, and Boundary and Thousand Springs from the group of springs below the lake. The Chaski Slide springs are high in sulfate, calcium, and magnesium compared to other springs above the lake. The springs occur in a large block of hydrothermally altered volcanic rock, and the alteration is a likely source of this anomaly. The plot of the chemistry of Boundary and Thousand Springs shows that they are similar (Fig. 4, upper) although located quite some distance apart (Fig. 1). These springs have noticeably higher concentrations of calcium and magnesium than other springs above and below the lake elevation.

The Cedar Springs area group includes Cedar Springs and four other springs southwest of the park boundary (Fig. 1, Table 1). The Mg value of 9.0 mg/L for Mare's Egg Spring is shown on Fig. 5, but the tie lines are not drawn because the point is an outlier as discussed above. The data in Fig. 5 show that as bicarbonate increases, the silica concentration actually starts to decrease. This is an indication that the amount of silica is being limited by solubil-

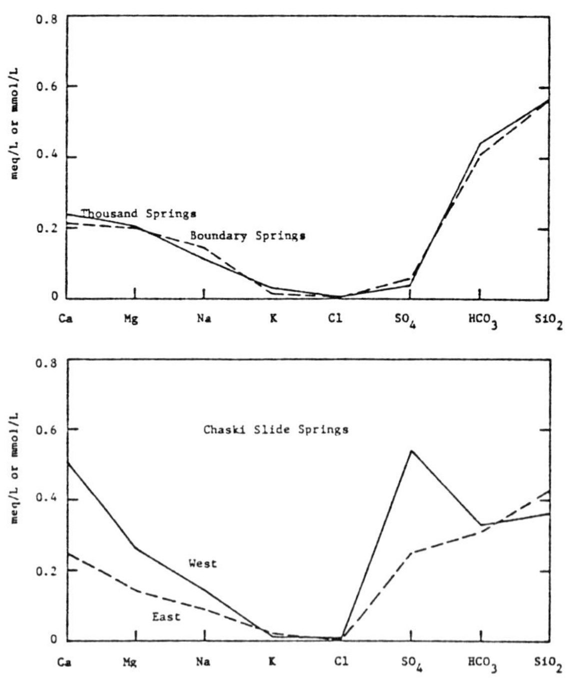

Figure 4. Modified Schoeller plots for springs above and below lake not shown in Fig. 3.

ity considerations as was also found for spring-water at high bicarbonate concentrations in the Lake Tahoe area (Nathenson 1989). Figure 6 shows a plot of specific conductance versus cation and anion concentrations for the Cedar Springs area group to assess data quality. The anions define a linear variation that can easily be passed through zero but the cations are quite variable. The highest conductance is for Mare's Egg Spring, and the outlier magnesium value is reflected in the large deviation of the cation value. The lowest conductance value is for Ranger Spring, and the large difference between the anions and cations may indicate that the calcium value is too high. Although the anion values closely follow a linear variation with conductance, the bias shown

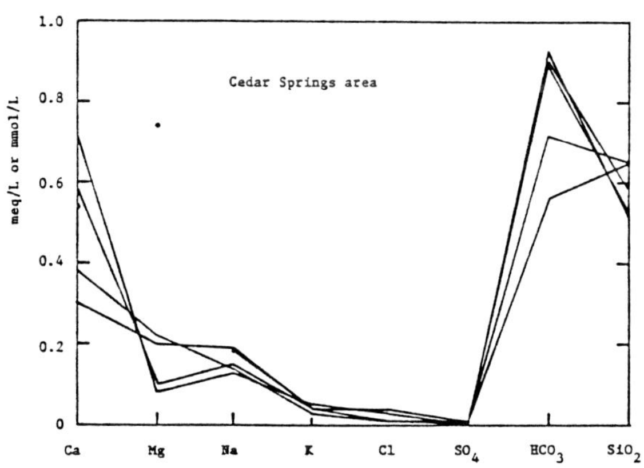

Figure 5. Modified Schoeller plots for Cedar Springs area springs from Table 1. Magnesium value for Mare's Egg Spring is shown as a dot but is not connected to other values for this spring.

in Fig. 2 prevents one from assuming that this is the true relation. Unlike the data in Fig. 2, the data in Fig. 6 do not indicate a consistent bias between anions and cations.

The systematic pattern of water chemistry of springs above and below the lake (Fig. 3) suggests that there is a common process causing this chemistry. In their study of waters in the Sierra Nevada, Garrels and MacKenzie (1967) suggested reactions of several minerals that could be important. The basic weathering process, as they outlined it, is that carbon dioxide dissolves in precipitation, forming an acid solution. This acid solution reacts with minerals in the rock to produce dissolved silica, bicarbonate, and major cations and residual clay. The major rock at Crater Lake is volcanic glass (which

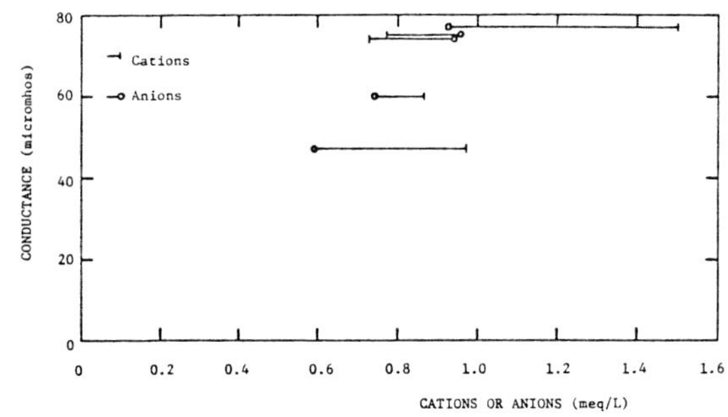

Figure 6. Specific conductance versus concentrations in meq/L for Cedar Springs area springs.

should be relatively easily dissolved), and the major minerals are plagioclase and pyroxenes (C. R. Bacon, oral comm., 1988). Based on the reported composition of the glass sample 81C-563G (Bruggman *et al.* 1987), we can convert the analysis to a chemical formula for the composition of glass. The plagioclase is intermediate in composition between albite and anorthite, and the clinopyroxene is approximately diopside in composition. Reactions for these constituents may be written assuming that the end product is kaolinite clay as was found in the Sierra Nevada:

(1) Glass

$$Na_{0.45}Ca_{0.082}K_{0.16}Mg_{0.037}Al_{0.77}Si_{3.21}O_8 + 0.85\ CO_2 + 1.195\ H_2O = 0.45\ Na^+ + 0.164 \times 1/2\ Ca^{+2} + 0.074 \times 1/2\ Mg^{+2} + 0.16\ K^+ + 0.85\ HCO_3^- + 2.44\ SiO_2 + 0.385\ \underset{\text{Kaolinite}}{Al_2Si_2O_5(OH)_4}$$

(2) Plagioclase

$$Na_{0.5}Ca_{0.5}Al_{1.5}Si_{2.5}O_8 + 1.5\ CO_2 + 2.25\ H_2O = 0.75\ \underset{\text{Kaolinite}}{Al_2Si_2O_5(OH)_4} + 0.5\ Na^+ + 1.0 \times 1/2\ Ca^{+2} + SiO_2 + 1.5\ HCO_3^-$$

(3) Clinopyroxene

$$CaMgSi_2O_6 + 4\ CO_2 + 2\ H_2O = 2 \times 1/2\ Ca^{+2} + 2 \times 1/2\ Mg^{+2} + 4\ HCO_3^- + 2\ SiO_2$$

The reactions are written with a factor 1/2 in front of the doubly charged species so that calculations in milliequivalents are easily done.

Based on these reactions, Table 2 shows calculations of water compositions compared to the calculated mean compositions from Table 1. The mean compositions are used as a convenient base for comparison. Concentrations for a number of constituents in precipitation collected at Medford, Oregon, (90 km southwest of Crater Lake) measured by

TABLE 2. CALCULATION OF WATER COMPOSITIONS (in meq/L for dissolved ions and mmol/L for silica) BASED ON REACTIONS (1) TO (3)

	Ca	Mg	Na	K	Cl	SO_4	HCO_3	SiO_2
Average of springs below lake	0.120	0.074	0.117	0.038	0.017	0.027	0.377	0.599
2 x Precipitation	0.052	0.016	0.014	0.006	0.012	0.034	0.032	0.004
0.23 x glass to kaolinite	0.038	0.017	0.104	0.037			0.196	0.561
0.018 x clinopyroxene	0.036	0.036					0.072	0.036
Calculated water composition	0.126	0.069	0.118	0.043	0.012	0.034	0.300	0.601
Average of Boundary and Thousand Springs	0.230	0.197	0.126	0.020	0.006	0.042	0.426	0.566
2 x Precipitation	0.052	0.016	0.014	0.006	0.012	0.034	0.032	0.004
0.23 x glass to kaolinite	0.038	0.017	0.104	0.037			0.196	0.561
0.070 x clinopyroxene	0.140	0.140					0.280	0.140
Calculated water composition	0.230	0.173	0.118	0.043	0.012	0.034	0.508	0.705
Average of Cedar Springs area springs	0.499	0.148	0.161	0.041	0.023	0.004	0.803	0.582
2 x Precipitation	0.052	0.016	0.014	0.006	0.012	0.034	0.032	0.004
0.3 x glass to kaolinite	0.049	0.022	0.135	0.048			0.255	0.732
0.050 x clinopyroxene	0.100	0.100					0.200	0.100
Calculated water composition	0.201	0.138	0.149	0.054	0.012	0.034	0.487	0.836
Residual Ca, HCO_3, and SiO_2	0.298						0.316	-0.254

Junge and Werby (1958) are given in Table 2. Junge and Werby (1958) did not measure all the major constituents, and I have estimated values for silica, magnesium, and bicarbonate based on a comparison with the values in the Sierra Nevada given in Feth *et al.* (1964). Crippen and Pavelka (1970) found for the Lake Tahoe area that about half of the precipitation on the land surface evaporated before infiltrating or flowing into streams; therefore the concentrations in precipitation are doubled in Table 2. For springs below the lake (springs above the lake have nearly the same mean composition), the contribution from the reaction glass to kaolinite was based on matching the amount of sodium in the average spring water. The remaining cations are calcium and magnesium, which can be obtained from the clinopyroxene reaction. The largest source of dissolved constituents is volcanic glass. The cations of the average composition of springs below lake level total 0.35 and the anions 0.42, so the lack of perfect agreement of all constituents is not surprising. If the cations in the average composition were higher, the calculated concentration of bicarbonate would be higher and the agreement would be closer. Conversely, if the anions were lower, the calculated contribution for bicarbonate would agree more closely. Considering all the uncertainties, the comparison of the calculated and average compositions is quite good.

No additional chloride or sulfate beyond that in precipitation is required to produce the composition of the springs below the surface elevation of Crater Lake (Table 2). The precision of these two constituents is such that small losses or additions cannot be detected in the data, but the data indicate that there is no large source of chloride or sulfate required in the weathering process. The maximum chloride in Crater Lake rocks is about 0.13% (Bruggman *et al.* 1987). Combining this value with the glass analysis used above, the mole ratio of chloride to sodium is about 0.023. Assuming that all the chloride would dissolve with the sodium, the added chloride is 0.023 x 0.104 = 0.0024 meq/L or 0.08 mg/L. Thus, the amount of chloride that could be added by a simple weathering process is much less than the amount from precipitation.

The calculated composition of Boundary and Thousand Springs (Table 2) differs from the calculated composition of the springs below the lake only in the dissolution of more clinopyroxene. The contribution from the reaction of glass to kaolinite is the same as for the springs below the lake, based on the nearly identical sodium concentration, and the contribution from the clinopyroxene reaction is calculated based on the calcium concentration. Again, the average analysis is not perfect, so that it is not surprising that the agreement is not perfect. The calculated composition for the Cedar Springs area springs (Table 2) has residual calcium and bicarbonate and missing silica compared to the actual composition. The silica could be brought into better agreement if the amount obtained from glass were reduced and dissolution of plagioclase were added, but this does not change the fundamental point that the chemical composition of the springs in the Cedar Springs area is not produced by a simple weathering process. The composition of these waters requires that solubility considerations become important. This same affect was found at the higher concentrations of bicarbonate in the study of the Tahoe area springs (Nathenson 1989).

In order to explore these relationships further, we can plot certain constituents versus others to see if the variation follows what is proposed in Table 2 and the reactions (1) to (3). Figure 7 shows silica versus sodium. The reaction of glass to kaolinite requires that the slope of the line should be 5.4:1. Considering all the uncertainties, the line is an excellent match to the data for the springs above and below the lake. The data for the Chaski Slide springs are not shown on the figure, because they involve addition of sulfate. For Boundary and Thousand Springs (plus symbols on Fig. 7), most of the silica is calculated to come from the glass reaction (Table 2), and the plotted points still approximately follow the line. The points for the Cedar Springs area samples (circles on Fig. 7) show a significant deviation from the line, which is to be expected based on the decrease in silica with increasing bicarbonate shown in Fig. 5 and the calculations in Table 2. The plot and calculations for the Lake Tahoe springs show a slope of between 2:1 and 1:1 (Fig. 7 in Nathenson 1989), showing that the local mineralogy has a large effect on what is dissolved in the weathering process.

The major parameter indicating the degree of re-

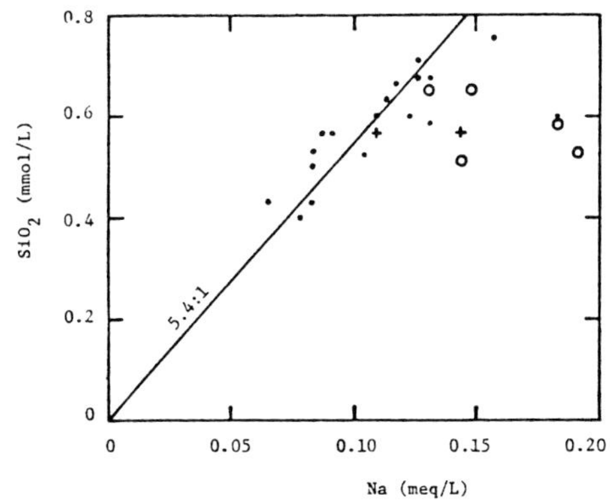

Figure 7. Silica versus sodium for springs above and below Crater Lake (points), Boundary and Thousand Springs (plus symbols), and Cedar Springs area springs (circles). Slope of 5.4:1 is based on reaction of glass to

action is the bicarbonate concentration, and Fig. 8 shows silica versus bicarbonate. The predicted slope from the reaction of glass to kaolinite is 2.9:1, and a line of this slope is shown on the figure. Because of the bias shown in Fig. 2, it is worthwhile to show this relationship in another way. In the reaction of glass to kaolinite, silica should also increase with a slope of 2.9:1 if it is plotted versus the sum of the cations, and this plot is shown in the bottom of Fig. 8. The bottom plot shows a more consistent variation than the top plot. For Boundary and Thousand Springs, most of the bicarbonate is generated by the clinopyroxene reaction rather than the glass reaction, and the amount of silica generated per amount of bicarbonate is smaller in the clinopyroxene reaction (Table 2). Thus, it is to be expected that these springs would plot away from the line. The data points from the Cedar Springs area springs deviate significantly from the proposed relationship, indicating that the silica concentration is probably limited by solubility considerations. The plot and calculations for the Lake Tahoe springs show a slope of between 1:1 and 0.5:1 (Fig. 8 in Nathenson 1989), so the concentration of silica added for a given bicarbonate concentration is much higher in the volcanic rocks of Crater Lake.

As the water from the springs flows into streams, it is likely that additional processes modify the chemistry. Figure 9 (top) shows the data from two samples of Annie Spring and a sample from Annie Creek at the south park boundary. There is a suggestion that calcium and bicarbonate have increased in Annie Creek, and a likely mechanism is the addition

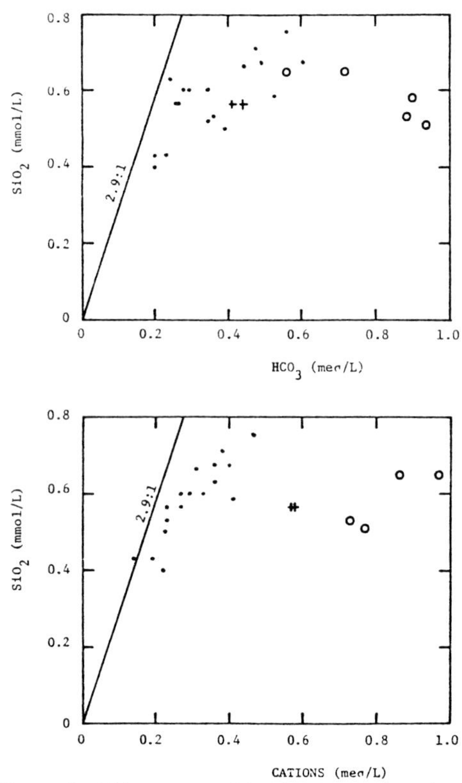

Figure 8. Silica versus bicarbonate and cations for springs above and below Crater Lake (points), Boundary and Thousand Springs (plus symbols), and Cedar Springs area springs (circles). Slope of 2.9:1 is based on reaction of glass to kaolinite.

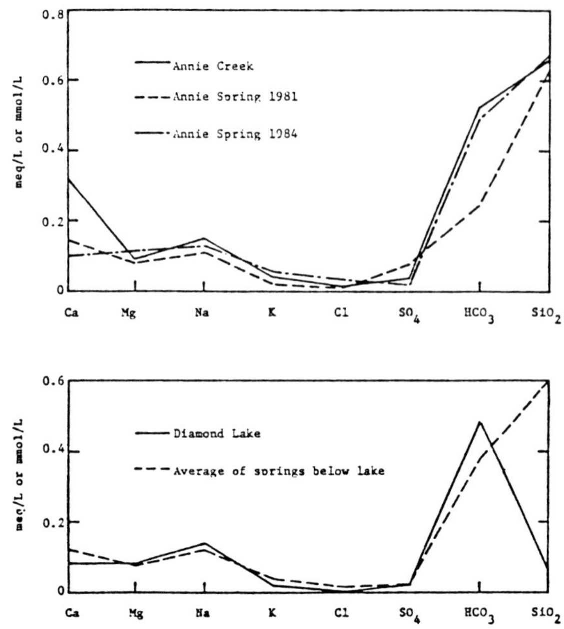

Figure 9. Modified Schoeller plots for samples listed.

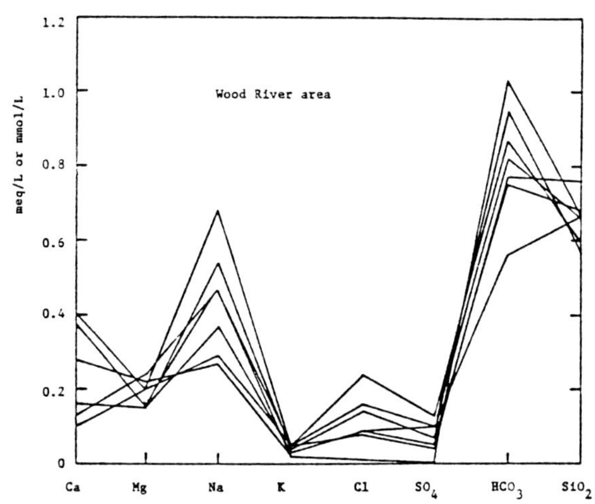

Figure 10. Modified Schoeller plots for Wood River area springs. Calcium value for Reservation Spring is shown as a dot but is not connected to other values for this spring.

of a water such as that found in the Cedar Springs area. The variation in other constituents is within the analytical error, and more data would be required to look for small effects. Figure 9 (bottom) shows the data for Diamond Lake and the average concentration of springs below the elevation of Crater Lake. To the accuracy of the analyses, Diamond Lake appears to reflect the composition of this average except for a large loss of silica, probably due to the action of diatoms in the lake. The lake is only 14 m deep and quite productive (Lauer *et al.* 1979), so the consumption of nearly all the silica is possible.

WOOD RIVER AREAS SPRINGS AND CRATER LAKE

The Wood River area springs (the springs that are the source of the Wood River and five other springs southeast of the park boundary) have a different chemistry than the springs discussed above. Chloride, sulfate, and sodium concentrations increase systematically with each other (Fig. 10). The concentrations of sodium, sulfate, and chloride are higher than in the waters discussed above (except for the sample from the Cattle Crossing Cafe which has low chloride and sulfate but high sodium). Figure 11 (top) shows the average composition of these springs compared to that of the Cedar Springs area samples. The concentrations of sodium, chloride, and sulfate are quite high compared to the Cedar

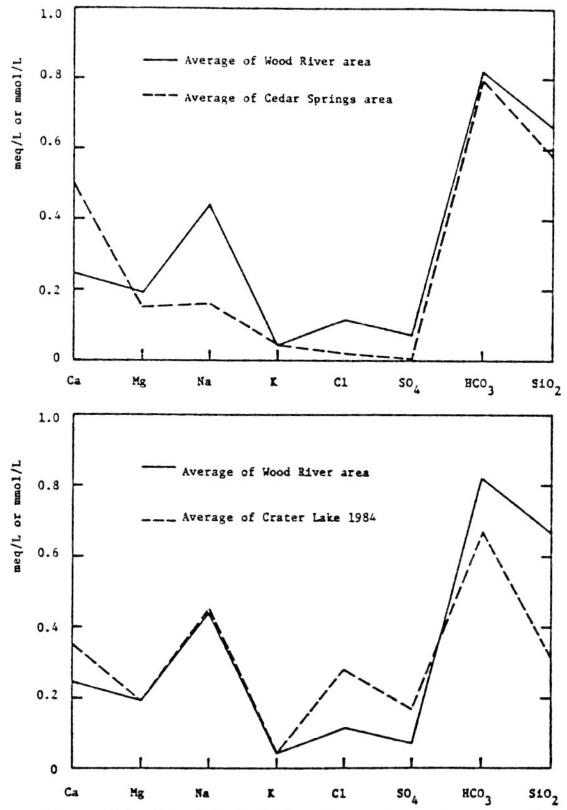

Figure 11. Modified Schoeller plots for samples listed.

Springs area data. The composition of the Wood River area springs cannot be explained as a simple weathering reaction of the volcanic rocks of Mount Mazama. The springs occur at a low elevation (about 4200 ft) compared to the lake elevation of about 6200 ft. It is possible that buried lake sediments are contributing the anomalous constituents (Thompson *et al.* 1987). The elevation of Wood River area springs corresponds approximately to the elevation of the bottom of Crater Lake. Figure 11 (bottom) compares the composition of the these springs to that of Crater Lake. Except for silica, the Wood River area springs and Crater Lake could be considered to be members of the same chemical group. Diatoms consume silica in Crater Lake, and the amount of silica in the lake is lowered by their deposition in the sediments (Nathenson 1990). Thompson *et al.* (1987) showed that the Wood River area springs are not a mixture of Crater Lake water with other water based on a plot of deuterium versus chloride. However, the comparison shown in Fig. 11 suggests that the same chemical process that provides the anomalous constituents in Crater Lake could provide the anomalous constituents in the Wood River area springs. Wells and Peck (1961) and Kienle *et al.* (1981) showed a fault in the area of the Wood River springs which could provide a source of the chloride and sulfate through deep circulation of water along the fault. In that case, the similarity of chemistry between Crater Lake and these springs would be a coincidence. Data needed to differentiate between these alternative explanations would be chemistry and stable isotopes for the water that feeds dissolved constituents into Crater Lake and the water that is the end member for the Wood River springs. The occurrence of chloride and sulfate may suggest a high-temperature process; however, geothermometer temperatures (Fournier 1981) for the average water of the Wood River area are 61°C for chalcedony and 46°C for Na-K-Ca (no magnesium correction is necessary). The systematic variation of chloride with sulfate would be consistent with these springs being a series of mixed waters with one end member having high chloride and sulfate that has not been sampled. The magnesium concentrations of these samples may be too high to be consistent with a hydrothermal origin; however, the magnesium could be added by continued reaction after mixing. Thompson *et al.* (1987) have proposed that high concentrations of boron and lithium reflect addition of thermal water to Crater Lake, and Fig. 12 shows data from their Table 4 for these two constituents. The boron data for the Wood River area springs (plus symbols) lie along a mixing line with Crater Lake, while the lithium may or may not define a mixing line. The Wood River area springs appear to be different from Crater Lake's chemistry in that they do not follow a chloride/lithium mixing relationship with Crater Lake; however, they are similar in that they contain significant lithium.

SPRINGS ON MINNEHAHA CREEK

Figure 13 shows the analyses for two springs on Minnehaha Creek. The two springs are high in bicarbonate relative to all other springs. The stable isotope data for the two springs plot on the meteoric water line but have quite different values (Table 1 and Thompson *et al.* 1987). The soda spring is dominantly a calcium-magnesium-bicarbonate spring while the other spring is a sodium-bicarbonate spring. Even though total ion contents differ by a factor of 5, amorphous silica temperatures (Fourn-

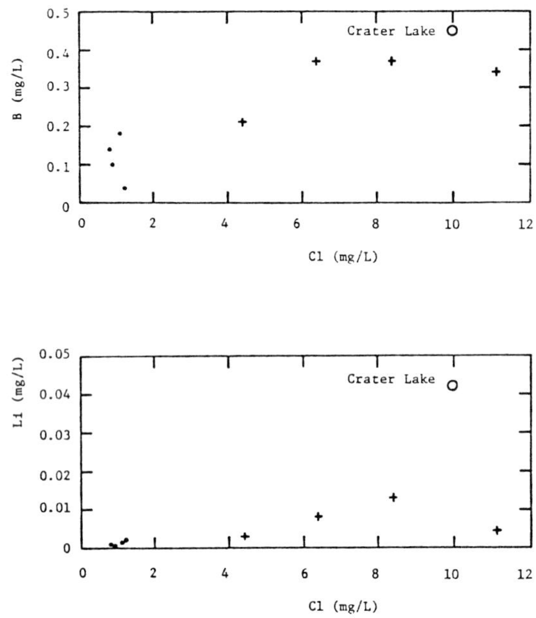

Figure 12. Boron and lithium versus chloride for Crater Lake, Wood River area springs (plus symbols) and other cold springs from Table 4 of Thompson *et al.* (1987).

ier 1981) are similar (1°C for the soda spring and 16°C for sample 84-11). Measured temperatures are 10°C, and the concentration of silica in solution may reflect dissolution of rock by large amounts of carbon dioxide in solution near the mean annual temperature. Carbon dioxide springs occur in the Western and High Cascades both north and south of these springs (Wagner 1959; Irwin and Barnes 1982). Umqua Hot Springs to the north is 46°C (Mariner *et al.* 1978), but other soda springs in the Cascades are less than 15°C and are not warm enough to be considered thermal (Wagner 1959). The relation of the springs on Minnehaha Creek to the input of constituents to Crater Lake is unclear. The chemical patterns shown on Fig. 13 are quite different from that for Crater Lake water (Fig. 11). The soda spring does have elevated chloride and sulfate compared to other spring waters (Table 1); however, its bicarbonate concentration is much higher than these other constituents. It seems unlikely that a water such as that found in the soda spring could be feeding Crater Lake and then dropping most of its bicarbonate to produce the pattern shown in Fig. 11. The high bicarbonate concentrations in these two springs, however, demonstrates that there is a significant source of carbon dioxide available locally in the subsurface. Depending on the rocks that it passes through, a water charged with carbon dioxide can produce a wide variety of water compositions.

CONCLUSIONS

The chemistry of most of the cold springs in the Crater Lake area can be explained by a simple weathering process in which volcanic glass and clinopyroxene react with water containing dissolved carbon dioxide to produce kaolinite clay. At higher concentrations of bicarbonate, the concentration of silica in solution is limited by the solubility of an unidentified phase. Springs that do not satisfy this simple model are the springs on Minnehaha Creek and the group of springs in the Wood River area. The soda springs are an unlikely source of the dissolved constituents in Crater Lake, but their chemistry demonstrates that, in some areas, there is a substantial amount of carbon dioxide in the subsurface available to drive chemical reactions. The springs in the Wood River area are chemically quite similar to Crater Lake water. The process that generates their chemistry could be the same as the process that generates the chemistry of the inflow to

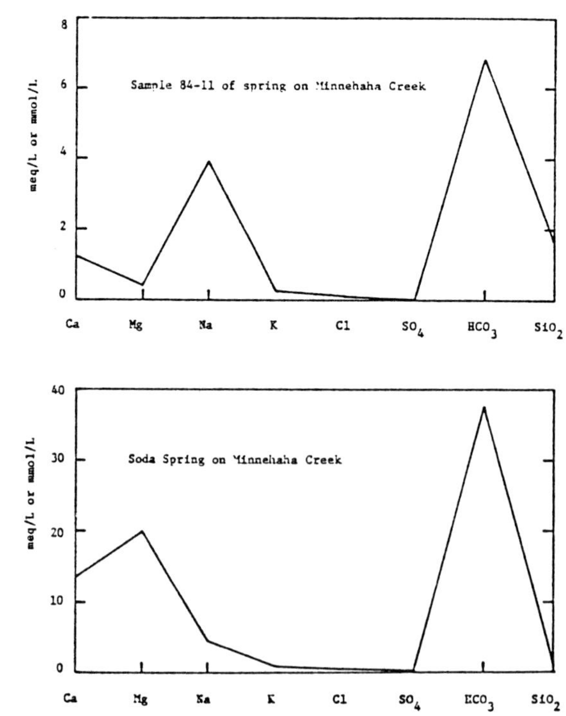

Figure 13. Modified Schoeller plots for springs on Minnehaha Creek.

Crater Lake, or the similarity of the chemistry could be a coincidence. Until the source water of both is characterized, it is not possible to settle this question.

Nathenson (1990) has used a chemical balance for Crater Lake to determine rates of addition of major ions to keep the lake at its present chemical composition. A major unknown in that work is whether the dissolved constituents are carried in a low flow of water with high concentrations or a high flow with low concentrations. However, the relative proportions of major elements in either case are similar to that for Crater Lake, so that the comparison in Fig. 11 with the springs in the Wood River area remains valid. If the flow into Crater Lake is low with high concentrations of major elements, the comparison in Fig. 11 would have to be based on diluting the flow to bring it to similar concentrations to those found in the Wood River area springs. The data in Fig. 10 show that the Wood River area springs are probably a mixture of two end-member compositions. The comparison with Crater Lake cannot be done in detail without knowing the composition of the high-chloride and -sulfate end-member component for the Wood River area springs.

ACKNOWLEDGMENTS

Robert Mariner, Robert Fournier, John Feth and Charles Bacon provided helpful reviews.

LITERATURE CITED

Bruggman, P. E., C. R. Bacon, P. J. Aruscavage, R. W. Lerner, L. J. Schwarz, and K. C. Stewart. 1987. Chemical analyses of rocks and glass separates from Crater Lake National Park and vicinity, Oregon. U. S. Geol. Surv. Open-File Rep. 87-57. 36 pp.

Crippen, J. R., and B. R. Pavelka. 1970. The Lake Tahoe basin, California-Nevada. U. S. Geol. Surv. Water-Supply Pap. 1972. 56 pp.

Feth, J. H., C. E. Roberson, and W. L. Polzer. 1964. Sources of mineral constituents in water from granitic rocks, Sierra Nevada, California and Nevada. U. S. Geol. Surv. Water-Supply Pap. 1535-I. 70 pp.

Fournier, R. O. 1981. Application of water geochemistry to geothermal exploration and reservoir engineering. Pages 109-143 *in* L. Rybach and L. J. P. Muffler, eds., Geothermal Systems: Principles and Case Histories. John Wiley, New York.

Garrels, R. M. 1967. Genesis of some ground waters from igneous rocks. Pages 405-420 *in* P. H. Abelson, ed., Researches in Geochemistry, vol. 2. John Wiley, New York.

Garrels, R. M. and F. T. MacKenzie. 1967. Origin of the chemical compositions of some springs and lakes. Pages 222-242 *in* Equilibrium Concepts in Natural Water Systems. 151st Meet. Amer. Chem. Soc., 1966. American Chemical Society, Washington, DC.

Hem, J. D. 1985. Study and interpretation of the chemical characteristics of natural water. U. S. Geol. Surv. Water-Supply Pap. 2254 (3rd ed.). 263 pp.

Irwin, W. P., and I. Barnes. 1982. Carbon dioxide rich springs and gas wells in the conterminous United States. U. S. Geol. Surv. Misc. Investigations Ser. Map I-1301, scale 1:5,000,000.

Junge, C. E., and R. T. Werby. 1958. The concentration of chloride, sodium, potassium, calcium, and sulfate in rain water over the United States. Jour. Meteor. 15:417-425.

Kienle, C. F., C. A. Nelson, and R. D. Lawrence. 1981. Faults and lineaments of the southern Cascades, Oregon. Oregon Dep. Geol. Mineral Industries, Spec. Pap. 13. 23 pp., 2 pls. (map, scale 1:250,000).

Lauer, W. L., G. S. Schuytema, W. D. Sanville, F. S. Stay, and C. F. Powers. 1979. The effects of decreased nutrient loading on the limnology of Diamond Lake, Oregon. U. S. Environ. Protection Agency Rep. EPA-600/8-79-017a. 59 pp.

Mariner, R. H., C. A. Brook, J. R. Swanson, and D. R. Mabey. 1978. Selected data for hydrothermal convection systems in the United States with estimated temperatures ≥ 90°C: Back-up data for U.S. Geol. Surv. Circ. 790. U. S. Geol. Surv. Open-File Rep. 78-858. 493 pp.

Nathenson, M. 1989. Chemistry of Lake Tahoe, California-Nevada, and nearby springs. U. S. Geol. Surv. Open-File Rep. 88-641. 27 pp.

Nathenson, M. 1990. Chemical balance for major elements in water in Crater Lake, Oregon. Pages 103-114 *in* E. T. Drake *et al.*, Crater Lake: An Ecosystem Study. Pacific Div., Amer. Assn. Advance. Sci., San Francisco, Calif.

Simpson, H. J., Jr. 1970. Closed basin lakes as a tool

in geochemistry. Ph.D. Thesis. Columbia Univ., New York. 325 pp.

Thompson, J. M., L. D. White, and M. Nathenson. 1987. Chemical analyses of waters from Crater Lake, Oregon, and nearby springs. U. S. Geol. Surv. Open-File Rep. 87-587. 26 pp.

Thompson, J. M., M. Nathenson, and L. D. White. 1990. Chemical and isotopic compositions of waters from Crater Lake, Oregon, and nearby vicinity. Pages 91-102 *in* E. T. Drake *et al.*, eds., Crater Lake: An Ecosystem Study. Pacific Div., Amer. Assn. Advance. Sci., San Francisco, Calif.

Wagner, N. S. 1959. Natural sources of carbon dioxide in Oregon. The Ore.-Bin 21(11):103-113.

Wells, F. G., and D. L. Peck. 1961. Geologic map of Oregon west of the 121st meridian. U. S. Geol. Surv. Misc. Investigations Map I-325, scale 1:500,000.

CRATER LAKE CLIMATE AND LAKE LEVEL VARIABILITY

Kelly T. Redmond[1]
Office of State Climatologist
Department of Atmospheric Sciences
Oregon State University
Corvallis, Oregon 97331

A consistent climatic record exists from the Crater Lake area since 1931, except for major gaps during World War II. The level of Crater Lake has varied by nearly five meters during the 20th Century. Climatic variability can account for these lake-level changes. The major elements of the hydrologic budget responsible for year-to-year level changes are precipitation, evaporation, and seepage. The unique setting of the lake and the availability of moderately long records of daily weather and lake level allow the evaluation of contributions from these components. The annual cycle of water level based on daily measurements since 1961 as well as the annual cycle of temperature, precipitation, snowfall, and snow depth, in terms of averages and probability distributions, are described. The simplified hydrology of the lake is primarily a consequence of the ratio of the lake area to its catchment basin area (approximately 0.8). The lake acts both as a large raingage and a large evaporation pan under appropriate circumstances. Preliminary invistigations suggest that the general assumption that evaporative water loss is less during cold seasons or cold episodes may not be correct..

Crater Lake was discovered in 1853 and the first observation of the water level was made in 1878. Since that time a variety of observers using different methods have contributed to the creation of a century-long record of lake level. Figure 1 shows the variations that have been observed since 1900. The most prominent feature in this record is the extended period of low water from 1930 to the late 1940s, when the water surface was as much as five meters below levels seen before and after.

An understanding of the climate of the lake and its environment is essential to explaining the observed water level fluctuations. Climate variability over a wide range of spatial and temporal scales also affects many other aspects of the physical behavior of the lake and of the plant and animal communities in the park. Climate influences such factors as available sunlight for photosynthesis; the life cycles of insects; growth of plant and animal pathogens; forest fires; plant phenology; wildlife mortality; transpiration; lake and stream chemistry and temperature; animal habitat; lake mixing; avalanche behavior; and forest blowdown from high wind. Weather and climate also affect human activities, including tourist visitation; outdoor recreation; seasonal labor costs; design of structures to withstand cold, heat, and wind and snow loads; road design, construction and maintenance; water treatment facilities; firefighting expenses; and import of pollutants from outside the park.

Crater Lake is unusual in that there is perhaps no other lake with the combination of relatively simple hydrological circumstances and the existence of a long period of climate and lake level observations. The lake can be thought of as acting in turns as a large natural (leaky) raingage and as an evaporation "pan."

The purpose of this paper is twofold: (1) to show that climate variability can account for the changes in lake level seen this century, and (2) to provide updated background information about typical val-

[1] Present address: Western Regional Climate Center, Desert Research Institute, P. O. Box 60220, Reno, Nevada 89506.

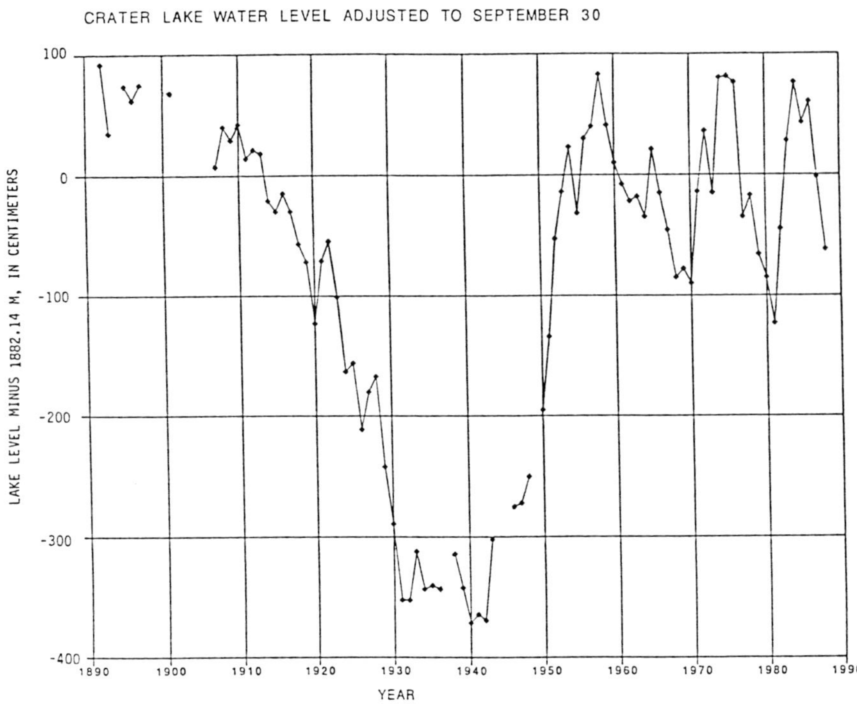

Figure 1. Crater Lake water level variations adjusted to 30 September. Water stage is referenced to an elevation of 1882.14 m (6175 feet) above mean sea level, and shown as departure in centimeters from that elevation. Adjustments are based on average rate of lake level change from the closest available date to 30 September, shown in Fig.12. In some years no measurements are available.

ues of important climatic elements, along with measures of dispersion about the central tendency, for use by other limnological and ecological studies.

PHYSICAL SETTING

Crater Lake sits astride the gentle crest of the Cascade Range, which lies at an average of approximately 1500 m in this portion of Oregon. The average elevation of the lake is 1882 m. The rim and even the lake itself thus lie well above the surrounding countryside. Steep slopes descend from the rim, which has an average elevation of about 2100 m, and a number of high points around the lake rise to about 2500 m. The area of the water surface is 53.2 km^2, and the drainage basin is 67.8 km^2 (Phillips 1968). The lake thus occupies 78.5 percent of its own drainage basin. No streams spill over the rim into the lake, and no streams drain the lake. The slopes facing away from the rim lie at an angle of about 15°. The only ways for water to leave the basin are through seepage and evaporation.

DATA

Daily weather measurements are made at park headquarters, about 2 km south of Rim Village, on the southwest side of the lake, at an elevation of 1973 m, or about 180 m below the elevation of the rim in this area. The station is the highest in the state of Oregon. A nearly continuous record exists from this site since 1931, with the exception of a few years during World War II when the park was closed and lengthy interruptions are present. Each day the maximum and minimum temperatures, precipitation, snowfall, and snow depth on the ground are measured and recorded. Prior to October 1958, readings were made in the afternoon; since that time they have been made each day at 8 a.m. The thermometer shelter is moved up a mast as the winter progresses to avoid becoming buried by snow. Prior to 1930 no consistent long-term measurements are available, although a few shorter records of 5 years' duration or less can be found in and near the park. Digital hourly precipitation measurements from a recording rain gage are available from this site since 1948.

Water levels have been measured at Cleetwood Cove in the northwest corner of the lake since 14 September 1961. The gage, operated by the U. S. Geological Survey, is capable of reporting changes in increments of 0.305 cm (0.01 foot) four times per day. For this study, only the measurements made at the close of the day near midnight have been used. Daily maximum and minimum surface water temperatures are also recorded here. Lake level measurements before the establishment of the Cleetwood Cove gage were made sporadically by visitors to the lake. At least six different gages have existed since the first one was installed in August of 1896. Several measurements are available in most years, usually during the warm season. In winter, the lake is practically inaccessible. Phillips (1968) has tabulated these earlier observations.

GENERAL CLIMATE DESCRIPTION

The main mid-latitude storm track in the atmosphere migrates southward in winter and northward in summer. The lake is, thus, alternately under (in winter) and to the south of (in summer) this band of active weather and exhibits the pronounced seasonality of precipitation common in the Pacific Northwest.

During the long transition from winter to summer, precipitation gradually becomes less frequent. Extended dry periods with abundant sunshine and no precipitation are common in summer. Near the end of July measurable precipitation occurs on less than 10% of the days, and July averages just 3 days with measurable rain (see Fig. 2). For the 30-day period commencing on July 15, there is a 20% probability of no measurable rain during this entire interval (based on 1931-1986 data). For the period from 1947-1986, all years have had at least a 15-day interval without measurable rain. About once a year a 5-week period without measurable rain occurs. Since 1931 there have been 10 periods with at least 50 consecutive days without rain, the longest being 97 days in 1951. Thunderstorms provide some rain during the warm season, but are not especially common. Thunder is recorded at Klamath Falls on an average of 14 days per year, and on 11 days at Medford, the nearest sites with reliable statistics (Changery 1981).

As autumn approaches, the hemispheric circulation of the atmosphere expands and intensifies, bringing a steadily increasing chance of precipitation to Crater Lake. Precipitation likelihood increases most rapidly in October, when daily probabilities of measurable precipitation rise from 25% at the beginning of the month to 45% at the end. During the passage of numerous cool season disturbances, moist maritime air is forced to rise from near sea level to the crest of the Cascade Range. As a result, the lake and its environment experience frequent moderate precipitation events during the winter months. The probability of heavier precipitation episodes is at a maximum from mid-November through late January. Light and moderate episodes occur most frequently from mid-November through mid-March. Winter conditions are usually fully established by early November. Measurable precipitation (at least 0.25 mm) occurs on about 60% of the days in the heart of winter (see Fig. 2), and on a total of about 140 days per year. The average length of the longest spell each year with measurable precipitation is about 16 days, with a maximum of 39 consecutive days in 1973.

The area is high enough that winter precipitation usually falls as snow at the weather station. The mean freezing level measured with balloons in the free air over nearby Medford, 84 km to the southwest, is at an average elevation of 2169 m from December through February. Freezing level statistics above Medford are shown in Table 1. It is of interest to note that historically both the freezing levels and the temperatures (shown below) remain nearly constant until mid-March. Over elevated surfaces, nighttime minimums will be cooler and daytime maximums (during the snow-free season) will be warmer than the free air temperature at the same height.

The average annual snowfall is 1306 cm, for the reference period in most widespread use at this writing (1951-1980). The long-term median snowfall for the period of record (1931/32-1987/88) is 1314 cm. Winter-centered snowfall (July-June) has ranged from 648 cm in 1976/77 to 2233 cm in 1932/33. The snowiest consecutive 12 months were from April 1948 through March 1949, when 2365 cm fell. Snow has occurred in all months but is quite uncommon in summer. Snow typically begins to accumulate at the weather station in early November, reaching a maximum depth at the beginning of April, when the maximum average depth (1931-1986) reaches about 325 cm. Figure 4 shows the

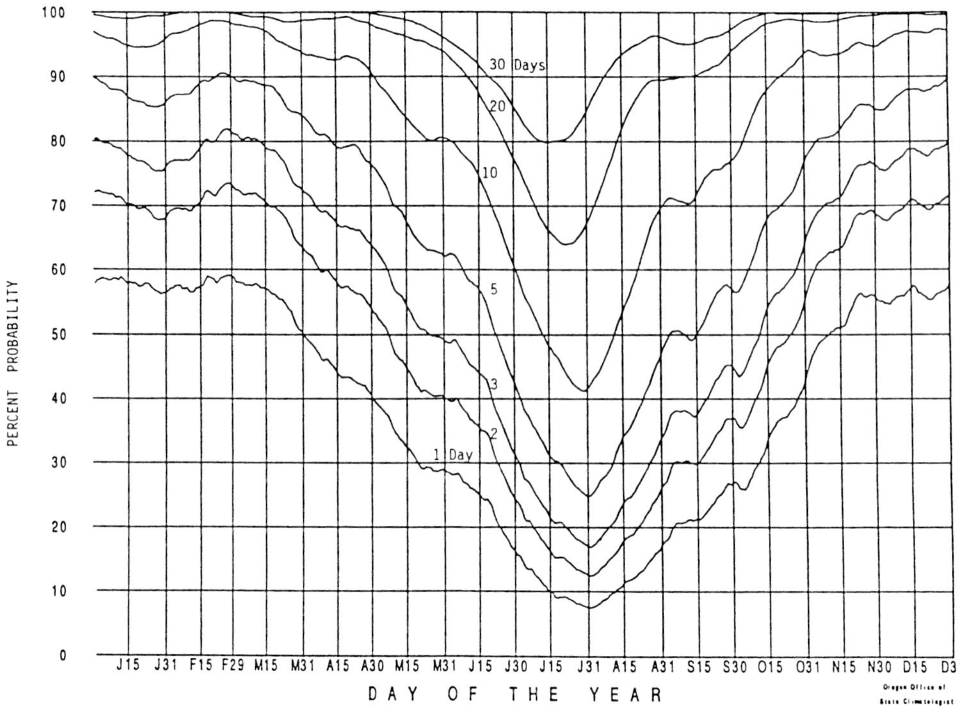

Figure 2. Probability of measurable precipitation (0.25 mm/0.01") for varying durations. Based upon daily observations from 1931-1986. Smoothed with a 29-day running mean filter. From the bottom, the curves show the probability of 0.25 mm of precipitation or more in the (1, 2, 3, 5, 10, 20, and 30) days starting on the plotted date.

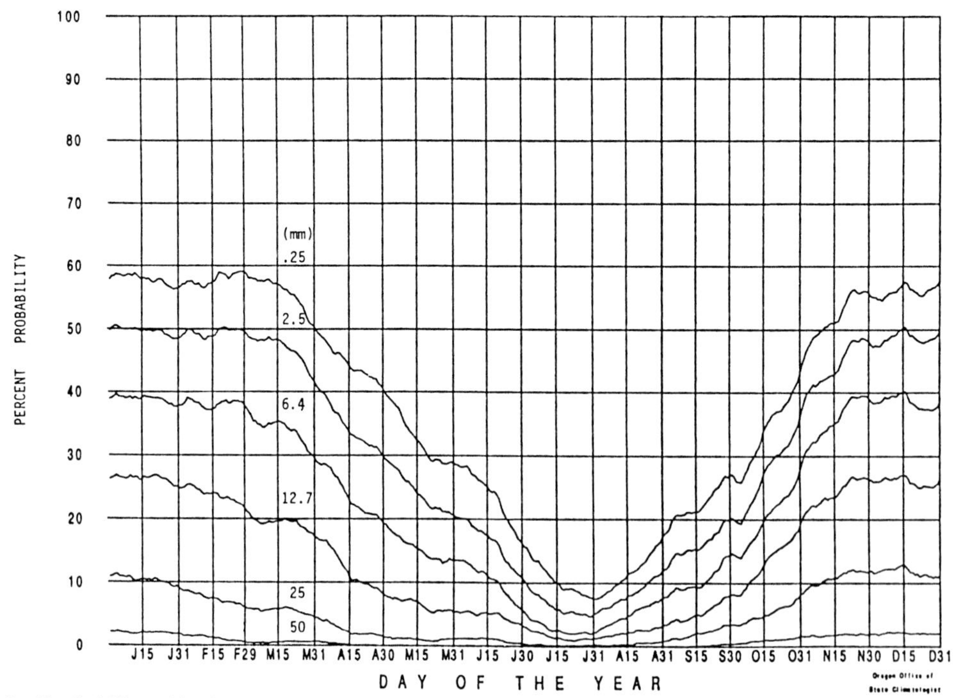

Figure 3. Probability of indicated 24-hour precipitation amount for each day of the year. Based upon daily observations from 1931-1986. Smoothed with 29-day running mean filter. Curves, from top, are for (0.25, 2.54, 6.35, 12.7, 25.4, and 50.8 mm; corresponding to English units, 0.01", 0.10", 0.25", 0.50", 1.00", 2.00").

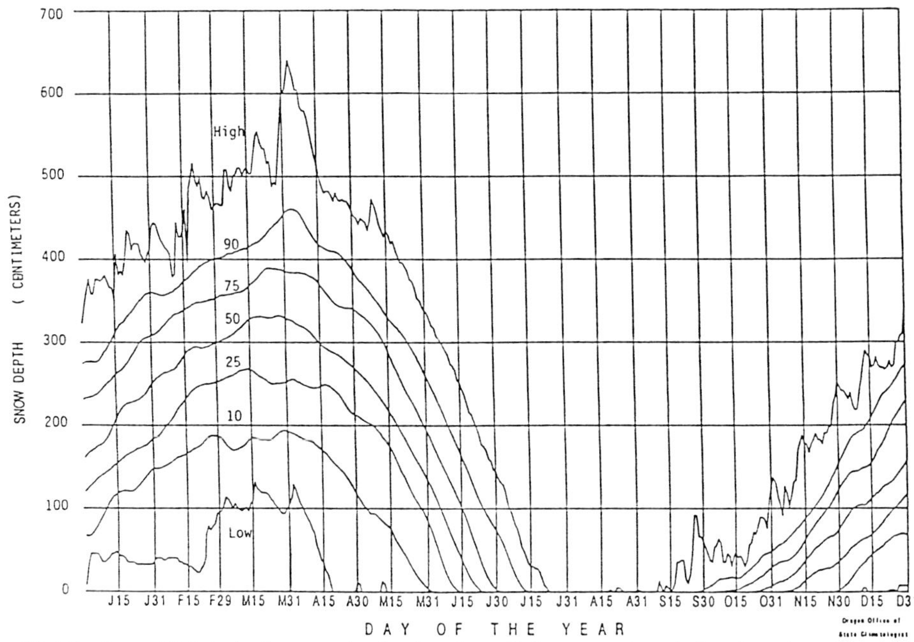

Figure 4. Distribution of historical snow depths by day of the year at park headquarters. Based on the period 1931-1986. Percentiles smoothed with 9-day running mean filter, daily extremes unsmoothed. From the bottom, the curves are: lowest observed, 10th, 25th, 50th, 75th, 90th percentiles, and the highest observed. Depths in cm.

daily distribution of snow depth observed over the past half century. Figure 5 shows the average snow depth and daily precipitation amount. The average extreme maximum depth in the spring (i.e., the most extreme value for the year) is 378 cm, and it has varied between 163 cm in 1977 and 640 cm in 1983. Snowfall statistics are not available for the rim, but snowfall on the southwest rim is certainly greater than at the weather station. It is not unusual, for example, for light summer snowfall events (a few centimeters) at park headquarters to be accompanied by 30-50 cm at rim level, just a few hundred meters higher.

During the passage of a typical cool-season storm system, warm southerly winds and moderate to heavy precipitation occur in advance of the passage

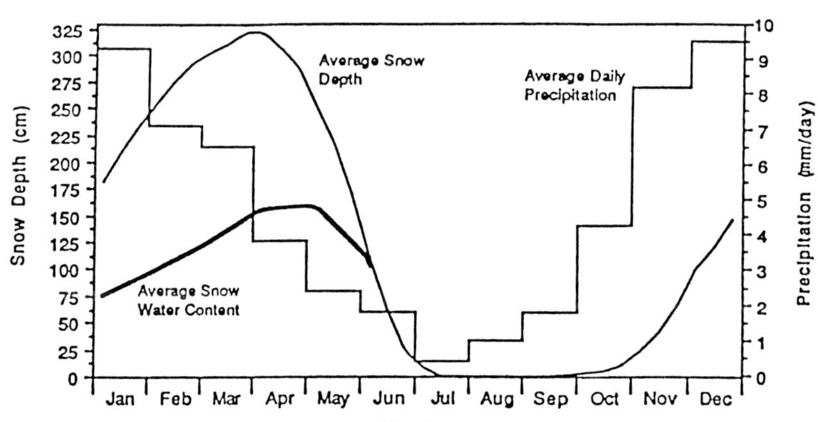

Figure 5. Crater Lake park headquarters. Daily average snow depth (1931-1986) plotted every fifth day; average month-end snow water content (1961-1985) from Snow Survey (1987); and monthly average of daily precipitation (1951-1980).

TABLE 1. FREEZING LEVEL STATISTICS ABOVE MEDFORD, JULY 1957 THROUGH JUNE 1967

If multiple freezing levels are present, the highest one is used. Park headquarters height is 1974 m. Rim height is about 2130 m. Original statistics were heights above ground level (401 m at Medford) and are here presented in meters above sea level. Because of radiative effects, minimum temperatures over elevated surfaces will usually be cooler than this table indicates, and maximum temperatures over elevated snow-free surfaces will usually be warmer than this table indicates. Source: Pacific Northwest River Basins Commission, 1975.

Period beginning	Average height	Standard deviation	% of time above 2002m	% of time above 2202m
Jan 1	2072m	923m	50%	44%
Jan 16	2125	818	53	42
Feb 1	2172	787	53	46
Feb 16	2000	716	46	38
Mar 1	1797	650	37	28
Mar 16	2085	734	43	36
Apr 1	2416	791	61	57
Apr 16	2200	740	51	44
May 1	2561	787	66	61
May 16	3017	759	86	82
Jun 1	3321	724	91	89
Jun 16	3821	730	97	95
Jul 1	3998	598	99	99
Jul 16	4381	408	100	100
Aug 1	4321	511	100	100
Aug 16	4038	688	99	97
Sep 1	4024	638	99	97
Sep 16	3853	798	99	94
Oct 1	3466	956	83	81
Oct 16	3433	896	84	84
Nov 1	2774	894	69	65
Nov 16	2345	846	55	46
Dec 1	2468	911	59	56
Dec 16	2179	849	53	41
Annual	2960	1134	70	65

of the cold front. High freezing levels allow rain to fall on many occasions. After passage of the cold front, freezing levels fall and precipitation becomes showery and locally intense in the unstable air.

Since the remnants of the former Mount Mazama stand above the surrounding landscape, the area has some of the temperature characteristics of a free air location, though modified by the presence of the large area of elevated land. Strong daytime solar heating causes the temperature at park headquarters to rise to an average of 20°C during mid-summer, while in mid-winter the average maximum is near 0°C. The extreme maximum of 33.3°C at the weather station is greater than the extreme maximum of 26.4° C observed at approximately the same elevation in the free air over Medford at 800 mb (average July height of the 800 mb level is 2037 m). The weather station lies within the valley of Munson Creek on a southwest-facing slope, and nighttime drainage winds prevent minimums from falling to the extremely low values which undoubtedly occur in flatter portions of the park. As a result, the lowest temperature on record is -29°C, a rather modest value given the heavy snow cover and high elevation. Probability distributions of mean daily maximum, mean, and minimum temperatures, and of the daily temperature range (maximum minus minimum), are shown in Figs. 6-9.

The weather station is below and away from the rim itself. The site is therefore not entirely representative of the conditions at either the rim or the lake. Logistical considerations greatly hinder the ability to make measurements away from areas with a permanent human presence. Pronounced east-west gradients in long-term annual precipitation (6-8 cm per km) are present at the crest of, and to the east of, the Cascade Range in Oregon. Crater Lake lies on this gradient. The southwest rim is wetter than the weather station, and the northeast rim is considerably drier. High capacity storage gages, designed to be read once per year, have been located on the rim above Cleetwood Cove since 1961 and have been recently installed near Rim Village and on Wizard Island. Consideration of previously published precipitation maps (Soil Conservation Serv. 1964; Froehlich *et al.* 1982) and the more reliable figures from the storage gages indicates that a marked difference in annual precipitation exists from one side of the lake to the other. Annual precipitation along the southwest rim appears to be 15% to 30% greater than at the weather station, and is nearly double the amount received along the northeast rim. Even as early as 1902, Diller, on the basis of snow depth and density measurements, recognized that "the average annual precipitation for that region [southwest rim] is nearer 80 [203 cm] than 70 [178 cm] inches." It is also quite likely that higher peaks all around the rim,

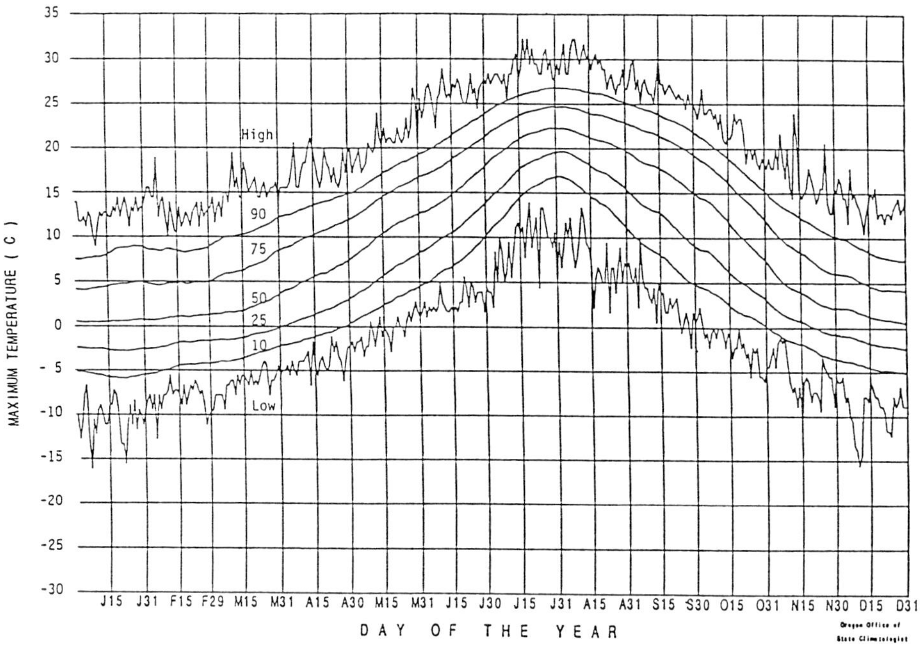

Figure 6. Distribution of individual daily maximum temperatures. Data from 1931-1986. Curves are, from bottom: lowest observed; 10th, 25th, 50th, 75th, 90th percentiles; and highest observed. Percentiles smoothed with 29-day running mean, extremes unfiltered.

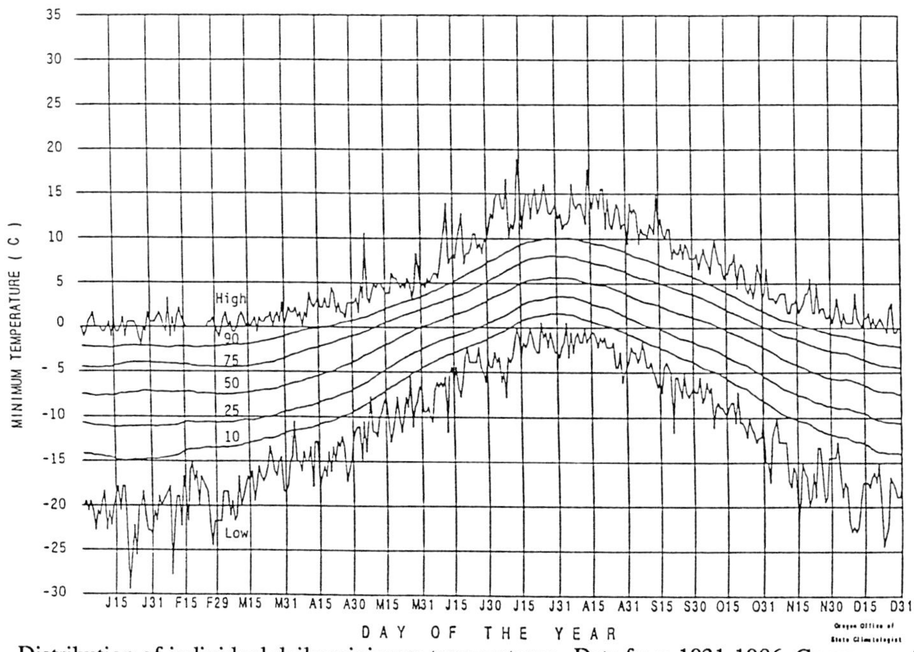

Figure 7. Distribution of individual daily minimum temperatures. Data from 1931-1986. Curves are, from bottom: lowest observed; 10th, 25th, 50th, 75th, 90th percentiles; and highest observed. Percentiles smoothed with 29-day running mean, extremes unfiltered.

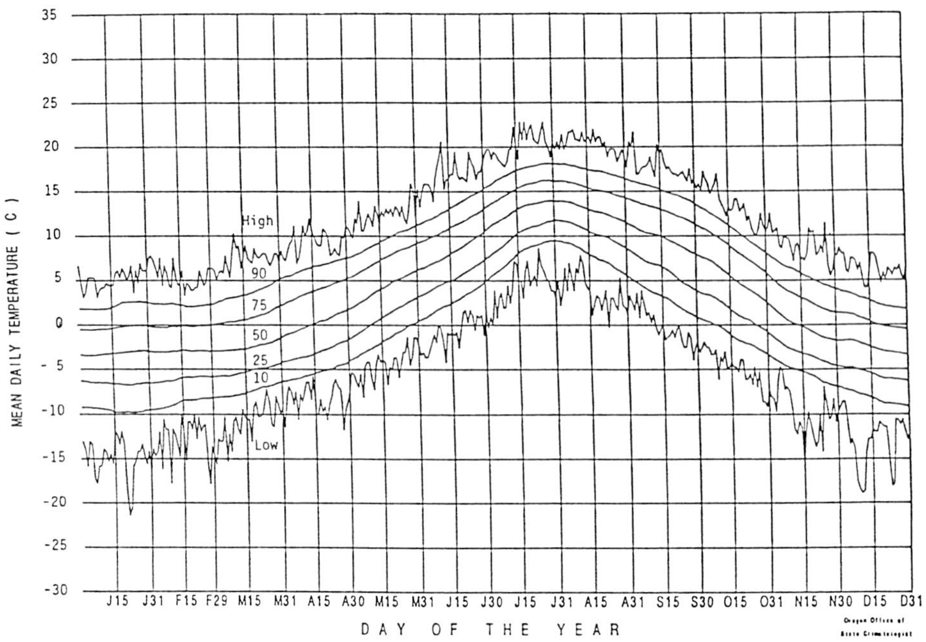

Figure 8. Distribution of individual daily mean temperatures. Means are (maximum + minimum)/2. Data from 1931-1986. Curves are, from bottom: lowest observed; 10th, 25th, 50th, 75th, 90th percentiles; and highest observed. Percentiles smoothed with 29-day running mean, extremes unfiltered.

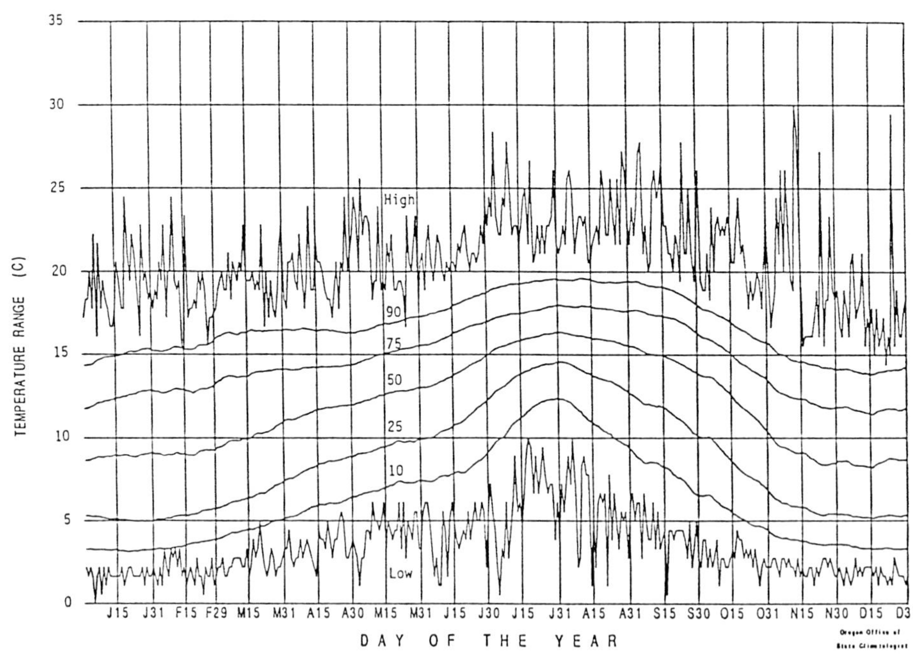

Figure 9. Distribution of individual daily mean temperature ranges (maximum minus minimum). Data from 1931-1986. Curves are, from bottom: lowest observed; 10th, 25th, 50th, 75th, 90th percentiles; and highest observed. Percentiles smoothed with 29-day running mean, extremes unfiltered.

and indeed any surface facing toward the southwest quadrant, and to a lesser extent the northwest, will cause local enhancements of precipitation. In its journey across the lake, air can pick up significant amounts of moisture and may deposit some of this as it ascends the "downwind" crater wall along the eastern side.

PRINCIPAL ELEMENTS OF THE WATER BUDGET

The lake gains water mass by two processes with greatly different time constants. Precipitation falling directly on the lake produces an immediate rise in lake level. A second component of input is runoff from the slopes of the caldera, primarily derived from the melting of snow in spring and early summer. The unfrozen lake and steep inner slopes act as a trap for wind-blown snow from the southwest. An unknown amount of water enters the basin this way. On average, snow disappears at the weather station about mid-June. South-facing slopes melt earlier, and north-facing slopes retain snow patches throughout many summers. No surface streams drain into or out of the caldera. A number of small rivulets trickle into the lake along the walls of the caldera. Many of these dry up in summer. These streams have been described by Diller (1902), who estimated the total flow of these streams to be 0.28 m^3/s, enough to raise the lake by 0.045 cm per day when they are flowing. Groundwater seepage above lake level into and out of the caldera have been assumed to be approximately in balance, after Phillips (1968). Water can only leave the lake in two ways, through evaporation and seepage from the bottom.

ANNUAL CYCLE OF WATER LEVEL

Figure 10 shows an example of the daily lake level fluctuations during a recent year (1985). Figure 11 shows the daily maximum and minimum temperature and accumulated precipitation at park headquarters during this same year. It is clear that the lake falls as a result of seepage and evaporation during periods of little or no precipitation, and rises during heavier precipitation events.

The lake can thus be considered to act in some senses as a giant raingage, albeit a leaky one. When precipitation (and inflow) exceeds the leakage rate, the lake rises. Otherwise, the lake falls or remains at constant elevation.

A group of years with complete lake level records has been used to construct a composite of the annual cycle of lake level, shown in Fig. 12. As the wet season sets in during autumn, the lake begins to rise. Precipitation thus exceeds seepage and evaporation. The lake typically continues to rise until early April.

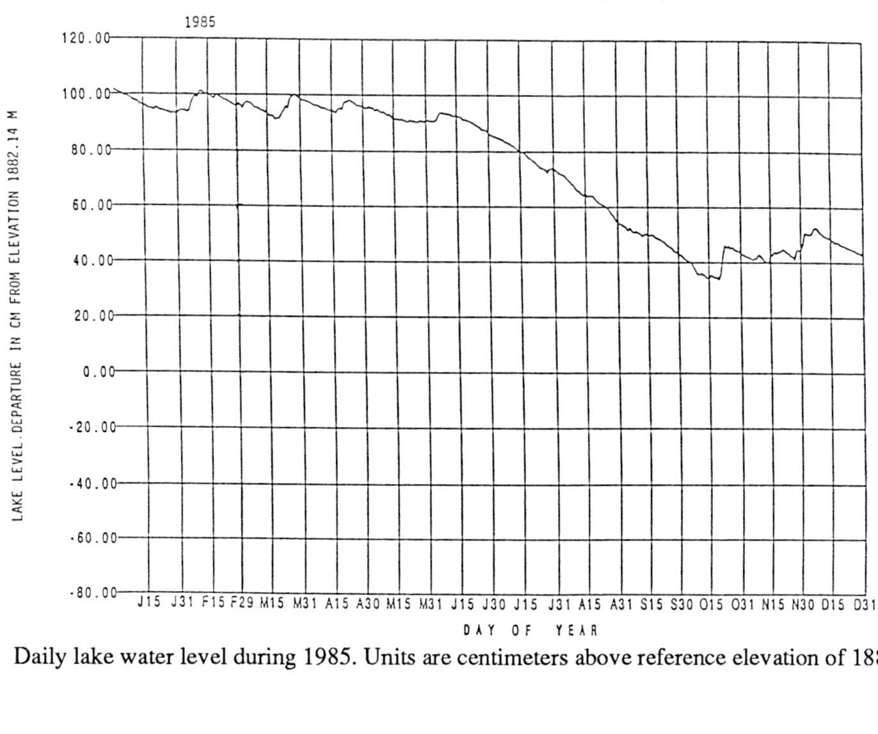

Figure 10. Daily lake water level during 1985. Units are centimeters above reference elevation of 1882.14 m.

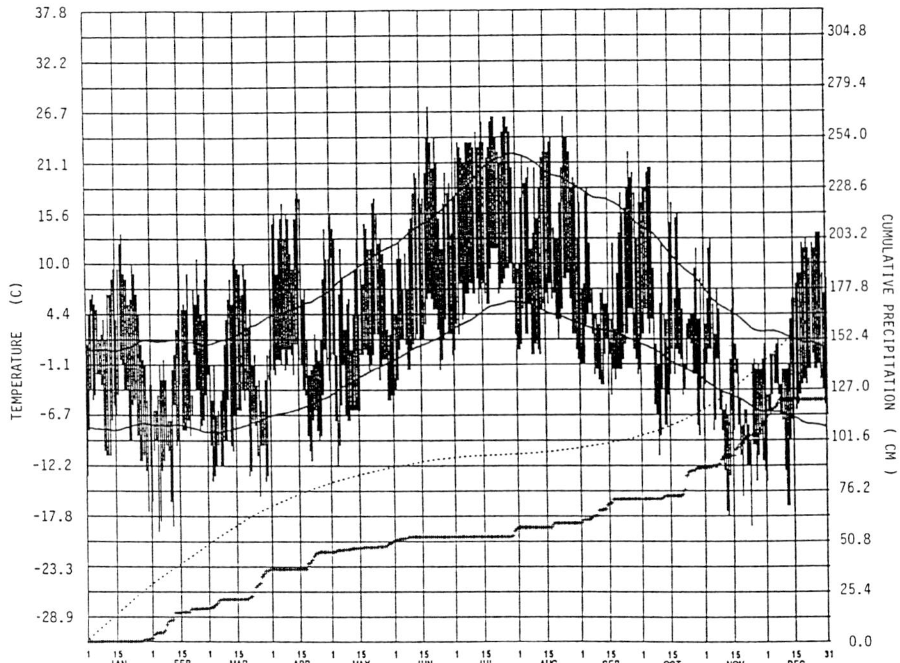

Figure 11. Daily maximum and minimum temperatures (vertical bars), and accumulated precipitation (bottom curve) at park headquarters since 1 January, for 1985, to correspond with Fig. 10. Long-term averages (solid lines: temperature; dotted lines: precipitation) are from 1951-1980. Metric equivalents of original English values.

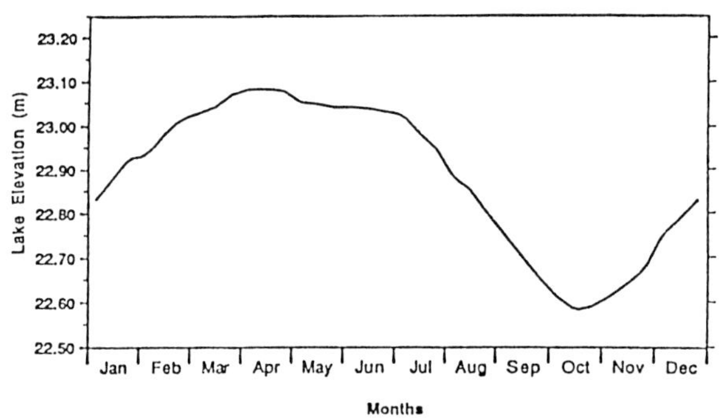

Figure 12. Annual cycle of water level. Fifteen-year composite of 1966-1971, 1975-1976, 1979-1985. Units: Elevation in meters minus 1859.28 m.

By this time the precipitation has begun to diminish and continues to decrease in frequency and amount as spring proceeds. Despite this, the lake remains high until the latter part of June. By this time the precipitation input has decreased substantially. However, as seen in Fig. 5, snow depths and snow water content are decreasing rapidly in May and June, so that the lake level is kept high by surface and groundwater runoff from the caldera slopes. After the snow has melted, the lake begins to fall steadily toward its early autumn minimum. Examination of individual years shows many interesting departures from this picture, however. The most recurrent feature is the relatively steady decline from July through September.

As can be seen, the lake usually reaches its low point sometime in October. This is very near the end of the "water year" used by hydrologists (1 October -30 September). Because the influence of snowmelt should be minimal by this time, and because the winter precipitation season has usually not yet begun by this date, water stage on 30 September was used to relate year-to-year changes in water level to climate.

RECONSTRUCTION OF LAKE LEVEL RECORD

Figure 13 shows the relationship between the annual change in lake level measured at the gaging station at Cleetwood Cove and annual precipitation at the weather station from 1961/62 to 1987/88. It is obvious that at this time scale lake level is closely linked with precipitation variability. The least-squares fit is:

$$LLC = -246.5 + 1.457\, P_{HQ} \qquad (1)$$

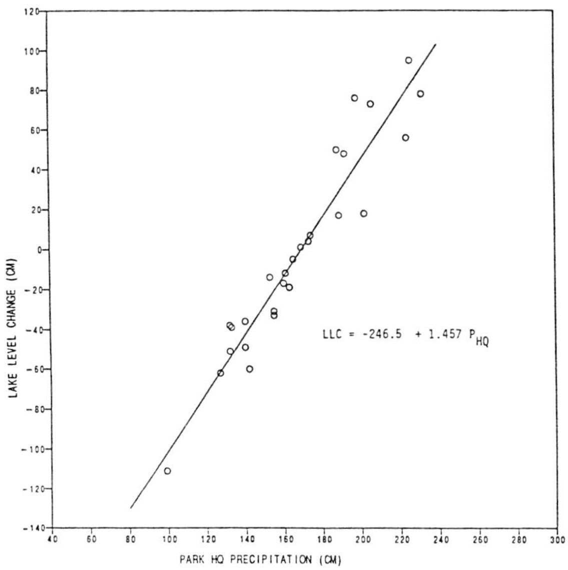

Figure 13. Annual change in lake level versus annual precipitation at park headquarters. Period is from 30 September to 30 September. 1961/62-1987/88. Least squares fit is shown.

where *LLC* is the lake level change (30 September of present year minus 30 September of previous year, cm), and P_{HQ} is the October-September precipitation (cm) measured at park headquarters. For this relationship, $r^2 = 0.915$.

Consistent precipitation records from a single site begin in 1931, at park headquarters. Longer precipitation records can be found in the valleys of western Oregon. Much of the precipitation in this region falls from stratiform clouds associated with synoptic scale cool-season disturbances. As a result, precipitation amounts show considerable spatial coherence. Monthly precipitation at Roseburg, 105 km to the west and 1800 m lower in elevation, is well correlated with that at Crater Lake. Correlation coefficients between Roseburg and Crater Lake monthly precipitation totals between 1931 and 1987 vary from 0.74 in June and July to 0.88 in October; for the water year $r = 0.79$. The record at Roseburg was used to reconstruct water year precipitation at Crater Lake prior to establishment of the headquarters station. Monthly values for Crater Lake were estimated from several nearby stations for the missing periods during World War II. The observed and reconstructed time series are shown in Fig. 14.

The observed and reconstructed time series of annual precipitation were then used to reconstruct the historical lake levels. We began with the known level on 30 September 1986 (60 cm above the arbitrary reference level 1882.14 m), and the known precipitation for the previous water year ending on this date. Using equ. (1) we estimate the lake level change and subtract from the current level to estimate the level on the previous 30 September. This backward stair-stepping method was then applied for each year using the precipitation value from the previous water year. Observed park headquarters precipitation data are used to 1931, and reconstructed precipitation values from Roseburg are used prior to this time.

The observed and reconstructed lake levels are shown in Fig. 15. It is quite apparent that precipitation variations can account for the extended period with low lake levels between 1920 and 1950. For those years prior to installation of the gaging station at Cleetwood Cove for which park headquarters precipitation and annual lake level changes are available, $r^2 = 0.78$ between observed and reconstructed annual lake level changes. For the earlier years when park headquarters precipitation was reconstructed from Roseburg, $r^2 = 0.444$, reflecting the imperfect relationship between annual precipitation at Roseburg and Crater Lake annual water level change.

Reconstructed lake levels agree well with observed levels from the 1940s onward. The rapid rise in lake level around 1950 is accounted for by a series of several very wet years. The lake rose about a meter from 1942 to 1946, a rise that is not adequately simulated. Reconstructed levels decrease during the decade of the 1930s, whereas the observed levels remained constantly low during this period. The missing precipitation values during the early 1940s may have been underestimated, or else the relation between lake-averaged precipitation and park headquarters precipitation differed in these years. Because the level each year is dependent upon the previous year, this method is sensitive to systematic errors and to large errors in one or two individual years, which will affect all remaining elements in the reconstructed time series. The magnitude and shape of the drop in reconstructed lake levels from 1910 to 1930 is quite well simulated. Either the lake levels are in error, or the relationship between Roseburg and Crater Lake precipitation changed around 1900. The reconstructed levels from Roseburg indi-

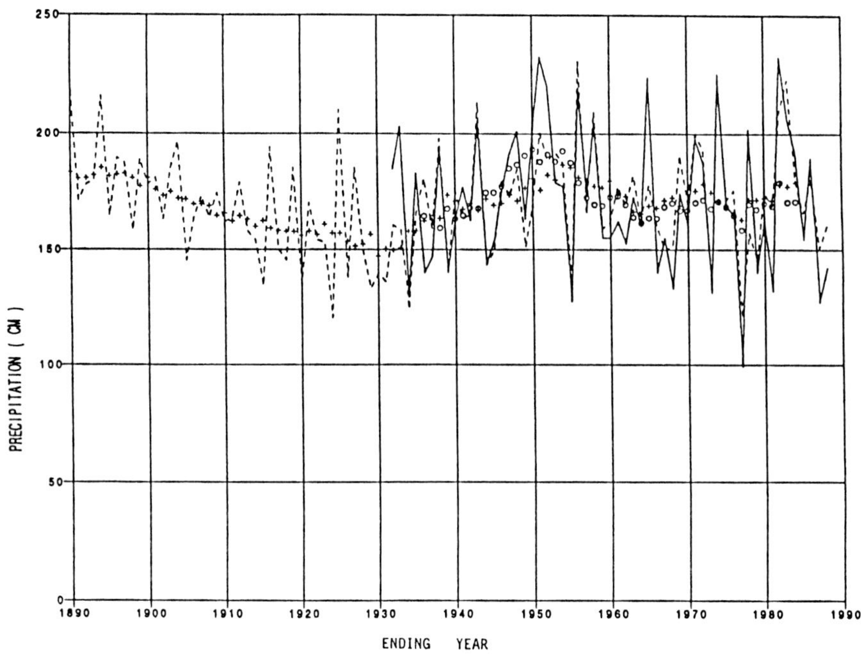

Figure 14. Observed (solid) and reconstructed (dashed) water year precipitation at Crater Lake. Reconstructed values from Roseburg. Nine-year running mean shown for observed (o) and reconstructed (+) time series. Regression coefficients based on 1931-1987.

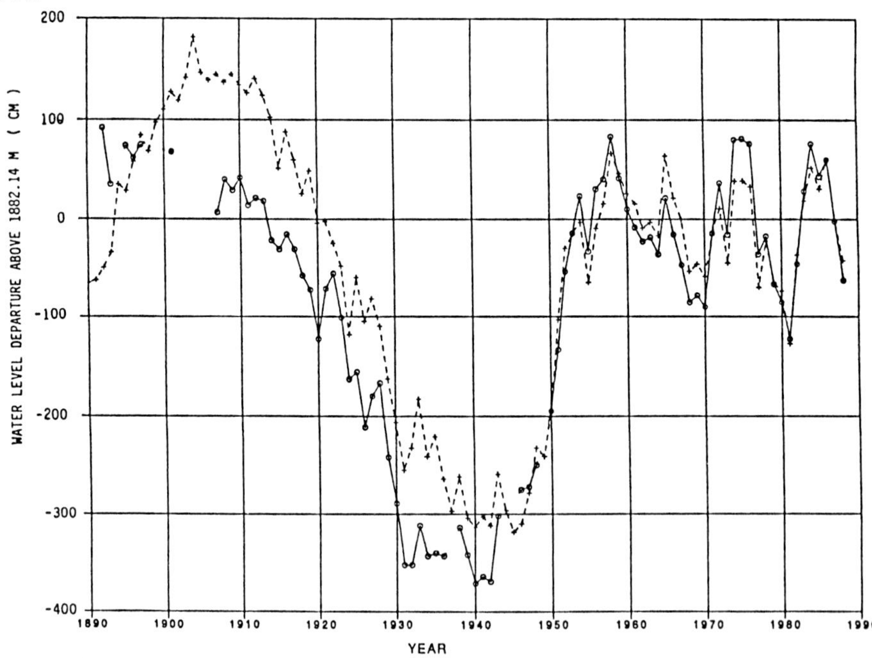

Figure 15. Observed (o) and simulated (+) lake levels, starting with the known value on 30 September 1986. Observed lake levels adjusted to 30 September. Reconstructed lake levels use Crater Lake park headquarters precipitation from 1985/86 backward to 1931/32, and use precipitation estimated from Roseburg prior to that time. Observed Crater Lake precipitation also used for forward extrapolation from 1986 to 1988. Water level expressed in centimeters above 1882.14 m.

cate that the lake should have been rising during the 1890s. A change in location and instrument exposure at Roseburg occurred in October 1905, about the time the reconstructed and observed levels began to deviate from each other (when working backward in time).

The dry conditions which led to the rapid fall in lake level in the 1920s are seen in other climatic records in the region. Note that it is the slope of the lake level curve, and not the level itself, that is proportional to precipitation. A constant lake level, such as in the 1930s, implies average precipitation. The period from the mid-1920s through the early 1940s brought minimum lake levels and stream discharges to much of eastern Oregon. Alexander *et al.* (1987a, 1987b) present data showing that Crescent Lake, 45 km to the north, reached its lowest level in October 1931, and the Rogue River just above Prospect, 30 km to the southwest, experienced its lowest flow in November 1931. The 50-meter-long Quinn River, which gushes from a spring 90 km to the north, showed no flow in a portion of 1941, and Upper Klamath Lake fell to its lowest elevation in 1944. Johnson and Dart (1982) refer to the generally dry conditions in eastern Oregon during this interval. They also present diagrams showing that precipitation in eastern Oregon is not closely correlated with precipitation in the wetter western portion of the state. Because Crater Lake lies at the sharp boundary between these two quite different regimes, it should not be surprising if it exhibits some of the characteristics of each.

CONCLUSIONS AND DISCUSSION

From equ. (1), the lake will be at the same level on 30 September from one year to the next if annual headquarters precipitation totals 170 cm. This is very close to the long-term average at this site. Indeed, after about a century of observations, the lake remains near the same level as when first observed.

Dividing the constant in equ. (1) by 365 days, the average loss rate of the lake is -0.675 cm/day. In 1985 Crater Lake was covered by ice for two extended periods, once in January (last 2-3 weeks) and again in December (last 3 weeks). Fortunately, almost no precipitation fell during the January episode and none at all fell in the December episode. Assuming that sublimation from the ice surface was negligible, the lake could only fall through seepage. The lake fell at an average rate of 0.347 cm/day. This leaves evaporation to account for the remaining -0.328 cm/day loss, or 120 cm per year.

The seepage rate estimated here is less than was estimated by Phillips (1968), who arrived at a value of 2.52 m^3/sec, equivalent to -0.410 cm/day change in lake level. Phillips cites earlier unpublished seepage estimates by F. F. Henshaw in 1913 of -0.382 cm/day and by S. T. Harding in 1953 of -0.267 cm/day. Most of these earlier estimates have relied on assumptions regarding runoff and evaporation. In particular, the most common has been the assumption that during cold periods in winter with no precipitation, runoff and evaporation were low, and therefore that observed water loss was due almost entirely to seepage.

One way to estimate evaporation is to make direct measurements for a full annual cycle, on an hourly basis, of each of the factors that influence this process, such as wind speed, humidity and temperature gradients, and exchange coefficients (see Mahrt and Ek 1984). This task is made difficult and even hazardous by the physical circumstances that prevail, especially in winter, with steep walls subject to frequent avalanches, heavy snow, high winds, and cold water. Rugged and dependable floating automatic sensors which do not need maintenance atten-

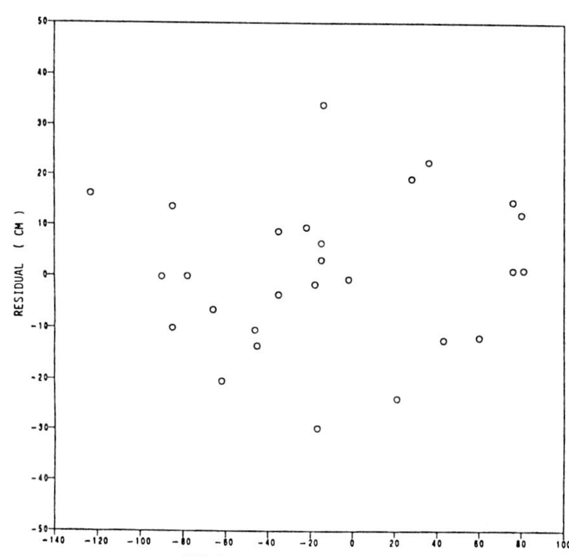

Figure 16. Scatter diagram of (a) departure of observed annual lake level change [30 September to 30 September] minus lake level change estimated from annual precipitation using equ. (1) against (b) lake stage, expressed in cm above 1882.14 m.

tion need to be deployed for this purpose. Until such parameters can be measured reliably, other methods based primarily on budget considerations must be used.

Phillips (1968), assuming that both the evaporation loss and the precipitation-inflow relationship are constant from year to year, derived a relation between lake stage and seepage rate (his Fig. 5). This shows seepage rate increasing substantially as lake level rises above 1882 m. However, this conclusion can only be regarded as speculative, since at least one of the required assumptions (constant annual evaporation) is likely not true. The coefficient of variation (standard deviation divided by precipitation) of water year precipitation is 18% at Crater Lake. The coefficient of variation for pan evaporation during the growing season ranges from 8% to 16% at various locations in the area. Although Crater Lake does not evaporate like a standard pan, there are almost certainly year-to-year fluctuations in evaporation from the lake surface. To test whether a stage-seepage relationship might exist, annual lake level change estimates obtained from equ. (1) were subtracted from observed annual lake level changes (30 September to 30 September). These residuals are plotted in Fig. 16 against the elevation of the lake. The random scatter indicates that there is no systematic bias as a function of stage. Therefore, either there is no association between seepage rate and stage, or else a relationship exists between annual precipitation and annual evaporation that fortuitously counterbalances any seepage-stage relationship. The latter possibility seems unlikely.

Simpson (1970) discussed in some detail the hydrology of Crater Lake, a knowledge of which was needed for a study of the tritium budget of the lake. Simpson reiterated the expectation that summer evaporation would exceed winter evaporation by at least a factor of 1.5 to 2, but he also noted that the lake fell at a similar rate during periods with no precipitation in both winter and summer. He took this to be evidence that seepage is the dominant mode of water loss. Using lake level from the four water years available at that time (1962/63-1965/66), the relation analogous to equ. (1) would have been flatter (slope 1.115, intercept -192.3). Simpson concluded that two-thirds of the total water loss (his estimate being -0.57 cm/day) occurred through seepage, somewhat less than the 72% estimated by Phillips. The value Simpson determined for seepage was thus -0.38 cm/day, close to the value of -0.347 cm/day found here. The fraction of the water loss accounted for by evaporation in this study is 52%, using a total loss rate of -0.675 cm/day. Diller (1902), using a floating pan in the lake (the only direct on-site measurements ever made), determined a 10-day average evaporation of -0.38 cm/day. However, whether this value is representative of annual conditions is open to question.

Preliminary investigations strongly hint, and physical reasoning suggests, that the general assumption that evaporative water loss is less during cold seasons and during cold episodes may not be correct. Further study of daily lake level behavior is in progress, and is expected to shed additional light on this important question.

ACKNOWLEDGMENTS

Thanks are extended to Gary Larson, Mark Buktenica, and Tom McDonaugh of the National Park Service for their helpful comments and observations, and to Jim Moffet and Larry Hubbard of the U. S. Geological Survey for lake data. Additional comments by Ted Strub and Kenneth Phillips were appreciated.

LITERATURE CITED

Alexander, C. W., R. L. Moffatt, P. R. Boucher and M. L. Smith. 1987a. Water resources data, Oregon, Water Year 1985, vol. 1. Eastern Oregon. U.S. Geol. Surv. Water-Data Rep. OR-85-1. 218 pp.

Alexander, C. W., R. L. Kraus, C. G. Kroll, R. L. Moffatt and M. L. Smith. 1987b. Water resources data, Oregon, Water Year 1985, vol. 1. Western Oregon. U. S. Geol. Surv. Water-Data Rep. OR-85-2. 396 pp.

Changery, M. 1981. National thunderstorm frequencies for the contiguous United States. NUREG/CR-2252. National Climatic Data Center, Asheville, North Carolina.

Diller, J. S. 1902. The geology and petrography of Crater Lake National Park. U. S. Geol. Surv. Prof. Pap. 3(pt. I). 62 pp.

Froelich, H. A., D. H. McNabb, and F. Gaweda. 1982. Average annual precipitation, 1960-1980, in southwest Oregon. Oregon State Univ. Ext. Serv., Corvallis, Ore., EM 82:20.

Harding, S. T. 1953. Water supply of Crater Lake, Oregon. (unpublished notes [available from Office of State Climatologist, Oregon State Univ., Corvallis, Ore. 97331.])

Johnson, D. M., and J. O. Dart. 1982. Variability of precipitation in the Pacific Northwest: Spatial and temporal characteristics. WRRI-77. 181 pp. (with Appendix, 144 pp.) Water Resources Res. Inst., Oregon State Univ., Corvallis, Ore.

Mahrt, L., and M. Ek. 1984. The influence of atmospheric stability on potential evaporation. Jour. Climate & Appl. Meteor. 23:222-234.

Pacific Northwest River Basins Commission. 1975. Lower atmosphere handbook, Columbia Basin States. Vol. 5. 201 pp.

Phillips, K. N. 1968. Hydrology of Crater, East, and Davis Lakes, Oregon. U. S. Geol. Surv. Water Supply Pap. 1859-E. 60 pp.

Simpson, H. J. 1970. Tritium in Crater Lake, Oregon. Jour. Geophys. Res. 75:5195-5207.

Snow Survey. 1988. Oregon annual data summary. Water Year 1987. Soil Conservation Serv., 1220 SW 3rd Ave., Portland, Ore. 28 pp.

Soil Conservation Service. 1964. Normal annual precipitation map, 1930-1957. 1 p. [available from Office of State Climatologist, Oregon State Univ., Corvallis, Ore.]

Sterns, G. L. 1963. Climate of Crater Lake National Park. Crater Lake Nat. Hist. Assoc. 12 pp.

SECCHI DISK, PHOTOMETRY, AND PHYTOPLANKTON DATA FROM CRATER LAKE: LONG-TERM TRENDS AND RELATIONSHIPS

Clifford N. Dahm[1], Douglas W. Larson[2], N. Stan Geiger[3], Lois K. Herrera[1]
[1] Department of Biology, University of New Mexico, Albuquerque, NM 87131
[2] Corps of Engineers, P.O. Box 2946, Portland, OR 97208
[3] Scientific Resources, Inc., 11830 SW Kerr Parkway, Suite 375, Lake Oswego, OR 97035

There has been a decrease in average Secchi disk transparency measurements in Crater Lake during the late 1970s and 1980s relative to earlier studies. The decrease is frequently most pronounced in the late summer. This period of time corresponds to epilimnetic water temperatures generally above 10°C and a bloom of the diatom, *Nitzschia gracilis*. These blooms have occasionally exceeded one million cells per liter. Depth integrated estimates of primary production rates in the epilimnetic zone (0-30 m) for measurements from 1980-1983 are 50% higher than those taken from 1967-1969. An inverse relationship between the log of total cell numbers of *Nitzschia gracilis* and Secchi depth was seen in the summer of 1983. Photometer data from July 16, 1969 and July 16, 1982 showed an increase in green and a decrease in blue and unfiltered light penetration in 1982 relative to 1969. The evidence from comparative measurements made recently and in the late 1960s points to an approximately 20% decrease in overall epilimnetic water clarity in Crater Lake.

The phenomenal clarity of Crater Lake sets this lake apart from most others (Smith and Tyler 1967; Larson 1972). A 20-cm Secchi disk measurement of 44 m made on July 16, 1969 stands as a record for lakes worldwide. The possibility that the outstanding clarity within Crater Lake might be diminishing brought about the establishment of a ten-year limnological monitoring program at Crater Lake National Park beginning in 1982 (Larson 1984). This program has as one of its main components a monitoring program to assess whether lake clarity has changed and to provide a baseline for future studies of the lake.

The historical data from Crater Lake are meager. A few measurements of Secchi depth were made by interested individuals in 1913 and 1937. The first concerted studies of Crater Lake were conducted from 1967-1969 by Jack Donaldson and coworkers (Hoffman and Donaldson 1968; Kibby *et al.* 1968; Malick 1971; Larson 1972). Secchi disk measurements from this work were the main basis upon which a concern for diminished water clarity was expressed in the early 1980s. These new measurements were taken by Doug Larson from 1978-1981 while he worked as a volunteer researcher at Crater Lake National Park.

Fortunately, in addition to the Secchi disk measurements, the work from 1967-1969 also included a number of other analyses that can be used to compare conditions in the surface waters of Crater Lake to recent measurements. These analyses include the rate of phytoplankton primary production using $^{14}CO_2$ uptake; photometry studies using unfiltered, blue, and green light; chlorophyll-α measurements; and the species and numbers of algae in the upper photic zone.

The purpose of this paper is to take a careful look at the relevant historical data from 1967-1969 and to compare these data to the more recent measurements. The primary question we wish to address is whether the exceptional clarity of Crater Lake has changed in recent times.

SITE DESCRIPTION

Crater Lake is located along the summit of the Cascade Range, approximately 105 km north of the Oregon/California border and 195 km inland from the Pacific Ocean. The precise location is 42°56′ N and 122°06′ W. Crater Lake occupies the collapsed caldera of the volcano, Mount Mazama. Crater Lake is the deepest lake in the United States at 589 m and ranks as the seventh deepest lake in the world (Hutchinson 1967). The lake formed in the collapsed crater left behind after the catastrophic eruption of Mount Mazama about 6600 years ago (Fryxell 1965). The surface area of the lake is 48 km^2 and the entire drainage area for the lake, including the lake, is 63 km^2. Most of the water enters the lake through direct precipitation on the lake's surface. Subsequent vulcanism since the catastrophic eruption has produced an emergent secondary cone, Wizard Island, and at least two other submerged cones. The walls of the caldera are steeply sloped and approximately half the lake bottom is 325 m or deeper (Byrne 1965). Morphometric features of Crater Lake are given in Table 1.

TABLE 1. MORPHOMETRIC CHARACTERISTICS OF CRATER LAKE, OREGON

Elevation, surface (m)	1882
Area (km^2)	48
Volume (km^3)	16
Depth, maximum (m)	589
Depth, mean (m)	325
Shoreline, length (km)	31

Crater Lake lies in an enclosed basin with no surface outlets (Nelson 1967). Seepage (73%) and evaporation (27%) are estimated to be the main modes of water loss. The total input of water supplied to the lake on an annual basis is approximately 0.7% of the lake's total volume (Phillips and Van Denburgh 1968). Flushing time is extremely slow. The level of the lake has fluctuated a maximum of about 5 m over a period of record since 1878 (U. S. Geol. Surv. 1983). The lake is ultraoligotrophic with exceptional clarity and lies fully within the confines of Crater Lake National Park.

MATERIALS AND METHODS

The logistics of sampling at Crater Lake present many difficulties. Severe winter weather, large accumulations of snow, and the steep, treacherous caldera walls make all but summer sampling very problematic. The analyses in this paper are all based on samples collected from June to September.

Secchi disk measurements were made with a standard black and white 20-cm disk. Measurements were compared only for those days when clear weather and calm surface conditions were noted. Sampling was done either at Station 13 or Station 23 (Fig. 1).

Phytoplankton primary production was measured on 125-ml samples which were inoculated with 1 ml ^{14}C-$NaHCO_3$ (5.0 $\mu Ci\ ml^{-1}$) and incubated *in situ* for 4 hr near midday. Dissolved inorganic carbon concentrations in the samples were calculated from measurements of pH and alkalinity (Wetzel and Likens 1979). Alkalinity was measured with a colorimetric determination using 0.018N H_2SO_4 as titrant and bromocresol green - methyl red indicator solution (APHA 1980). Field measurements of pH were done with an Altex meter standardized with pH 7.00 and pH 10.00 buffer solutions. Light and dark bottles were deployed at each depth. The samples were retrieved and immediately filtered through a Millipore filter (0.45 µm, HA-type) and placed in a desiccator to dry. Sample filters were counted by liquid scintillation (Beckman LS 9000 counter) using a standard counting cocktail. Counts were automatically corrected for blanks and counting efficiency.

Water samples for chlorophyll-α analysis were dispensed into 1-liter plastic bottles and treated with saturated $MgCO_3$. Samples were filtered through Millipore HA-type filters (0.45 µm) within 4-6 hours of collection. The method of Strickland and Parsons (1972) was followed for the extraction and measurement of the chlorophyll-*a*. A Bausch and Lomb Spectronic 70 spectrophotometer was used for the absorbance measurements and a standard algal chlorophyll-*a* (Sigma Chemical Co.) was used for standardization.

Water samples for phytoplankton determinations were treated as described in Geiger and Larson (1990). Briefly, samples were emptied into a 10-µm net and the material retained in the net was washed into plastic bottles and fixed with 3% formalin. An aliquot of 500 ml of the water passing through the net was also collected and fixed in the same manner.

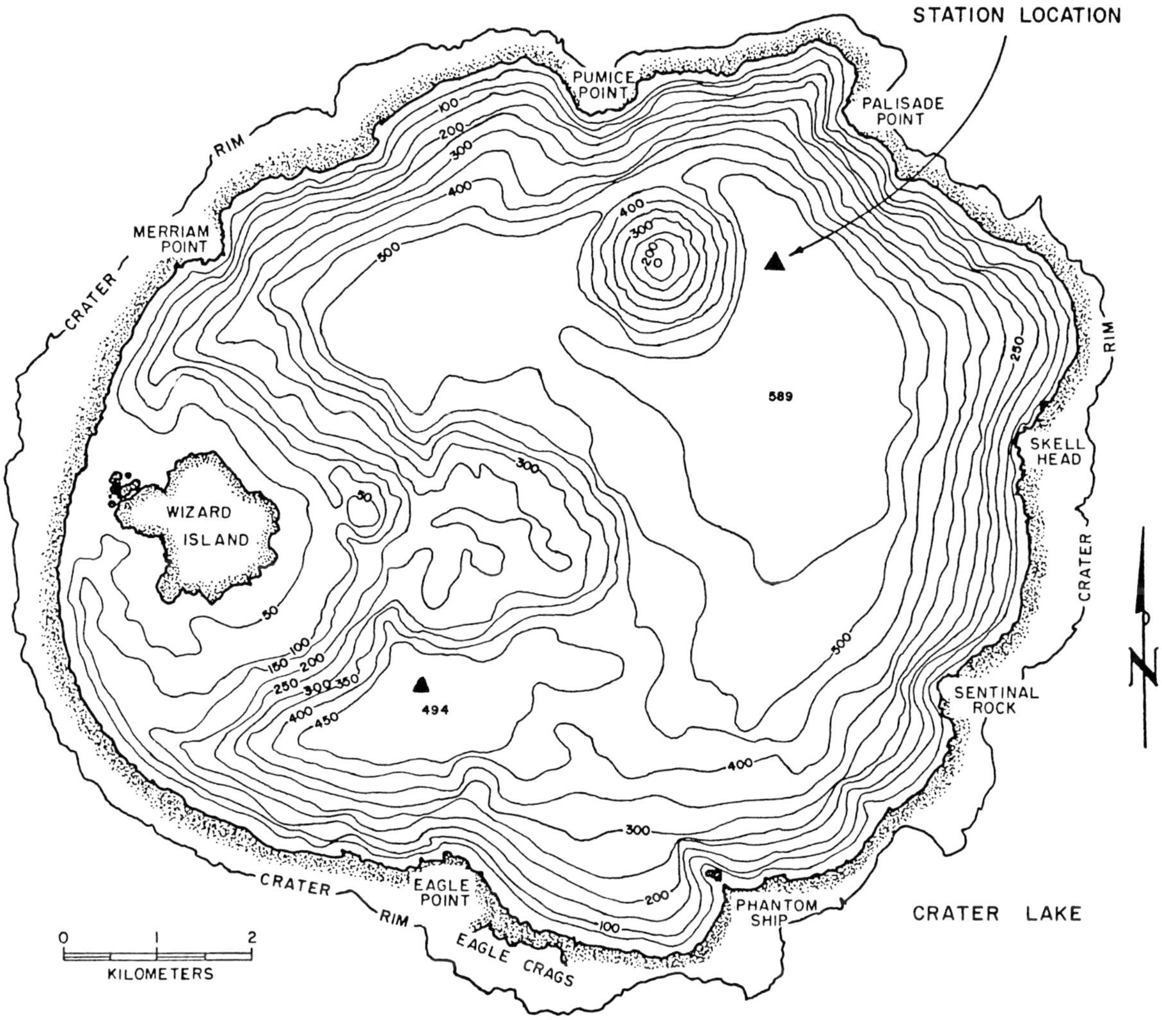

Figure 1. Bathymetric map of Crater Lake, Oregon, with the two main station locations marked by the solid triangles.

Subsamples of the material retained on the 10-µm net and passing through the net were gently filtered through 0.45-µm MF-Millipore membrane filters. Filters were made transparent by placing them on immersion oil on microscope slides and warming to draw oil into the filter (Lind 1974). The filters were observed using phase and brightfield microscopy at a magnification of 1000x. Counts were made of at least 100 discrete algal cells or colonies (Greeson *et al.* 1977).

Photometry measurements were carried out using a Kahl photometer. A Kahl model 268-WA310 was used during the 1967-1969 measurements and a Kahl model 268-WA350 was used in the early 1980s. The two instruments were intercalibrated to assure consistency of the two data sets. Measurements were made of unfiltered light, green light, blue light, and red light. A deck cell was used to monitor for changes in light reaching the surface during the lowering of the photometer. Measurements were made from the surface to a depth of 90 to 100 m.

RESULTS

A comparison of the Secchi disk measurements made from before 1970 with those made from 1978 to 1986 is shown in Table 2. The mean depth for Secchi disk measurements in the period before 1970 was 36.5 m. Only eight measurements were made during this period. The mean depth for Secchi disk measurements in the period from 1978-1986 was 29.1 m. This mean was calculated based on 83 measurements. A Student *t*-test was used to analyze for a significant difference in the two means. A significant difference existed with a probability of error of less than 0.001.

TABLE 2. SECCHI DISK COMPARISON FOR THE YEARS FROM 1937-1969 VERSUS 1978-1986

A significant difference in the means exists at the $p < .001$ level

Years	Average	Std Dev	N
1937-1969	36.5	3.4	8
1978-1986	29.1	3.1	83

The Secchi disk readings from 1982-1984 are shown in Fig. 2. The maximum depth measured during this period was 32.0 m. The minimum depth that was recorded was 21.9 m. A seasonal pattern was seen each year. The greatest clarity normally occurred in measurements made during the early summer. A gradual decrease was seen as the summer progressed. Minimum values were found in August or September.

Primary production measurements were made seven times in the period from 1967-1969 and eight times from 1980-1983 (Table 3). Measurements were made at midday (1100-1500 hr). The average total primary production throughout the photosynthetically active portion of the water column (ca. 0 to 200 m) was 24.0 mg C $m^{-2}hr^{-1}$ from 1967-1969 and 21.4 mg C $m^{-2}hr^{-1}$ from 1980-1983. The slightly higher overall rates of primary production during the late 1960s were not significantly different from the values in the early 1980s. Photosynthetic activity in the surface water of Crater Lake, which roughly corresponded to average Secchi disk measurements (ca. 0-30 m), was 2.4 mg C $m^{-2}hr^{-1}$ for the 1967-1969 measurements and 3.6 mg C $m^{-2}hr^{-1}$ for the 1980-1983 measurements. The rates of photosynthesis in the surface waters of Crater Lake were on average 50% higher in the early 1980s than the late 1960s. The difference, however, was not statistically significant.

There was a seasonal pattern in the rate of photosynthesis within the surface waters of Crater Lake. Measurements made in the early summer in all years were lower in total primary production (Figs. 3-5). As the summer progressed, higher rates of primary production were found in the years 1968, 1969, and 1983 when repeated assays were made. The minimum level of primary production measured in the surface waters occurred on July 16, 1969 (0.5 mg C $m^{-2}hr^{-1}$) and the maximum level occurred on August 24, 1981 (6.5 mg C $m^{-2}hr^{-1}$). The mean of measurements made in all years for August and September was 4.0 mg C $m^{-2}hr^{-1}$, while the mean of measurements made in June and July was 2.2 mg C $m^{-2}hr^{-1}$. The difference between the August/September and June/July means is significantly different at a probability of error of .012 using either a Student *t*-test or Mann-Whitney *U*-test.

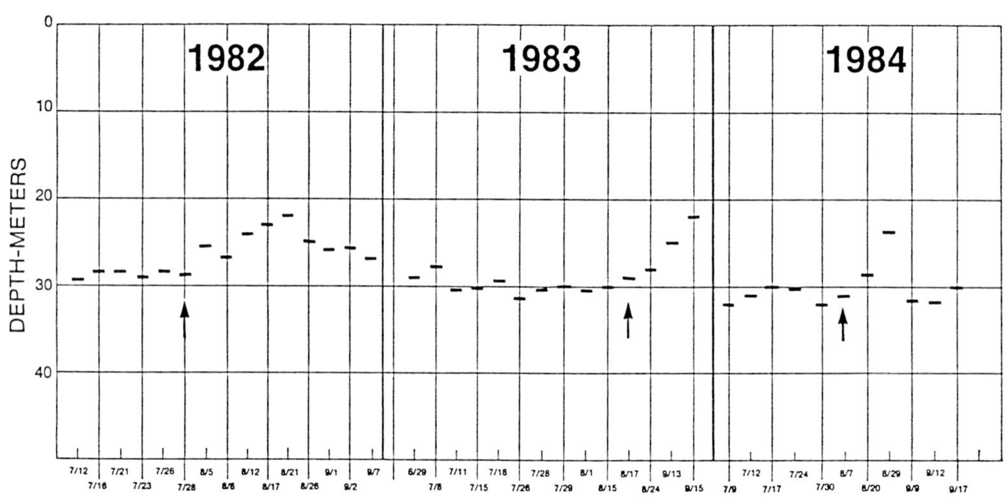

Figure 2. Secchi disk readings for Crater Lake from 1982 - 1984.

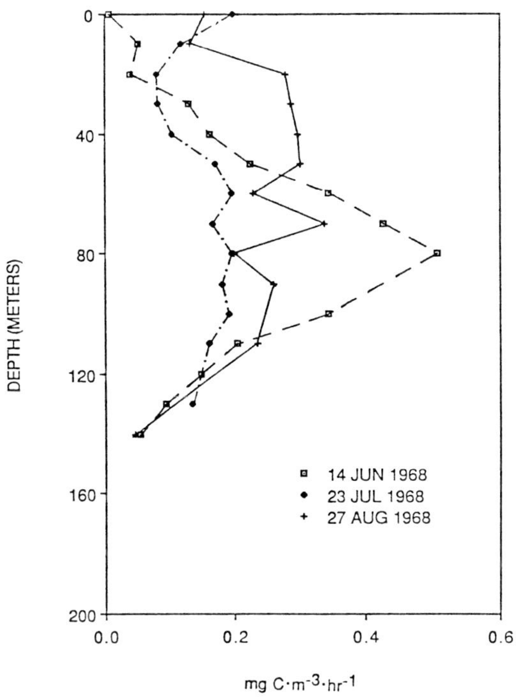

Figure 3. Phytoplankton primary production rates measured by $^{14}CO_2$ uptake on three dates in 1968.

The chlorophyll-*a* data from before 1970 are very limited. A chlorophyll-*a* maximum was noted at 100 m on data collected August 5, 1969 (Larson 1972). Near-surface values ranged from .086 to .159 mg m^{-3}. Chlorophyll-*a* data for 1983 ranged from .000 to .930 mg m^{-3}. In general, the amount of chlorophyll-*a* measured in the 1983 data set showed a great deal of variation both temporally and vertically within the water column. The highest levels of chlorophyll-*a* were measured on August 18 and September 2, 1983 in the 0-20 m portion of the water column.

Photometry measurements have also been made both in the period from 1967-1969 and from 1979-1985. Three profiles were made in the 1960s and 14 profiles have been carried out from 1979-1985. Two of the photometry profiles were made exactly 13 years apart on July 16, 1969 and July 16, 1982. The profiles of % transmittance versus depth for unfiltered light, blue light, and green light for these two dates are shown in Figs. 6-8. Unfiltered light and blue light did not penetrate as deeply into the lake in 1982 as in 1969. The depth of 50% transmittance for unfiltered light was 11 m in 1969 and 7 m in 1982. The depth of 10% transmittance was 54 m in 1969

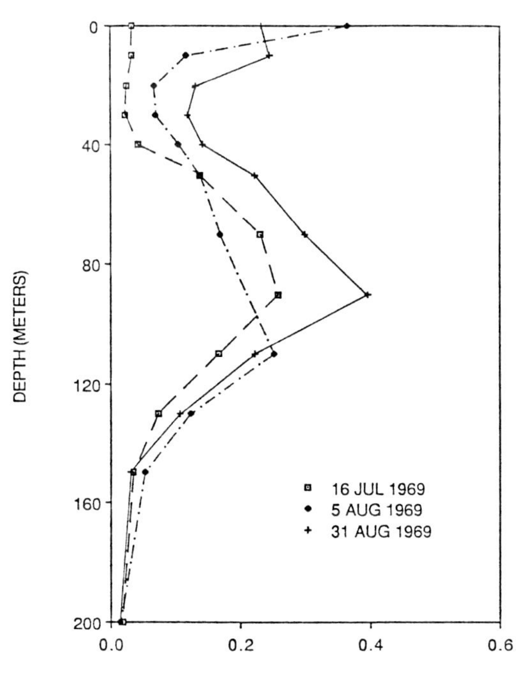

Figure 4. Phytoplankton primary production rates measured by $^{14}CO_2$ uptake on three dates in 1969.

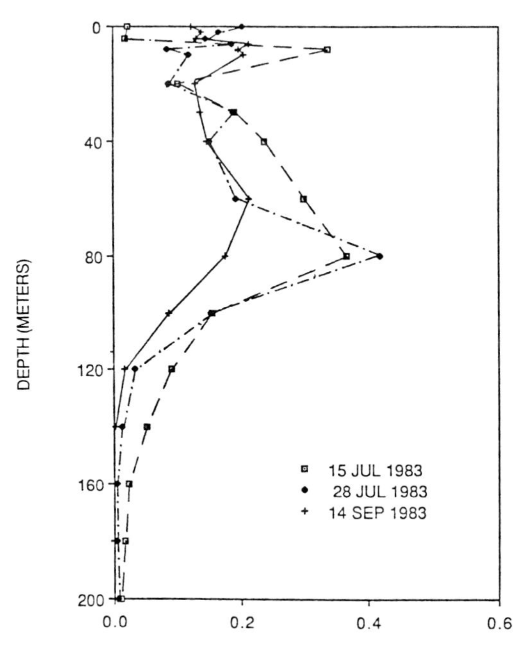

Figure 5. Phytoplankton primary production rates measured by $^{14}CO_2$ uptake on three dates in 1983.

and 34 m in 1982 (Fig. 6). The depth of 50% transmittance for blue light was 30 m in 1969 and 12 m in 1982. The depth of 10% transmittance was 69 m in 1969 and 55 m in 1982 (Fig. 7).

The opposite trend existed for the measurements of % transmittance of green light on July 16 in 1969 and 1982 (Fig. 8). Green light penetrated further on this date in 1982 than in 1969. The depth of 50% green light transmittance was 9 m in 1969 and 14 m in 1982. The depth of 10% transmittance was 36 m in 1969 and 44 m in 1982.

TABLE 3. CRATER LAKE PRIMARY PRODUCTION RATES DURING MIDDAY UNDER OPTIMAL WEATHER CONDITIONS TAKEN FROM 1967-1983.

Uptake of $^{14}CO_2$ was used to estimate primary production.

Date	mg C $m^{-2}hr^{-1}$ 0-200m	mg C $m^{-2}hr^{-1}$ 0-30m
26 Jul 67	18.3	2.7
14 Jun 68	27.3	1.0
23 Jul 68	19.7	2.2
27 Aug 68	28.1	4.1
16 Jul 69	21.4	0.5
5 Aug 69	20.7	2.6
31 Aug 69	32.7	3.5
27 Aug 80	29.4	2.8
24 Aug 81	23.2	6.5
14 Jul 83		1.9
15 Jul 83	27.2	2.5
28 Jul 83	23.7	4.1
29 Jul 83	17.3	2.6
9 Sep 83	11.8	4.0
14 Sep 83	17.5	4.6
	Mean & Std Dev.	
Depth	1967-1969	1980-1983
(0-200 m)	24.0±5.4	21.4±6.2
(0-30 m)	2.4±1.3	3.6±1.5

DISCUSSION

Has the exceptional clarity of Crater Lake changed in recent times relative to conditions in the past? The limited data base which exists suggests that clarity has indeed declined. The Secchi disk measurements made from 1978-1986 average more than 7 m less than those in earlier periods. The difference is statistically significant (36.5 m versus 29.1 m). The photometry data show a decrease in the penetration of unfiltered light and blue light with an increase in penetration of green light when 1960s data are compared to 1980s data. This is in keeping with an increase of fine particulates in the water column. Surface water primary production rates also show a trend towards higher values in recent years. The regrettable lack of a rigorous long-term data base for a national treasure such as Crater Lake makes our conclusion somewhat tentative, but the evidence points towards a diminished clarity on the order of about 20%.

Secchi disk measurements are easily made, but elements such as the visual acuity of the observer, surface water conditions, light conditions, time of day, and the size of the disk produce variability in the reported depth. The measurements analyzed in this paper were limited to those values where the measurements were made near midday, under calm conditions, with bright sunshine, and using a 20-cm disk. Given the many limitations on the absolute accuracy of the method, Secchi disk measurements do still provide a good measure of the transparency of water to light (Tyler 1968; Wetzel 1983). Secchi disk transparency generally represents a depth of 1 to 15% transmission with the norm being approximately 10% of surface light (Wetzel 1983). The phenomenal clarity of Crater Lake, with the need for outstanding eyesight to see a small disk at a long distance, makes the light transmission values usually somewhat above this 10% value (see Figs. 6-8). Although the Secchi disk measurements do not represent the most ideal data for comparison, the consistency of the two data sets and a careful analysis of the conditions under which measurements were made provide some confidence that a real difference exists.

Photometry profiles are less subjective and more rigorous means than the Secchi disk to analyze lake clarity, although the number of profiles, particularly before 1970, is quite limited. The photometry data support a conclusion of decreased clarity in the period of 1979-1986 versus the period from 1968-1969. Larson (1972) published three profiles made with a photometer that looked at unfiltered, blue, green, and red portions of the light spectrum. The measurements were made on June 14, 1968, July 16, 1969, and August 9, 1969. The average depth for 10% transmittance of unfiltered light was 54 m with a range from 49-58 m. Blue light had an average depth of 10% transmittance of 67 m with a range from 64-69 m. Green light had an average depth of 10% transmittance of 37 m with a range of 36-40 m. These values can be compared to more recent profiles such as the July 16, 1982 measurements (Figs. 6-8). The depths for 10% light transmittance in Crater Lake on July 16, 1982 were 34 m, 55 m, and 44 m for unfiltered, blue, and green light, respectively. Other profiles made with a photometer in the period from 1979-1986 yield similar results. The pattern is consistent with an increase in fine particulates in the water in recent years, which decreases overall light and blue light penetration but increases green light penetration due to scattering by the small particles.

Smith *et al.* (1973) measured maximum penetration of light into Crater Lake at a wavelength of 469 nm in the early 1970s. This wavelength is very strongly in the blue portion of the spectrum (blue is 465.0 nm and green is 526.2 nm as presented in Tam and Patel 1979). The indication from the recent photometry data is that maximum penetration has shifted somewhat towards greener wavelengths. A

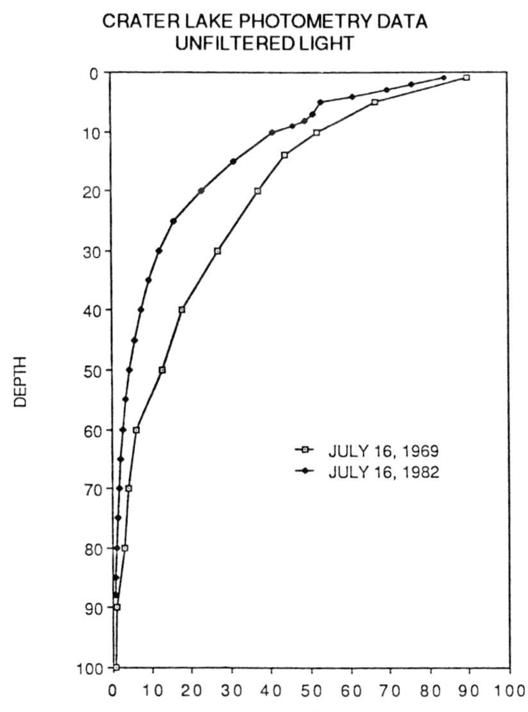

Figure 6. Comparison of unfiltered light transmittance profiles in Crater Lake on July 16, 1969 and July 16, 1982.

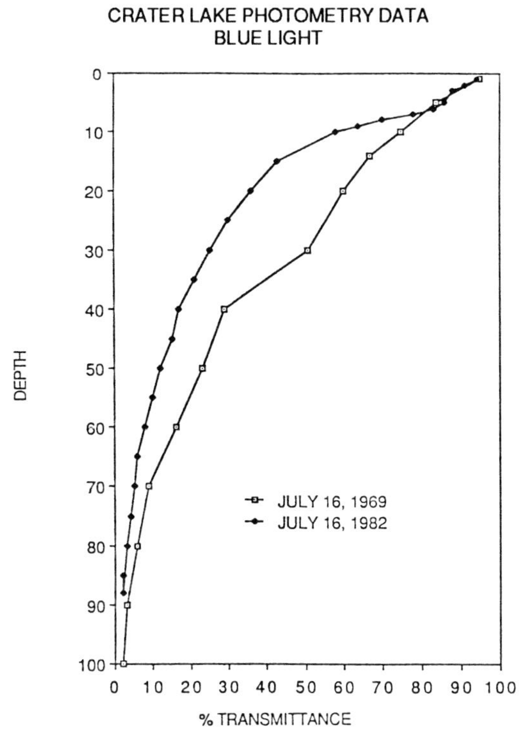

Figure 7. Comparison of blue light transmittance profiles in Crater Lake on July 16, 1969 and July 16, 1982.

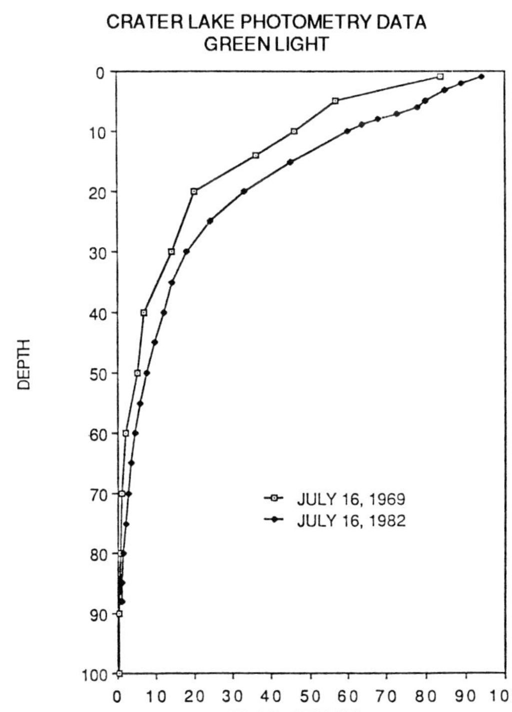

Figure 8. Comparison of green light transmittance profiles in Crater Lake, July 16, 1969 and July 16, 1982.

repeat survey of light penetration with equipment capable of such narrow band resolution of light transmittance would be an excellent method by which to analyze in more detail the question of clarity changes within Crater Lake.

The decline in clarity of surface waters within Crater Lake is due to small particles present in the upper 30-40 m of the lake. The exact nature of these particles remains unknown, but a link to phytoplankton cells can be seen in at least some of the periods of diminished Secchi disk readings. A good example of the possible link to phytoplankton numbers was seen in the 1983 data. A diatom, *Nitzschia gracilis*, had been noted to occur in large numbers in the surface waters of Crater Lake (see Larson *et al.* 1987). This observation was consistent with the general increase in the rate of photosynthesis in the surface waters of the lake, which was seen in both the 1968-1969 and 1983 studies. Secchi disk measurements in 1983 were at a maximum around 30-31 m in July. Total cell counts for *Nitzschia gracilis* in the 0-20 m zone of the water column in July were also low during this period (Fig. 9). *Nitzschia gracilis* began a logarithmic growth phase in late July, which extended through the middle of August. By the middle of August, *Nitzschia gracilis* cell count estimates had reached a total of between 10^9-10^{10} cells within a 20 m^3 volume extending from the surface to 20 m. These levels persisted through the period of measurement that lasted until mid-September. Secchi disk measurements began to decrease during the period of logarithmic growth of the *Nitzschia gracilis*. Minimum Secchi disk measurements, including three values less than 25 m, occurred in September when algal cell numbers in the surface waters were near their maximum. The bloom of this diatom in the late summer appears to have decreased the Secchi disk reading in 1983; comparable patterns are seen in 1982 and 1984 (Fig. 2).

The seasonal bloom of *Nitzschia gracilis* occurs in waters that are extremely impoverished in nitrogen and trace minerals (Larson *et al.* 1987; Dymond and Collier 1986; Collier and Dymond 1988). The concentration of nitrate-N in the surface waters of Crater Lake during the period from July-September of 1983 never exceeded the detection limit of 1 $\mu g\ l^{-1}$. The concentration of ammonium-N was also less than 1 $\mu g\ l^{-1}$ on July 8, July 15, and August 10. Ammonium-N values in the range of 0-5 $\mu g\ l^{-1}$ occurred in the surface water on August 17, September 7, and September 14. This diatom proliferates in very oligotrophic waters with extremely low levels of key nutrients. It is interesting to note that earlier studies of Crater Lake did not find a large bloom of this phytoplankton species (Utterback *et al.* 1942, Sovereign 1958; Thomasson 1962; Larson 1970; Larson 1972). It is not known whether the absence of reports concerning *Nitzschia gracilis* is due to the limited nature of past sampling or a much lower abundance, historically, for the organism. In any case, this species of phytoplankton presently grows to great numbers (sometimes exceeding 10^6 cells l^{-1}) in the epilimnion during the period of warmer surface waters, usually above 10°C, which is normally present in Crater Lake during the middle to late summer.

The cause or causes for the recent change in clarity within the surface waters of Crater Lake are not known. We believe that this change is at least in part

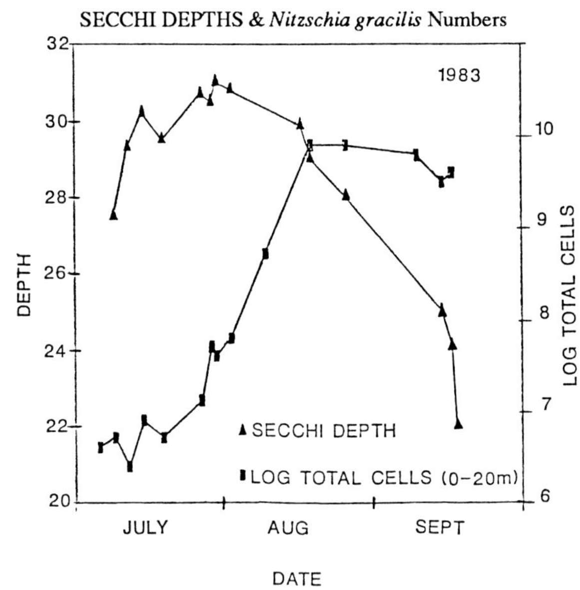

Figure 9. Secchi disk depths and the log of the summation of total *Nitzschia gracilis* algal cells in the 0-20 m stratum of Crater Lake during the summer of 1983.

associated with greater algal primary production in the upper portions of the photic zone. Inorganic particles and increased bacterial numbers may also play a role. A number of plausible hypotheses can be put forth to explain the difference in clarity measured in the 1980s relative to before 1970.

Some of these include:

(1) Sewage contamination has been entering the lake through groundwater and surface springs and has produced a gradual buildup in a chemical constituent, probably nitrogen, previously limiting primary production.

(2) Changes have occurred within the airshed in the area of Crater Lake that have altered the composition of the precipitation and dry fallout entering the lake.

(3) A shift has occurred in the structure of the zooplankton community that results in greater overall algal numbers in the upper photic zones during this past decade compared to previous periods of study.

(4) The algal component of Crater Lake, particularly the diatom, *Nitzschia gracilis*, has increased in size and vigor due to adaptation and/or recruitment into the near-surface waters where algal populations had been historically restricted in numbers.

(5) Physical processes within the caldera walls of Crater Lake have somehow become more active in recent years, adding to the number of inorganic particles within the lake.

It is impossible at present to sort out the relative merits of these or other hypotheses. Bits and pieces of data exist to give some credence to a number of the above hypotheses. The presence of a spring below the Rim Village area that may be contaminated by sewage remains worrisome. Air quality within the caldera needs more study, and quantification of nitrogen and trace metals inputs to the lake from airborne sources should be a high priority. The long-term changes in the structure of the zooplankton community require more years of monitoring. In any case, the recent work of Jack Dymond and Robert Collier with sediment traps in Crater Lake does provide strong evidence that the cycling of carbon and nutrients within the photic zone is a very efficient process. Very little of the net annual primary production within Crater Lake actually leaves the photic zone (Dymond and Collier, pers. comm.). If a long-term slow increase of a limiting nutrient to the photic zone of the lake has been occurring, the recent biotic response within the phytoplankton community is not surprising, and the diminished lake clarity is a by-product of this change in activity.

CONCLUSIONS

The available data from Crater Lake support a conclusion of recent decline in clarity. The extent of this decline is about 20% averaged throughout the summer months of the study. Secchi disk measurements and photometer profiles are the best data sets on which to estimate the extent of change in clarity. Primary production rates in the surface waters and the bloom of a diatom, *Nitzschia gracilis*, also point to changes in the upper photic zone in recent years. The cause or causes of change in lake clarity still remain a mystery, although a number of potential mechanisms have been postulated.

ACKNOWLEDGMENTS

We thank the National Park Service for partial support of our research. J. Salinas, H. Tanski, and M. Gillmore assisted in the collection of samples. E. Drake and G. Larson encouraged our participation in the symposium and in preparation of this manuscript for this book. NSF project BSR-8407429 provided support for C. Dahm during both the data collection and data analysis phase of the research. R. Graham lent editorial assistance, and J. Barnett was instrumental in the preparation of figures for the manuscript and for assistance in data analyses.

LITERATURE CITED

American Public Health Association. 1980. Standard Methods for Examination of Water and Wastewater. 15th ed. APHA.

Byrne, J. V. 1965. Morphometry of Crater Lake, Oregon. Limnol. Oceanogr. 10:462-465.

Collier, R. W. and J. Dymond. 1988. Studies of hydrothermal processes in Crater Lake. Oregon State Univ., Coll. Oceanogr. Ref. Rep. 88-5. 49 pp.

Dymond, J. and R. Collier. 1986. Geochemistry and limnology of Crater Lake. *In* G. Larson, ed., Crater Lake Limnological Studies. 1985 Ann. Rep. U. S. National Park Service, CA 9000-3-0003. 72 pp.

Fryxell, R. 1965. Mazama and Glacier Peak volcanic ash layers, relative ages. Science 147:1288-1290.

Geiger, N. S. and D. W. Larson. 1990. The initial survey of phytoplankton species distribution in Crater Lake, Oregon 1978-1980. Pages 153-165 *in* E. T. Drake *et al.*, eds., Crater Lake: An Ecosystem Study. Pacific Div., Amer. Assn. Advance. Sci., San Francisco, Calif.

Greeson, P. E., T. A. Ehlke, G. A. Irwin, B. W.

Lium, and K. V. Slack. 1977. Methods for collection and analysis of aquatic biological and microbiological samples. Chapter 4A, Book 5, Laboratory Analysis. Techniques of the U. S. Geological Survey.

Hoffman, F. O. and J. R. Donaldson. 1968. Zooplankton population dynamics, Crater Lake. Crater Lake Rep. 2, U. S. Nat'l Park Serv. 12 pp.

Hutchinson, G. E. 1967. A Treatise on Limnology, Vol. 2. J. Wiley & Sons, New York. 1115 pp.

Kibby, H. V., J. R. Donaldson, and C. E. Bond. 1968. Temperature and current observations in Crater Lake, Oregon. Limnol. Oceanogr.13:363-366.

Larson, D. W. 1970. On reconciling lake classification with the evolution of four oligotrophic lakes in Oregon. Ph.D. Thesis. Oregon State Univ., Corvallis, Ore. 145 pp.

Larson, D. W. 1972. Temperature, transparency, and phytoplankton productivity in Crater Lake, Oregon. Limnol. Oceanogr. 17:410-417.

Larson, D. W. 1984. The Crater Lake study: detection of possible optical deterioration of a rare, unusually deep caldera lake in Oregon, U.S.A. Verh. Internat'l. Verein. Limnol. 22:513-517.

Larson, D. W., C. N. Dahm, and N. S. Geiger. 1987. Vertical partitioning of the phytoplankton assemblage in ultraoligotrophic Crater Lake, Oregon, U.S.A. Freshwater Biol. 18:429-442.

Lind, O. 1974. Handbook of Common Methods in Limnology. Mosby Co., St. Louis, Mo.

Malick, J. G. 1971. Population dynamics of selected zooplankton in three oligotrophic lakes. M.S. Thesis. Oregon State Univ., Corvallis, Ore. 112 pp.

Nelson, C. H. 1967. Sediments of Crater Lake, Oregon. Geol. Soc. Amer. Bull. 78:833-848.

Phillips, K. N. and A. S. Van Denburgh. 1968. Hydrology of Crater, East, and Davis lakes, Oregon. U. S. Geol. Surv. Water-Supply Pap. 1859-E. 60 pp.

Smith, R. C. and J. E. Tyler. 1967. Optical properties of clear natural water. Jour. Optical Soc. Amer. 57:589-595.

Smith, R. C., J. E. Tyler, and C. R. Goldman. 1973. Optical properties and color of Lake Tahoe and Crater Lake. Limnol. Oceanogr. 18:189-199.

Sovereign, H.E. 1958. The diatoms of Crater Lake, Oregon. Trans. Amer. Micros. Soc. 77:96-124.

Strickland, J. D. H. and T. R. Parsons. 1972. A practical handbook of seawater analysis. Fisheries Res. Bd. Can., Bull. 167. 310 pp.

Tam, A. C. and C. K. N. Patel. 1979. Optical absorptions of light and heavy water by laser optoacoustic spectroscopy. Applied Optics 18:3348-3358.

Thomasson, K. 1962. Planktological notes from western North America. Arkiv Bot. 4:437-463.

Tyler, J. E. 1968. The Secchi disc. Limnol. Oceanogr. 13:1-6.

U. S. Geological Survey. 1983. Water Resources Data, Oregon, Water Year 1982, Vol. 2: Western Oregon. U. S. Geol. Surv. Water-Data Rep. OR-82-2.

Utterback, C. L., L. D. Phifer, and R. J. Robinson. 1942. Some planktonic and optical characteristics of Crater Lake. Ecology 23:97-103.

Wetzel, R. G. 1983. Limnology, 2nd ed. Saunders Coll. Publ. Co., Philadelphia, Pa. 767 pp.

Wetzel, R.G. and G.E. Likens. 1979. Limnological analyses. W. B. Saunders Co., Philadelphia, Pa. 357 pp.

PHYTOPLANKTON SPECIES DISTRIBUTION IN CRATER LAKE, OREGON, 1978-1980

N. Stan Geiger[1] and Douglas W. Larson[2]
[1] Scientific Resources, Inc., 11830 SW. Kerr Parkway, Suite 375, Lake Oswego, OR 97035
[2] U. S. Army Corps of Engineers, P. O. Box 2946, Portland, OR 97208

The vertical distribution of phytoplankton in Crater Lake was examined by obtaining samples during 1978, 1979 and 1980 at the surface and at 20 m intervals to 200 m at a single station in the deepest basin of the lake. Samples were partitioned into material retained by and passed through a 10 micrometer mesh net. The fraction passed through the net, composed predominantly of diatoms, consistently comprised the highest percentage of the total densities. Recurring vertical distribution patterns of dominant species in combined fractions were observed through the three years with *Nitzschia gracilis* commonly more abundant at the surface in late summer, *Tribonema affine* more frequently found at mid-depths sampled, and *Stephanodiscus hantzschii* more frequently found at the lowest depths sampled. Similar distributions were observed in sampling four widely-spaced stations in the lake on the same day.

Until 1978 no systematic survey and analysis of Crater Lake phytoplankton had been performed. Notes on a few cursory examinations of attached and planktonic algae samples that were restricted to summer sampling (Brode 1938; Sovereign 1958; Kemmerer *et al.* 1924; Utterback *et al.* 1942; Thomasson 1962) prompted the authors to undertake an initial characterization of planktonic algae that would provide a basis for more detailed future studies. Sampling of the lake 1978-1980 was conducted to obtain information on algae at depths similar to those studied by Larson in 1969 (Larson 1970, 1972).

Specific objectives of the three-year sampling effort were to: (1) characterize the species of phytoplankton and their occurrence during summertime sampling; (2) examine the vertical distribution of algae species at 20 m intervals to 200 m.; (3) sample the large and small species more extensively by partitioning sampler contents into two fractions, those retained by and passed through a net with mesh openings 10x10 micrometers; and (4) describe horizontal distribution of the phytoplankton.

MATERIALS AND METHODS

Samples of water from Crater Lake for phytoplankton analysis were obtained during the summers of 1978, 1979 and 1980 during the months when the lake was accessible. In 1978, 101 samples were obtained on 13 occasions from July 11 through August 29. In 1979, 54 samples were obtainted on 6 occasions from May 23 through August 30. In 1980, 61 samples were obtained on six occasions from July 10 through October 10. On one date in 1980 four stations were sampled at five depths each.

Water samples were collected from selected depths from surface to 200 m with a 2.5 liter VanDorn PVC bottle. Samples were retrieved at a station located over the northeast and deepest basin of the lake (Fig. 1). The maximum depth sampled was determined on the basis of light penetration in the lake previously determined to be 1% of surface light intensity at 100 m (Larson 1972). This depth was doubled in an attempt to bracket the photic zone. Eleven sampling depths at intervals of 20 m were routinely sampled.

Contents of a VanDorn bottle were emptied into a 10 micrometer aperture mesh net. The material retained in the net was trasferred to a plastic bottle and fixed with 3% formalin. An aliquot of the water passing the net (500 ml) was collected and fixed in the same manner.

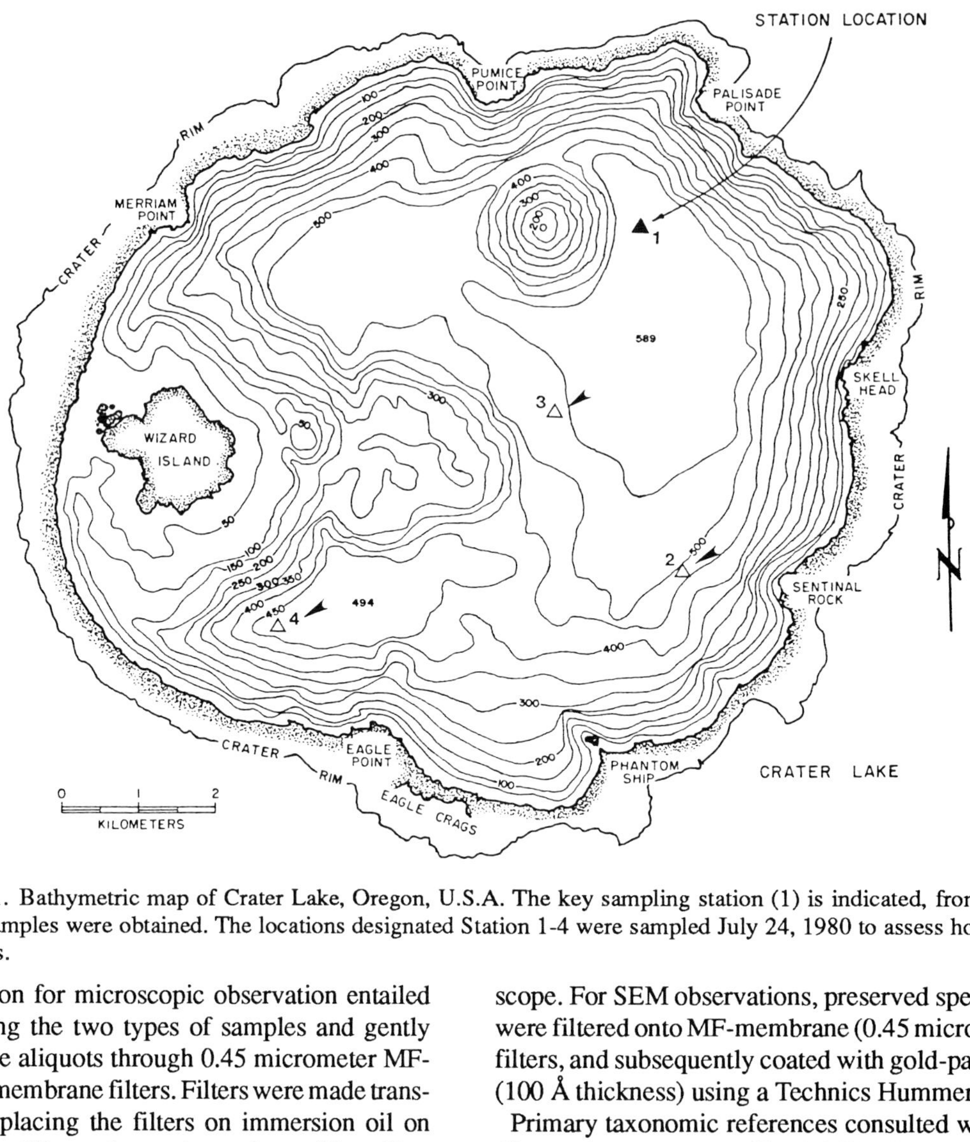

Figure 1. Bathymetric map of Crater Lake, Oregon, U.S.A. The key sampling station (1) is indicated, from which nearly all samples were obtained. The locations designated Station 1-4 were sampled July 24, 1980 to assess horizontal distributions.

Preparation for microscopic observation entailed subsampling the two types of samples and gently filtering the aliquots through 0.45 micrometer MF-Millipore membrane filters. Filters were made transparent by placing the filters on immersion oil on microscope slides and warming to draw oil into filter by water evaporation (Lind 1974).

Mounted filters were observed using phase and brightfield microscopy (American Optical H20). Counts were made of at least 100 discrete algal particles (cells or colonies) having distinct chromatophores at a magnification of 1000x (Greeson *et al.* 1977). This method of preparation provided permanent slides for future examination of preserved material. For supplemental identifications of fragile species, observations were made with an inverted microscope (Wild M40). Scanning electron photomicrographs of selected small diatoms were made using a JEOL JSM-35 scanning electron microscope. For SEM observations, preserved specimens were filtered onto MF-membrane (0.45 micrometer) filters, and subsequently coated with gold-paladium (100 Å thickness) using a Technics Hummer II.

Primary taxonomic references consulted were for Chrysophyceae, Bourrelly (1968); for Bacillariophyceae, Patrick and Reimer (1966, 1975), Sovereign (1958), Huber-Pestalozzi and Hustedt (1942), Archibald (1972); for Chlorophyceae, Prescott (1962, 1970), Smith (1950); and for Cryptophyceae and Dinophyceae, Huber-Pestalozzi (1968).

RESULTS

Phytoplankton Taxonomy

The phytoplankton of Crater Lake during the periods sampled in 1978-1980 consisted of 140 species of algae (Chrysophyta: Bacillariophyceae 102 spp. and Chrysophyceae 15 spp; Chlorophyta 11 spp;

Cryptophyta 3 spp. and Pyrrophyta 9 spp). Diatoms comprised 73% of the species. A complete listing of species observed for these three years is provided as an Appendix. A number of these species were observed only once or twice through the three full sampling seasons.

Some species that were prominent are not well-defined taxonomically due to the absence of observations of reproduction (e.g., *Mougeotia* sp. and *Tribonema affine*), scarcity of taxonomic features (typical of the small spherical cells of an apparent chrysophyte and a chlorophyte, each frequently abundant), present taxonomic uncertainty with the genus (e.g., *Nitzschia* and *Stephanodiscus*), or the hesitancy to assign specific names to certain members of a genus that has not been well-described in the Northwest (e.g., *Kephyrion, Pseudokephryion, Ochromonas* and *Chromulina*).

One of the prominent species throughout the study was the diatom identified as *Stephanodiscus hantzschii* (Fig. 2). This centric diatom was called *S. hantzschii* on the advice in 1979 of taxonomist Dr. Gary Collins of the USEPA laboratory at Cincinnati, Ohio, who made the first SEM observations of our Crater Lake material. Subsequent reviews of the identity of this diatom by other taxonomists suggest that the name *S. hantzschii* may require revision

Figure 2. SEM photomicrograph of *Stephanodiscus hantzschii* (length of bar equivalent to 1.0 micrometer).

following more detailed comparisons with other similar forms of *Stephanodiscus*.

Another prominent species over the three years was the diatom named *Nitzschia gracilis* (Archibald 1972) (Fig. 3). Sovereign (1958) had named a new species of *Nitzschia* [*exilis*] from attached material collected on the shores of Wizard Island that was similar to the *N. gracilis* from the plankton collections during 1978-1980.

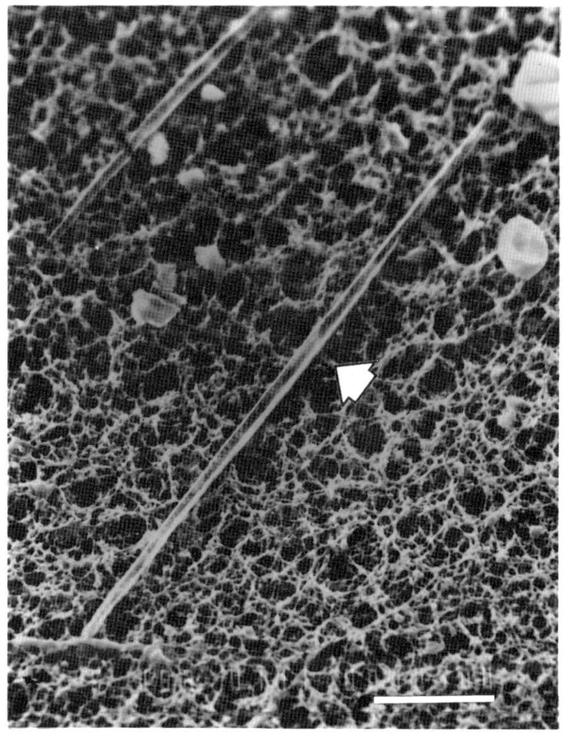

Figure 3. SEM photomicrograph of *Nitzschia gracilis* (length of bar equivalent to 10 micrometers).

One of the largest planktonic algae species observed was the species *Tribonema affine* (Fig. 4). Biovolumes among the dominant species varied from 29,000 cubic micrometers for *Tribonema affine* to 21 cubic micrometers for *Ankistrodesmus spiralis* (Table 1).

Net Retained Versus Passed Algae

A total of 71 species was observed in 1978; 71 were also observed in 1979. Table 2 shows the partitioning of species retained by and passed through the net. Only 17 of the total number of species observed in either year occurred at densities greater than 5% of the total number observed in both fractions. As

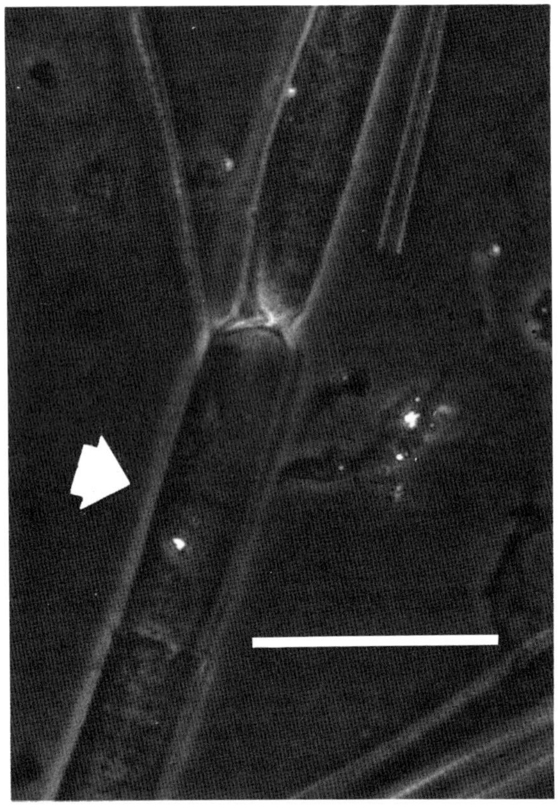

Figure 4. Phase contrast photomicrograph of *Tribonema affine* (length of bar equivalent to 10 micrometers).

noted in Table 2, species of *Synedra* and *Nitzschia* were prominent.

The needle-like and linear cylindrical shape of the largest prominent species (e.g., *Tribonema affine*, *Synedra delicatissima*, or *Nitzschia gracilis*) allowed them to pass through the net. While both the largest and smallest species were retained by and passed through the net, actual densities of these species on any day predictably showed size selection in the partitioning. For example, on August 29, 1978 *Stephanodiscus hantzschii* densities (units/l) were very much different in the two fractions (Table 3). Likewise, the two fractions were very much different for the largest prominent alga *Tribonema affine* in samples from August 14, 1979 (Table 4).

For 1978 the range of total algae densities retained in the net was from a high in early summer of 41,000 units/l to a low of 500 units/l in August. Densities of algae passing the net were nearly always higher, ranging from a low in early summer of 7,000 units/l to a high of 324,000 units/l in August. The profile of high surface and subsurface densities with increased densities at depths of 120-180 m was especially characteristic of the fraction passed through the net. In 1979 and 1980 total densities of algae in both net

TABLE 1. AVERAGE BIOVOLUMES OF TYPICAL DISCRETE UNITS OF DOMINANT PHYTOPLANKTON SPECIES IN CRATER LAKE, OREGON

SPECIES	BIOVOLUME (cu. micron.)
Ankistrodesmus spiralis (T.)Lemm.	21
Selenastrum minutum (Naeg.) Coll	24
Nitzschia gracilis Hantzsch	62
Stephanodiscus hantzschii Grun.	71
Rhodomonas minuta Skuja	132
Synedra delicatissima W. Smith	1584
Asterionella formosa Hass	2087
Tribonema affine G. S. West	29119

TABLE 2. PROMINENT SPECIES RETAINED BY (R) AND/OR PASSED THROUGH THE 10 X 10 MICROMETER APERTURE MESH NET FOR SAMPLE PARTITIONING 1978 AND 1979

SPECIES	1978		1979	
	R	P	R	P
Tribonema affine	R	P	R	P
Melosira distans v. *alpigena*				P
Stephanodiscus hantzschii	R	P	R	P
Asterionella formosa	R		R	
Synedra delicatissima	R		R	P
Synedra rumpens	R	P		
Synedra vaucheriae	R	P		
Synedra mazamaensis		P		P
Synedra tenera				P
Achnanthes minutissima				P
Nitzschia demota	R	P		
Nitzschia serpenticula	R	P		
Nitzschia gracilis	R	P	R	P
Epithemia sorex	R			
Navicula cryptocephala v. *min.*	R	P		P
Kephyrion spirale			R	
Ankistrodesmus spiralis		P		P
Spherical chlorophyte		P		P

TABLE 3. *STEPHANODISCUS HANTZSCHII* DENSITIES (UNITS/L) IN FRACTIONS RETAINED BY AND PASSED THROUGH THE NET AUGUST 29, 1978

DEPTH (m)	PASSED	RETAINED
0	4419	0
20	7164	0
40	9117	130
60	4892	28
80	1434	0
100	4410	0
120	51987	36
140	93635	766
160	119418	59
180	54858	170
200	43217	195

TABLE 4. DENSITIES OF *TRIBONEMA AFFINE* IN SAMPLES FROM AUGUST 14, 1979 IN FRACTIONS RETAINED BY AND PASSED THROUGH THE NET

DEPTH (m)	PASSED	RETAINED
0	0	1986
20	3558	5810
40	0	10904
60	380	4600
80	208	10668
100	0	7313
120	0	2468
140	0	2656
160	668	3972
180	0	4050
200	0	2521

retained and passed varied, but the proportional distribution of densities vertically remained the same.

Vertical Distributions

Total phytoplankton densities were highest in the 0-20 m depths, although densities were nearly as great at depths below 140 m (Fig. 5). In 1980 maximum phytopolankton densities were found at the lake's surface (Fig. 5). Densities in the mid-depth zone sampled (60-120 m) where both production and chlorophyll maxima occurred (Larson, Dahm and Geiger 1987) were considerably less than either the shallow- or deep-water assemblages (Fig. 5). How-ever, vertical distribution of total algae biovolume suggests biovolume (biomass) may correlate better with production and chlorophyll data. An example of this was the vertical distribution of total density (both fractions) and total dominant species biovolume August 26, 1980 (Table 5).

TABLE 5. THE VERTICAL DISTRIBUTION OF TOTAL DENSITY (BOTH FRACTIONS) AND TOTAL DOMINANT SPECIES BIOVOLUME, AUGUST 26, 1980

DEPTH	TOTAL DENSITY (unit/l x 10^4)	TOTAL BIOVOLUME (cu. micron./l x 10^7)
0	7.17	1.26
20	7.60	3.16
80	5.17	9.89
100	2.39	6.61
140	15.50	13.70
200	2.50	5.67

The vertical distribution of the lake's three dominant species followed a characteristic pattern in each of the three years as illustrated in Fig. 6. *Nitzschia gracilis* was the most abundant alga in the 0-20 m depth. Conversely, *Stephanodiscus hantzschii* was the dominant alga at depths of 140-200 m where light was <0.1% of surface light. However, both species occurred in great abundance at shallow depths May 1979 (Fig. 6). At mid-depth, *Tribonema affine* usually outnumbered all other species present (Fig. 6).

Horizontal Distribution

Results of sampling the lake during the same day, July 24 1980, at four widely-spaced locations (Stations 1-4, Fig. 1) show a rather narrow range of densities over all samples from 60,542 to 348,433 units/l (Table 6). Table 6 provides densities on species observed in samples that were >10% of the total densities. These species were *Nitzschia gracilis, Tribonema affine, Selenastrum minutum, Ankistrodesmus spiralis, Rhodomonas minuta* and *Stephanodiscus hantzschii.*

The pattern of recurring positioning of *Nitzschia gracilis* and *Stephanodiscus hantzschii* is apparent at each of the stations. Highest densities of *Tribonema affine* were observed in the mid-depth

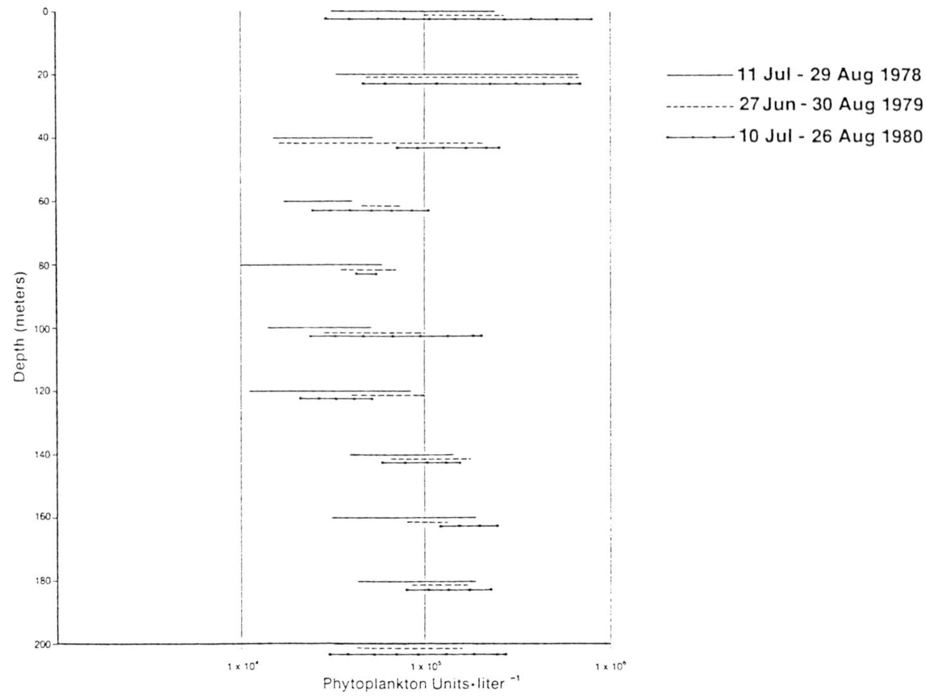

Figure 5. Ranges shown for total phytoplankton counts per sampling depth, Crater Lake, Oregon 1978-1980.

range of 40-100 m. The two green algae, *Senelastrum minutum* and *Ankistrodesmus spiralis* were widely distributed in the water colum. The absence of these two species in samples from the surface suggests densities of *Nitzschia gracilis* may have been sufficiently high to prevent detection of these lower density taxa with the counting method used.

DISCUSSION

Comparison with Previous Phytoplankton Studies

The 1913 plankton survey by Kemmerer *et al.* (1924) was probably the first of its kind for the lake. Although these investigators reported finding only two species of phytoplankton (*Mougeotia* sp. and the diatom *Asterionella* sp.) they did discover that both the phytoplankton and zooplankton were distributed to great depths, reaching maximum abundance between 60 and 200 m. The study is of limited value since no information was provided about how the plankton was sampled.

Studies of the lake in 1940 by Utterback, Phifer and Robinson (1942) indicated that: (1) phytoplankton was most abundant at a depth of 75-150 meters; (2) virtually no phytoplankton existed in the surface-to-20 m stratum, or in the deepest sample taken at 425 m; (3) most phytoplankton consisted of filamentous, blue-green algae (*Anabaena* sp.), and (4) diatoms constituted only about 15% of the total phytoplankton collected. Samples were obtained by hauling a No. 20 mesh plankton net (mesh aperture=79 micrometers) vertically through the water column, or by casting a Kemmerer bottle to discrete depths, retrieving the sample and then centrifuging the water to extract the phytoplankton. Population densities reportedly ranged from $1x10^3$ to $3x10^6$ cells per liter, but no indication is given as to the meaning of the term "cells." The work provided a brief taxonomic list, including, in addition to *Anabaena* sp., the diatoms *Nitzschia* sp., *Asterionella* sp., *Navicula* sp. and the filamentous alga *Mougeotia* sp.

Thomasson (1962) listed about a dozen species, one of which (*Ceratium hirundinella*) suggested that the samples were collected from sheltered shoreline locations rather than from the pelagic region of the lake. Since sampling locations were not given, the

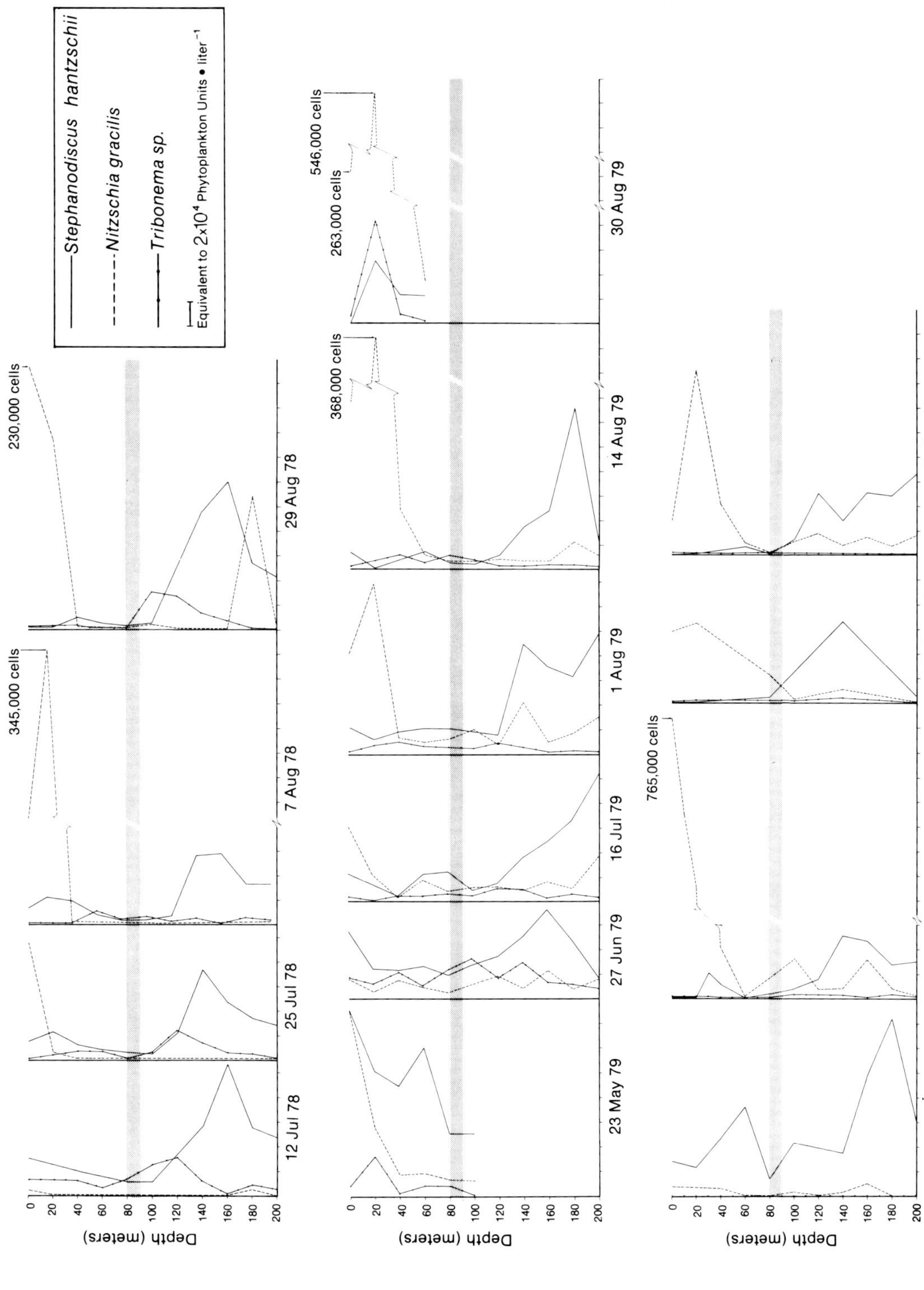

Figure 6. Vertical distributions of the three dominant phytoplankton species in Crater Lake, Oregon (*Nitzschia gracilis*, *Tribonema affine*, and *Stephanodiscus hantzschii*). "Cells" equivalent to cells/l. Shaded horizontal band (80-90m) indicates usual zone of 1% light transmission during summer.

TABLE 6. HORIZONTAL AND VERTICAL DISTRIBUTIONS OF TOTAL DENSITY AND DENSITIES OF SELECTED DOMINANT SPECIES, JULY 24, 1980, CRATER LAKE, OR.[1]

DEPTH (M)	STA 1	STA 2	STA 3	STA 3
Total densities (units/l)				
0	218,707	205,348	176,858	213,319
40	230,360	164,918	348,587	272,911
100	184,714	273,612	168,507	113,483
160	223,619	110,974	154,062	328,950
200	256,802	137,389	60,542	348,433
Nitzschia gracilis densities (units/l)				
0	192,462	166,331	168,015	165,684
40	0	0	0	0
100	0	0	0	7,865
160	0	0	0	6,579
200	0	0	0	58,647
Tribonema affine densities (units/l)				
0	2,187	0	0	2,071
40	18,428	11,544	20,306	0
100	3,694	21,888	8,179	17,977
160	2,236	2,219	3,081	0
200	2,568	4,121	3,057	4,064
Selenastrum minutum densities (units/l)				
0	2,187	0	0	0
40	46,072	3,298	23,690	168,562
100	31,718	5,527	44,604	1,123
160	4,384	2,197	1,540	0
200	22,883	4,163	4,114	40
Ankistrodesmus spiralis densities (units/l)				
0	0	0	0	0
40	36,857	29,685	67,686	32,107
100	1,865	44,220	8,260	22,471
160	52,616	13,185	12,324	49,342
200	58,479	33,306	4,280	48,297
Rhodomonas minuta densities (units/l)				
0	0	0	0	0
40	0	1,649	0	0
100	63,437	85,676	23,128	0
160	32,885	1,098	12,324	3,289
200	40,681	2,775	0	0
Stephanodiscus hantzschii densities (units/l)				
0	2,187	0	3,537	12,426
40	2,303	13,193	27,074	8,026
100	5,597	11,055	0	0
160	46,039	49,443	64,706	118,422
200	27,968	41,633	34,857	68,996

[1]See Fig. 1 for station locations.

study is of limited value. Thomasson also reported that the plankton was "very sparse" on the day he visited the lake (14 July 1959), and identified a rather abundant filamentous alga as *Tribonema* sp.

Coville (1897) reported of Crater Lake that "The lake itself is wholly devoid of aquatic vegetation. No algae, no mosses, and no aquatic flowering plants were found in its water." Apparent increases in phytoplankton in Crater Lake since 1897 can be attributed largely to improved methods of analysis that are able to document the presence of the small algae that dominate the lake. The ability to study attached algae (and mosses) more carefully on the bottom of the lake (e.g., Loeb and Reuter 1981) has also produced a more complete picture of the diverse and extensive flora than was previously perceived.

The presence of *Anabaena* as a dominant in the samples of Utterback, Robinson and Phifer in 1940 and the virtual absence of *Anabaena* sp. in the sampling reported here remains a puzzle. Species that would presumably be attached algae are common in the plankton (e.g., *Synedra mazamaensis* and six species of *Gomphonema*). The prominence of *Nostoc* sp. on the lake bottom (Loeb and Reuter 1981) suggests that sampling at a location where wave action would erode the colonies into suspension could produce the impression that "*Anabaena*" was prominent.

CONCLUSIONS

Changes in total phytoplankton densities differed between the fraction retained by and that passed through the net during summer. In early summer the fraction retained by the net and that passed through were more equally distributed vertically in the water column. Small algae cells were always more numerous than large cells, reaching density maxima near surface and 120-180 m strata and increased in number from early to late summer. Large cells became less dense as summer progressed. Increases in total densities were greatest near the surface in each of the three summers.

The phytoplankton in 1978-1980 was dominated by diatoms in numbers of species and individuals. In terms of biovolume or biomass the filamentous chrysophyte *Tribonema affine* comprised an appreciable amount of the total phytoplankton biomass on most sampling dates.

The recurrent vertical distribution patterns of *Nitzschii gracilis*, *Tribonema affine* and

Stephanodiscus hantzschii suggest a positioning in the water column by these species at optimum environmental conditions. Relationships between environmental parameters and the phytoplankton have been reported in Geiger and Larson 1981, Larson, Dahm and Geiger 1987, and Dahm, Larson, Geiger, and Herrera 1990.

LITERATURE CITED

Archibald, R.E.M. 1972. A preliminary key to the fresh and brackish water species of the genus *Nitzschia* in South Africa. Limno. Soc. S. Africa 18:33-55.

Bourrelly, P. 1968. Les Algue d'eu douce. Initiation à la systematique Tone II: Les algue, jaunes et brunes. Chrysophycées, Pheophycées, Xanthophycées, et Diatomées. Boubée à Cié, Paris.

Brode, J. Stanley. 1938. The denizens of Crater Lake. Northwest Sci. 12(3):50-57.

Coville, F. V. 1897. The August vegetation of Mount Mazama, Oregon. Mazama 1(2):170-203.

Dahm, Clifford N., Douglas W. Larson, N. Stan Geiger and Lois K. Herrera. 1990. Secchi disk, photometry, and phytoplankton data from Crater Lake: Long-term trends and relationships. Pages 143-151 *in* E. T. Drake *et al.*, eds., Crater Lake: An Ecosystem Study. Pacific Div., Amer. Assn. Advance. Sci., San Francisco, Calif.

Geiger, N. Stan and Douglas W. Larson. 1981. Crater Lake: Its planktonic algae. Mazama 63(13):54-59

Greeson, P. E., T. A. Ehlke, G. A. Irwin, B. W. Lium, K. V. Slack, eds. 1977. Methods for collection and analysis of aquatic biological and microbiological samples. Chapter A4, Book 5 Laboratory Analysis. Techniques of Water-Resources Investigations of the United States Geological Survey.

Huber-Pestalozzi, G. 1968. Das Phytoplankton des Süsswassers. 3 Teil Cryptophyceae, Chloromonadophyceae, Dinophyceae. 2 Auflage mit Neubearbeitung von B. Fott. E. Schweizerbart'sche Verlagsbuchhandlung, Stuttgart. 322 pp.

Huber-Pestalozzi, G. and F. Hustedt. 1942. Die Kieselalgen. *In* A. Thienemann (ed.), Das Phytoplankton des Süsswassers, Die Binnengewasser. Band XVI, Teil 2, Halfte II. E. Schweizerbart'sche Verlagsbuchhandlung, Stuttgart. 549 pp.

Kemmerer, G., F. Bovard, and W. R. Boorman. 1924. Northwestern lakes of the United States: Biological and chemical studies with reference to possibilities in production of fish. Bull. U. S. Bur. Fish. 39:51-140.

Larson, D. W. 1970. On reconciling lake classification with the evolution of four oligotrophic lakes in Oregon. Ph.D. Thesis, Oregon State Univ., Corvallis, Ore. 145 pp.

Larson, D. W. 1972. Temperature, transparency, and phytoplankton productivity in Crater Lake, Oregon. Limnol. Oceanogr. 17:410-417.

Larson, D. W., C. N. Dahm and N. S. Geiger. 1987. Vertical partitioning of the phytoplankton assemblage in ultraoligotrophic Crater Lake, Oregon, U.S.A. Freshwater Biol. 18:429-442.

Lind, O. 1974. Handbook of common methods in Limnology. Mosby & Co., St. Louis, Mo.

Loeb, Stanford L. and John E. Reuter. 1981. The epilithic periphyton community: A five-lake comparative study of community productivity, nitrogen metabolism and depth-distribution of standing crop. Verh. Internat'l Verein. Limnol. 21:346-352

Patrick, R. and C. W. Reimer. 1966. The diatoms of the United States. Vol. 1. Acad. Nat. Sci. Phila., Monogr. 13. 688 pp.

Patrick, R. and C. W. Reimer. 1975. The diatoms of the United States. Volume 2, Part 1. Acad. Nat. Sci. Phila., Monogr. 13. 213 pp.

Prescott, G. W. 1962. The algae of the Western Great Lakes Area. Wm. C. Brown Co., Dubuque, Iowa. 719 pp.

Prescott, G. W. 1970. The freshwater algae of the United States. Wm. C. Brown Co., Dubuque, Iowa. 348 pp.

Smith, G. M. 1950. The freshwater algae of the United States. McGraw-Hill, New York. 719pp.

Sovereign, H. E. 1958. The diatoms of Crater Lake, Oregon. Trans. Amer. Micr. Soc. 77:96-134.

Thomasson, K. 1962. Planktological notes from western North America. Arkiv Botanik 4:437-463.

Utterback, C. L., L. D. Phifer, and R. J. Robinson. 1942. Some planktonic and optical characteristics of Crater Lake. Ecology 23:97-103.

APPENDIX

PHYTOPLANKTON SPECIES OBSERVED 1978–1980

DIV*	SPECIES	AUTHORITY	1978	1979	1980
BAC	ACHNANTHES CL1		X		
BAC	ACHNANTHES DEFLEXA	REIM.	X		X
BAC	ACHNANTHES LANCEOLATA	(BREB.) GRUN.	X	X	X
BAC	ACHNANTHES LINEARIS	(W. SMITH) GRUN.	X		X
BAC	ACHNANTHES MICROCEPHALA	(KUETZ.) GRUN.			X
BAC	ACHNANTHES MINUTISSIMA	KUETZ.	X	X	X
BAC	ACHNANTHES PINNATA	HUST.			
BAC	AMPHORA OVALIS	(KUETZ.) KUETZ.		X	
BAC	AMPHORA PERPUSILLA	(GRUN.) GRUN.	X		X
BAC	ASTERIONELLA FORMOSA	HASS.	X	X	X
BAC	CALONEIS HYALINA	HUST.	X	X	X
BAC	COCCONEIS PEDICULUS	EHR.	X		
BAC	COCCONEIS PLACENTULA	EHR.	X		X
BAC	COSCINODISCUS LACUSTRIS	GRUN.	X		
BAC	CYCLOTELLA KUTZINGIANA	THWAITES			
BAC	CYCLOTELLA MENEGHINIANA	KUETZ.	X	X	
BAC	CYCLOTELLA STELLIGERA	CL. AND GRUN.			X
BAC	CYMBELLA MICROCEPHALA	GRUN.		X	
BAC	CYMBELLA MINUTA	HILSE EX RABH.			
BAC	CYMBELLA MUELLERI	HUST.			X
BAC	CYMBELLA NAVICULIFORMES	AUERSW. EX. EHI.	X		
BAC	CYMBELLA SINUATA	GREG.		X	
BAC	CYMBELLA TUMIDA	(BREB. EX KUETZ.)V.H	X		
BAC	CYMBELLA TURGIDULA	GRUN.			X
BAC	DIATOMA HIEMALE V. MESODON	(EHR.) GRUN.		X	X
BAC	DIATOMELLA BALFOURIANA	GREV.	X		X
BAC	DIPLONEIS PUELLA	(SCHUM.) CL.	X		
BAC	EPITHEMIA ADNATA	(KUETZ.) BREB.	X	X	
BAC	EPITHEMIA SOREX	KUETZ.	X	X	X
BAC	EPITHEMIA TURGIDA V. WESTER.	(EHR.) GRUN.			X
BAC	FRAGILARIA BREVISTRIATA	GRUN.		X	
BAC	FRAGILARIA CAPUCINA	DESM.		X	X
BAC	FRAGILARIA CONSTRUENS	(EHR.) GRUN.	X	X	X
BAC	FRAGILARIA CROTONENSIS	KITTON	X	X	X
BAC	FRAGILARIA LEPTOSTAURON	(EHR.) HUST.	X		
BAC	FRAGILARIA LEPTOSTAURON V. DUBIA	(GRUN.) HUST.	X		
BAC	GOMPHONEIS HERCULEANA	(EHR.) CL.			X
BAC	GOMPHONEMA ANGUSTATUM V. SARCOPH	(GREG.) GRUN.			X
BAC	GOMPHONEMA CL1		X	X	X
BAC	GOMPHONEMA GRACILE	EHR. EMEND. V.H.	X		
BAC	GOMPHONEMA OLIVACEOIDES	HUST.			X
BAC	GOMPHONEMA OLIVACEUM	(LYNGB.) KUETZ.	X		
BAC	GOMPHONEMA PARVULUM	KUETZ.	X	X	X

DIVISION PREFIX = CHL CHLORO; CRY CRYPTO; CHR CHRYSO; BAC BACILLARIO; PYR DINOFLAG

APPENDIX (continued)
PHYTOPLANKTON SPECIES OBSERVED 1978–1980

DIV*	SPECIES	AUTHORITY	1978	1979	1980
BAC	GOMPHONEMA SUBCLAVATUM	(GRUN.) GRUN.		X	
BAC	GOMPHONEMA VENTRICOSUM	GREG.		X	
BAC	HANNAEA ARCUS	(EHR.) PATR.	X		
BAC	MELOSIRA AMBIGUA	(GRUN) O. MUELL.	X		
BAC	MELOSIRA AMERICANA	KUETZ.			X
BAC	MELOSIRA DISTANS	(EHR.)KUETZ.	X	X	X
BAC	MELOSIRA ITALICA	(EHR.)KUETZ.		X	X
BAC	MERIDION CIRCULARE	(GREV.) AG.			
BAC	NAVICULA CANALIS	PATR.	X		
BAC	NAVICULA CRYPTOCEPHALA	KUETZ.	X		X
BAC	NAVICULA CRYPTOCEPHALA V. VENETA	(KUETZ.)RABH.	X	X	
BAC	NAVICULA DECUSSIS	OSTR.	X		
BAC	NAVICULA HEUFLERI	GRUN.			X
BAC	NAVICULA LATENS	KRASSKE		X	
BAC	NAVICULA MENISCULUS V. UP.	(GRUN.) GRUN.			X
BAC	NAVICULA MINIMA	GRUN.	X	X	X
BAC	NAVICULA NOTHA	WALLACE			X
BAC	NAVICULA PSUEDOREINHARDTII	PATR.	X		
BAC	NAVICULA RADIOSA	KUETZ.	X		
BAC	NEIDIUM AFFINE	(EHR.) PFITZ.			
BAC	NITZSCHIA ACICULARIS	W. SMITH	X	X	X
BAC	NITZSCHIA ACUTA	HANTZSCH.	X		
BAC	NITZSCHIA ALLANSONI	ARCH.		X	
BAC	NITZSCHIA AMPHIBIA	GRUN.		X	
BAC	NITZSCHIA ANGUSTATA	(W. SMITH) GRUN.			X
BAC	NITZSCHIA BACATA	HUST.		X	
BAC	NITZSCHIA COMMUNIS	RABH.		X	
BAC	NITZSCHIA DEMOTA	ARCH.	X	X	
BAC	NITZSCHIA DISSIPATA	(KUETZ.) GRUN.	X	X	X
BAC	NITZSCHIA FONTICOLA	GRUN.	X	X	
BAC	NITZSCHIA FRUSTULUM	KUETZ.	X	X	X
BAC	NITZSCHIA GRACILIS	HANTZSCH.	X	X	X
BAC	NITZSCHIA INNOMINATA	SOV.	X		
BAC	NITZSCHIA LATENS	ARCH.		X	
BAC	NITZSCHIA MEDIOCRIS	HUST.	X		
BAC	NITZSCHIA OBSOLETA	HUST.			X
BAC	NITZSCHIA OREGONA	SOV.	X	X	
BAC	NITZSCHIA PALEA	(KUETZ) W. SMITH			X
BAC	NITZSCHIA PERMINUTA	GRUN	X	X	X
BAC	NITZSCHIA RECTA	HANTZSCH	X	X	X
BAC	NITZSCHIA SERPENTICULA	ARCH.	X	X	X
BAC	NITZSCHIA SILICA	ARCH.	X	X	
BAC	NITZSCHIA SUBACICULARIS	HUST.			X

DIVISION PREFIX = CHL CHLORO; CRY CRYPTO; CHR CHRYSO; BAC BACILLARIO; PYR DINOFLAG

APPENDIX (continued)

PHYTOPLANKTON SPECIES OBSERVED 1978–1980

DIV*	SPECIES	AUTHORITY	1978	1979	1980
BAC	NITZSCHIA SUBPUNCTATA	ARCH.		X	
BAC	RHOICOSPHENIA CURVATA	(KUETZ.) GR. EX.RAB.	X	X	
BAC	RHOPALODIA GIBBA	(EHR.) O. MULL.			X
BAC	STAURONEIS KRIEGERI	PATR.	X		X
BAC	STEPHANODISCUS ASTRAEA	(EHR.) GRUN.	X	X	X
BAC	STEPHANODISCUS ASTRAEA V. MIN	(KG.) GRUN.	X		
BAC	STEPHANODISCUS HANTZSCHII	GRUN.	X	X	
BAC	SYNEDRA DELICATISSIMA	W. SMITH	X	X	X
BAC	SYNEDRA MAZAMAENSIS	SOV.	X	X	X
BAC	SYNEDRA PULCHELLA	RALFS EX KUETZ.			X
BAC	SYNEDRA RADIANS	KUETZ.		X	
BAC	SYNEDRA RUMPENS	KUETZ.	X	X	X
BAC	SYNEDRA TENERA	W. SMITH		X	
BAC	SYNEDRA ULNA	(NITZ.) EHR.	X	X	X
BAC	SYNEDRA VAUCHERIAE	KUETZ.	X	X	X
BAC	TABELLARIA FLOCCULOSA	(ROTH) KUETZ.		X	X
CHL	ANKISTRODESMUS FALCATUS	(CORDA) RALFS	X	X	X
CHL	ANKISTRODESMUS SPIRALIS	(TURNER) LEMM.	X	X	X
CHL	CHODATELLA WRATISLAWIENSIS	(SCHROEDER) LEY.			X
CHL	CLOSTERIOPSIS LONGISSIMA	LEMMA			X
CHL	MOUGEOTIA SP.		X	X	X
CHL	OOCYSTIS LACUSTRIS	CHODAT			X
CHL	OOCYSTIS PUSILLA	HANSGIRG			X
CHL	SELENASTRUM MINUTUM	(NAEG.) COLL.		X	X
CHL	SPHERICAL CHLOROPHYTE #1		X	X	X
CHL	STAURASTRUM GRACILE	RALFS.	X		
CHL	TETRAEDRON ARTHRODESMIFORME		X		
CHR	CHROMULINA-LIKE SP1			X	X
CHR	CHRYSOLYKOS PLANCTONICUS	MACK			X
CHR	CHRYSOPHYTE SPORE		X	X	
CHR	CHRYSOPHYTE STATOSPORE CL1			X	
CHR	CHRYSOPHYTE STATOSPORE #11		X	X	X
CHR	DINOBRYON SERTULARIA	EHR.		X	X
CHR	KEPHYRION CINCTUM	(LACK.) BOUR.		X	
CHR	KEPHYRION GRACILIS	(HILLARD) NOV. COMB.	X	X	X
CHR	KEPHYRION OVALE	(LACK.) HUB.-PEST.			X
CHR	KEPHYRION SPIRALE	(LACK.) CONRAD		X	X
CHR	OCHROMONAS SP.1				X
CHR	OCHROMONAS SP.2				X
CHR	PSEUDOKEPHYRION CONICUM	(SCHILL) SCHM.			X
CHR	SPHERICAL CHRYSOPHYTE #1			X	
CHR	TRIBONEMA AFFINE	G.S. WEST	X	X	X
CRY	CRYPTOMONAS EROSA	EHR.			X

DIVISION PREFIX = CHL CHLORO; CRY CRYPTO; CHR CHRYSO; BAC BACILLARIO; PYR DINOFLAG

APPENDIX (continued)
PHYTOPLANKTON SPECIES OBSERVED 1978–1980

DIV*	SPECIES	AUTHORITY	1978	1979	1980
CRY	RHODOMONAS LACUSTRIS	PASCH. T. RUTT.			X
CRY	RHODOMONAS MINUTA	SKUJA			X
PYR	DINOFLAGELLATE SP. (?)		X	X	
PYR	GLENODINIUM CL1			X	X
PYR	GLENODINIUM CL5			X	
PYR	GLENODINIUM PENARDIFORME	(LINDEM.) SCHILLER			X
PYR	GYMNODINIUM FUSCUM	STEIN			
PYR	GYMNODINIUM SP.1				X
PYR	PERIDINIUM ACICULIFERUM	(LEMM.) LEMM.			X
PYR	PERIDINIUM CL1			X	
PYR	PERIDINIUM INCONSPICUUM	LEMM.			X
	TOTAL TAXA 140		71	71	81

DIVISION PREFIX = CHL CHLORO; CRY CRYPTO; CHR CHRYSO; BAC BACILLARIO; PYR DINOFLAG

SPATIAL AND TEMPORAL PATTERNS IN THE PHYTOPLANKTON OF CRATER LAKE (1985 - 1987)

Mary K. Debacon and C. David McIntire
Department of Botany and Plant Pathology,
Oregon State University, Corvallis, Oregon 97331

The temporal and spatial distribution and abundance of phytoplankton populations in Crater Lake were investigated between June 1985 and April 1987. Phytoplankton samples processed during this study contained a total of 132 taxa, which included 49 diatoms, 45 chrysophytes, 2 xanthophytes, 19 chlorophytes, 3 blue-green algae, 10 dinoflagellates, and 4 cryptomonads. Total cell biovolume integrated to a depth of 250 m ranged from a minimum of 3,151 mm^{3}m^{-2} in January 1987 to a maximum of 33,706 mm^{3}m^{-2} in April 1987. From a community perspective, the phytoplankton of Crater Lake can be described as a sparse but diverse assemblage, which is spatially and temporally modified periodically by local variations in the abundance of the more dominant species. Lowest species diversity and highest dominance were found in the upper 10 m of the water column during the summer months when *Nitzschia gracilis* was the dominant organism. In contrast, species diversity was relatively high throughout the water column to a depth of 200 m during the winter and spring months when there was no evidence of thermal stratification. The dominant taxa at this time of year included *Stephanodiscus hantzschii* and *Gymnodinium inversum*.

Prior to the late 1960s, most of the phytoplankton research conducted at Crater Lake consisted of a qualitative evaluation of dominant taxa. While sporadic attempts were made to investigate the lake before 1970, restricted seasonal access and the rugged terrain surrounding the lake made more detailed studies difficult. Early surveys of the flora were conducted between 1913 and 1940 (Kemmerer *et al.* 1924; Brode 1938; Hasler 1938; Utterback *et al.* 1942). In these studies, phytoplankton samples were collected by net tows, either in conjunction with physical and chemical data or unsupported by limnological information. Because of the manner in which the early samples were collected and subsequently handled, no historical data exist that support the hypothesis that the species composition of the phytoplankton has changed significantly in Crater Lake during the last 70 years. Moreover, the phytoplankton data base currently available is still insufficient for management purposes (Larson 1987).

The research presented here describes the temporal and spatial distribution and abundance of phytoplankton populations in Crater Lake for a period between June 26, 1985 and April 14, 1987. More specifically, the paper presents a list of species, total cell biovolumes integrated to various depths in the water column, the distributional patterns of dominant taxa, and numerical expressions of community structure.

METHODS

During the period covered by this study, phytoplankton samples were obtained with 4-l Van Dorn bottles from different water depths at a location approximately 3 km south of Cleetwood Cove (National Park Service Station 13). Samples relevant to the analysis presented in the next section were collected near the lake surface and at depths of 5, 10, 20, 40, 60, 80, 100, 120, 140, 160, 180, 200, 225, and 250 m on 6/26/85, 7/23/85, 8/20/85, 9/18/85, 3/5/86, 5/28/86, 6/25/86, 7/23/86, 8/20/86, 9/17/86, 1/18/87, and 4/14/87. For the quantitative analysis of species composition, 1-liter subsamples of each plankton collection were fixed with Lugol's solution and concentrated by allowing the seston to settle for at least 70 hr. After decanting the supernatant water, a concentrated subsample of 50 ml was transferred to a plexiglass settling chamber. The chamber then was mounted on a Wild inverted microscope, and approximately 300 algal units were identified and counted at a magnification of 1750X (Lund *et al.* 1958). An algal unit was an individual cell or diatom valve, if the taxon was a unicellular form, or an individual filament in the case of multicellular taxa. To aid in the identification of some diatom taxa, it was sometimes necessary to use the modified Utermohl method to help clear the cells of organic material (Taylor *et al.* 1986).

Cell densities were estimated from the formula:

$$\text{algal units/liter} = (\text{count} \times [A/(W \times L)] / [V/1000 \text{ ml}]) \times CF$$

where A is the area of the chamber (cm^2), W is the field width (cm), L is the total length of the transects (cm), V is the volume of the chamber (ml) times one liter, and CF is the volume of the concentrated sample divided by the volume of the original sample. Biovolume conversion factors were determined for each taxon using appropriate geometric formulae. The densities of algal units were multiplied by these conversion factors to obtain estimates of biovolumes expressed as $\mu m^3\ l^{-1}$. The summation of the biovolumes of all taxa in each sample also was calculated, and these values were used to estimate total cell biovolume in the water column integrated to depths of 20, 40, 80, 120, and 250 m.

Raw data obtained from the microscopic examination of plankton samples were recorded on computer coding sheets according to a standard format required by the programs selected for data analysis. Species diversity was expressed by Shannon's information measure:

$$H' = \sum_{i=1}^{s} (n_i/N) \log_e (n_i/N)$$

where n_i is the number of organisms belonging to the i-th species in a sample of N individuals, and S is the number of species in the sample. As an indicator of dominance in the phytoplankton samples, the information-based measure of redundancy was calculated from

$$R = (H'_{max} - H')/(H'_{max} - H'_{min}).$$

In this case, H'_{max} and H'_{min} represent the maximum and minimum possible values for a sample of size N with S species (McIntire and Overton 1971). Shannon's index is a measure of heterogeneity (i.e., its value is determined by both species richness and relative abundance of each taxon), while the redundancy index can vary from 0 when taxa are equally common to 1 when there is one dominant taxon and all others are represented by a single individual.

RESULTS

The phytoplankton samples processed during this study contained a total of 132 taxa which included 49 diatoms (Bacillariophyceae), 45 chrysophytes (Chrysophyceae), 2 xanthophytes (Xanthophyceae), 19 chlorophytes (Chlorophyta), 3 blue-green algae (Cyanophyta), 10 dinoflagellates (Pyrrhophyta), and 4 cryptomonads (Cryptophyta). A list of 88 taxa identified to the genus, species, or variety level is presented in Table 1. When such an identification was not possible, the organism was given a unique number so that it could be recognized and counted in other samples. To facilitate comparisons with other studies, Table 1 also includes size class categories and a multiplier that converts cell densities to cell biovolume for each taxon.

Seasonal Abundance of Phytoplankton

Total biovolume to a depth of 250 m ranged from a minimum of 3,151 mm^3m^{-2} in January 1987 to a maximum of 33,706 mm^3m^{-2} in April 1987 (Fig. 1). Maximum biovolume in the epilimnion (upper 20 m) occurred in August 1985 (5,082 mm^3m^{-2}) and July 1986 (9,853 mm^3m^{-2}) when *Nitzschia gracilis* was the dominant organism. In June 1985, May 1986, September 1986, and April 1987 the integrated maxima to a depth of 250 m were the result of relatively high concentrations of organisms between 80 m and 250 m, and were not associated with maxima in the epilimnion.

Vertical Distribution of Selected Taxa

In this section, the vertical distributions and abundances of four phytoplankton taxa (*Nitzschia gracilis, Stephanodiscus hantzschii, Ankistrodesmus spiralis, Tribonema* sp., and *Gymnodinium inversum*) are described in terms of cell biovolume (Figs. 2-5). These taxa were among the ten most

TABLE 1. LIST OF TAXA FOUND IN PHYTOPLANKTON SAMPLES FROM CRATER LAKE DURING THE PERIOD FROM JUNE 26, 1985 THROUGH APRIL 14, 1987. THE TABLE ALSO INCLUDES SIZE CLASS ACRONYMS AND A FACTOR FOR CONVERTING ALGAL UNITS PER LITER TO BIOVOLUME EXPRESSED AS CUBIC MICROMETERS PER LITER

KEY
Size Class: NA (Nannoplankton); NE (Netplankton)
GLAD (Greatest linear axis diameter) Size Categories: 1 µm - 10 µm (1); 10 µm - 20 µm (2); 20 µm - 50 µm (3); 50 µm - 70 µm (4); 70 µm - 90 µm (5); 90 µm - 150 µm (6); >150 µm (7)

Taxon	*Size Class*	*Biovol. Factor*
DIVISION CHRYSOPHYTA		
Class Bacillariophyceae		
Achnanthes lanceolata (Bréb.) Grun.	NA3	675
Achnanthes lanceolata var. *dubia* Grun.	NA3	294
Achnanthes minutissima Kütz.	NA3	276
Asterionella formosa Hass	NE7	920
Cocconeis rugosa Sov.	NE4	1021
Cyclotella kutzingiana Thwaites	NA2	339
Cymbella turgida Greg.	NE6	2160
Epithemia sorex Kütz.	NE7	7800
Fragilaria construens (Ehr.) Grun.	NA2	540
Fragilaria construens var. *veneta* (Ehr.) Grun.	NA1	260
Fragilaria crotenensis var. *oregona* Sov.	NE6	1229
Fragilaria leptostauron (Ehr.) Hust.	NA3	820
Fragilaria pinnata Ehr.	NA2	300
Fragilaria vaucheriae (Kütz.) Peters.	NA3	640
Fragilaria vaucheriae var. *capitellata* (Grun.) Patr.	NA4	1100
Gomphonema heidinii Hust.	NA3	367
Gomphonema olivaceum var. *calcarea* (Cl.) Cl.	NA3	480
Gomphonema parvulum Kütz.	NA3	600
Navicula cryptocephala var. *veneta* (Kütz.) Rabh.	NA3	780
Navicula seminulum Grun.	NA1	176
Nitzschia acicularis W. Smith	NE4	144
Nitzschia acuta Hantzsch	NE6	1905
Nitzschia bacata Hust.	NE5	380
Nitzschia closterium (Ehr.) W. Smith	NE5	144
Nitzschia dissipata (Kütz.) Grun.	NA1	29
Nitzschia fonticola Grun.	NA3	120
Nitzschia frustulum Kütz.	NA3	276
Nitzschia gracilis Hantzsch	NE5	473
Nitzschia innominata Sov.	NA3	462
Nitzschia linearis W. Smith	NE7	9342
Nitzschia palea (Kütz.) W. Smith	NA5	1300
Nitzschia perminuta Grun.	NA3	315
Nitzschia serpenticula Arch.	NE6	2820
Nitzschia tryblionella Hantzsch	NE7	42000
Nitzschia vermicularis (Kütz.) Grun.	NE7	4000
Rhoicosphenia curvata (Kütz.) Grun. ex Rabh.	NE4	840
Stephanodiscus hantzschii Grun.	NA1	226
Synedra acus Kütz.	NE7	3114
Synedra delicatissima W. Smith	NE7	5000
Synedra mazamaensis Sov.	NA3	495
Synedra radians Kütz.	NE6	1210
Synedra rumpens Kütz.	NE4	998
Synedra rumpens var. *familiaris* (Kütz.) Hust.	NE5	1460
Synedra tenera W. Smith	NE6	1050
Class Chrysophyceae		
Bicoeca petiolatum (Stein) Pringsheim	NA1	137
Calycomonas sp.	NA2	73
Chromulina grandis Dolf.	NA2	268
Chromulina minor Pasch.	NA1	29
Chromulina spectabilis Scherffel	NA1	72
Chrysocapsa planctonica Pascher	NA1	103
Chrysochromulina sp.	NA1	92
Chrysolykos planctonicus Mack.	NA1	50
Dinobryon bavaricum Imhoff.	NA2	77
Dinobryon sertularia Ehr.	NA2	399
Dinobryon sociale Ehr.	NA2	124
Diplomitella socialis (Kent) Silva	NA1	172
Kephyrion asper (Lack) Bourr.	NA1	36
Kephyrion cupriloforme Conr.	NA1	14
Kephyrion spirale (Lack.) Conr.	NA1	35
Ochromonas elegans Dolf.	NA1	226
Ochromonas granulosa H. Meyer	NA2	113
Ochromonas miniscula Conr.	NA1	33
Ochromonas ovalis Dolf.	NA1	100
Ochromonas verrucosa Skuja	NA1	29
Pseudochromulina asymmetrica Dolf.	NA1	9
Pseudokephyrion conicum (Schill) Schm.	NA1	65
Pseudopedinella sp.	NA2	924
Class Xanthophyceae		
Tribonema affine G. S. West	NE5	1853
DIVISION CHLOROPHYTA		
Ankistrodesmus falcatus var. *acicularis* (A. Braun) G. S.	NA2	110
Ankistrodesmus spiralis (Turner) Lemm.	NA2	30
Crucigenia quadrata Morren	NA1	70
Kirchneriella contorta (Schmidle) Bohlin	NA1	9
Mougeotia sp.	NE7	33757
Oocystis pusilla Hansgrig	NA1	327
Planktosphaeria glatinosa G. M. Smith	NA1	144
Scenedesmus bijuga (Turp.) Lagerh.	NA1	88
Selenatrum minuta (Naeg.) Collins	NA1	22
DIVISION CRYPTOPHYTA		
Rhodomonas lacustris Pascher and Ruttner	NA2	1042
Rhodomonas minuta Skuja	NA2	823
Rhodomonas minuta var. *nannoplantica* Skuja	NA1	254

TABLE 1 (continued). LIST OF TAXA FOUND IN PHYTOPLANKTON SAMPLES FROM CRATER LAKE DURING THE PERIOD FROM JUNE 26, 1985 THROUGH APRIL 14, 1987

Taxon	*Size Class*	*Biovol. Factor*
DIVISION CYANOPHYTA		
Anabaena sp.	NA1	72
Spirulina major Kütz.	NA1	9

Taxon	*Size Class*	*Biovol. Factor*
DIVISION PYRRHOPHYTA		
Amphidium lutem Skuja	NA1	346
Cryptochrysis polychrysis Pasher	NA2	963
Gymnodinium fuscum (Ehr.) Stein	NE4	33000
Gymnodinium inversum Nygaard	NA3	5440
Peridinium aciculiferum (Lemm.) Lemm.	NA3	9425
Peridinium inconspicuum Lemm.	NA3	6750

abundant organisms (based on biovolume) found during the study, and were indicative of differences between late summer patterns when the epilimnion was well established and patterns during the spring months when thermal stratification was not present. Figures 2-5 also illustrate comparable seasonal data for two different years.

In August 1985 and August 1986, *Nitzschia gracilis* was dominant in the upper 20 m, while *Gymnodinium inversum* exhibited a more uniform biovolume throughout the water column down to depths between 120 m and 140 m (Figs. 2-3). The August data for both years also indicated that *Stephanodiscus hantzschii* reached its maximum cell biovolume at 140 m and was relatively rare at depths above 100 m. *Tribonema* sp. was present between 40 m and 100 m in the August samples of both years, but in lower biovolumes than either *N. gracilis* or *G. inversum.*

While the total cell biovolume in the water column was greater in April 1987 than in March 1986, both sets of samples indicated that the vertical distributions of the two dominant taxa (*Stephanodiscus hantzschii* and *Gymnodinium inversum*) were relatively uniform in the water column from the near-surface waters to depths between 200 m and 250 m (Figs. 1, 4, 5). The biovolume of *G. inversum* was greater in the mid-spring samples (April 1987) than in the early spring samples (March 1986), while *S. hantzschii* had similar biovolumes in both groups of samples. *Nitzschia gracilis* also was present in the spring samples of both years, but its biovolume was relatively small throughout the upper 200 m of the water column.

Community Properties

Mean species diversity (H') for all samples obtained from the water column on a particular day ranged from 1.38 (6/25/86) to 2.34 (1/18/87); the corresponding range for the redundancy index (R) was 0.32 (6/26/85 and 1/18/87) to 0.58 (5/28/86 and 6/25/86). Lowest species diversity and highest dominance was found in the upper 10 m of the water column during the summer months when *Nitzschia gracilis* was the dominant organism (Figs. 2, 3, 6 and 7). The greatest vertical variations in species diversity also occurred during late summer (8/20/85 and 7/23/86) when *N. gracilis* exhibited a maximum cell density and biovolume in a well-developed epilimnion. In contrast, species diversity and dominance was relatively uniform from the water surface to a depth of 200 m during the winter and spring months when there was no evidence of thermal stratification.

DISCUSSION

The phytoplankton assemblage in Crater Lake is a relatively diverse flora that exhibits seasonal and spatial variations in species composition. Distributional patterns of three dominant taxa, namely *Nitzschia gracilis, Stephanodiscus hantzschii,* and *Tribonema* sp., observed during the summers of 1985 and 1986 were similar to patterns observed by Larson (1984) and Larson *et al.* (1987) during 1978-1980. These similarities indicated consistency in the summer dynamics of these taxa over a 10-year period. However, there is no evidence from earlier studies that the same three taxa exhibited the same patterns or degree of dominance before 1970 (Kemmerer *et al.* 1924; Brode 1938; Utterback *et al.* 1942). In fact, many of the taxa listed in the early papers were more typical of benthic, littoral habitats than of the limnetic zone. Moreover, inconsistencies in the various reports of species composition for the lake were probably related in part to differences in sampling procedures.

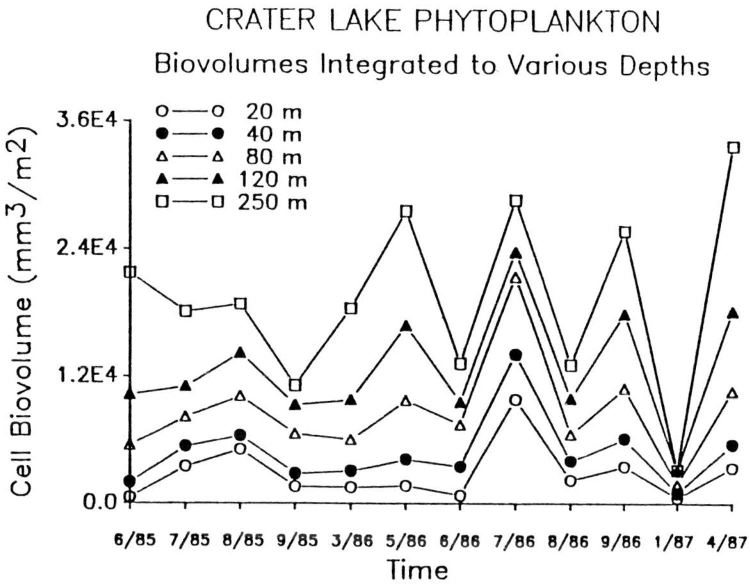

Figure 1. Total cell biovolume integrated from the water surface to various depths in the water column of Crater Lake. Each value is expressed as mm^3/m^2, and represents taxa abundant enough to occur in a sample count of 300 algal units.

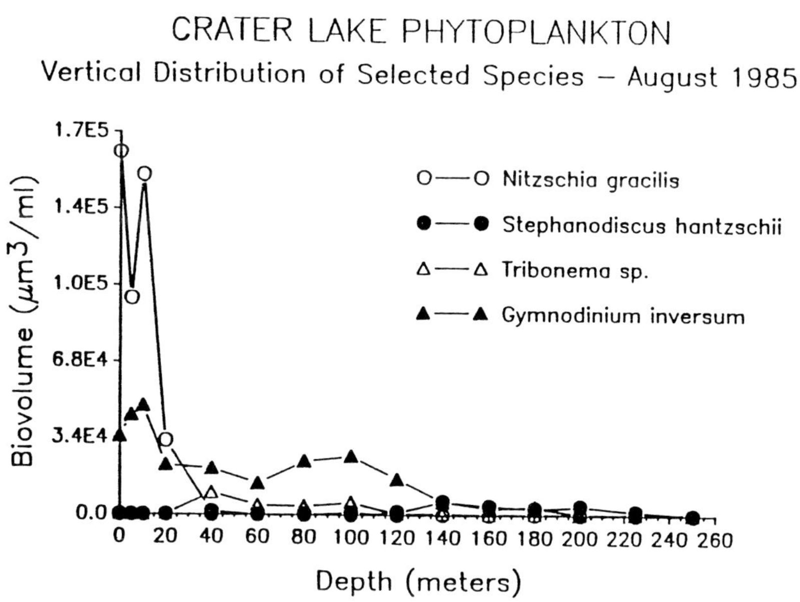

Figure 2. Vertical distribution of some dominant species of phytoplankton found in samples obtained in August 1985. Abundance is expressed as cell biovolume ($\mu m^3/l$).

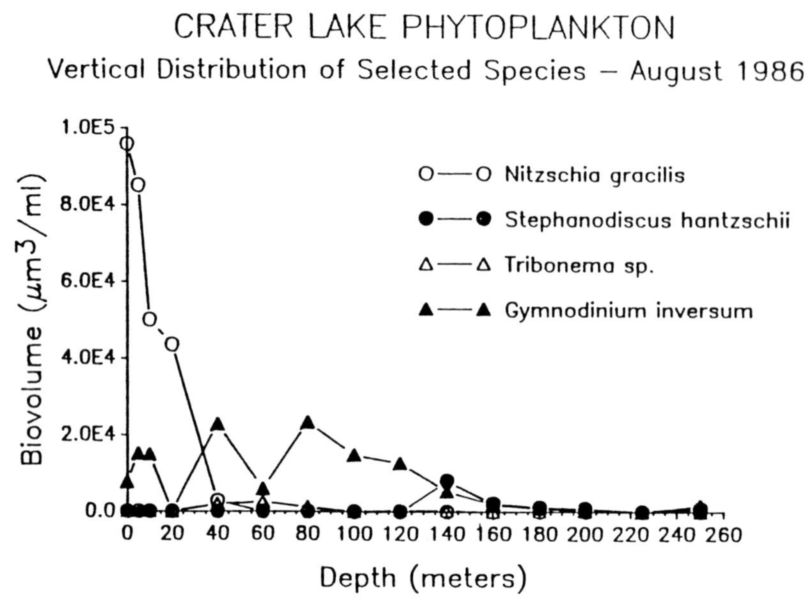

Figure 3. Vertical distribution of some dominant species of phytoplankton found in samples obtained in August 1986. Abundance is expressed as cell biovolume (μm^3/l).

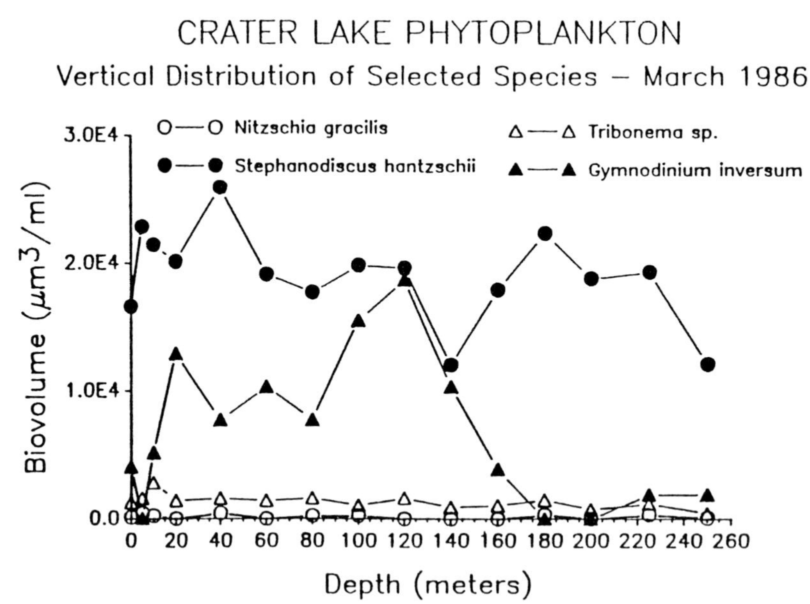

Figure 4. Vertical distribution of some dominant species of phytoplankton found in samples obtained in March 1986. Abundance is expressed as cell biovolume (μm^3/l).

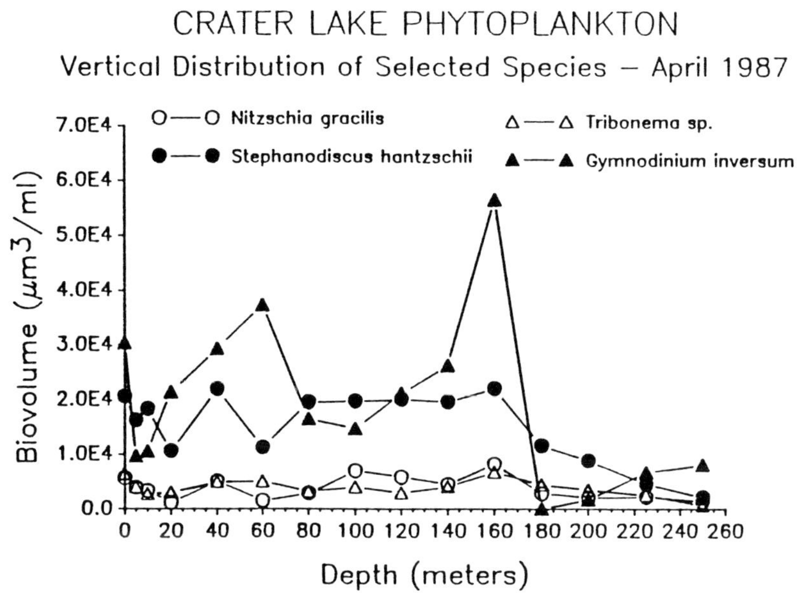

Figure 5. Vertical distribution of some dominant species of phytoplankton found in samples obtained in April 1987. Abundance is expressed as cell biovolume ($\mu m^3/l$).

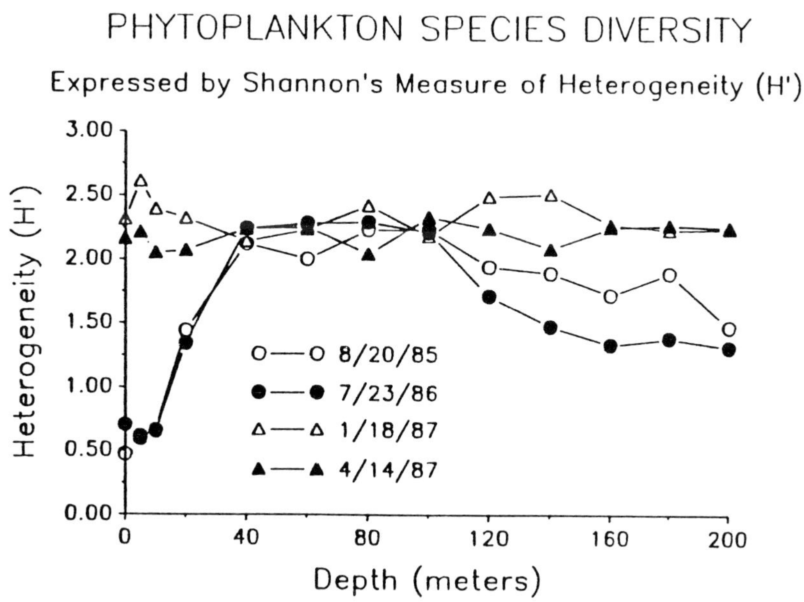

Figure 6. Phytoplankton species diversity at various depths in Crater Lake on August 20, 1985, July 23, 1986, January 18, 1987, and April 14, 1987. Species diversity is expressed by Shannon's information measure (H').

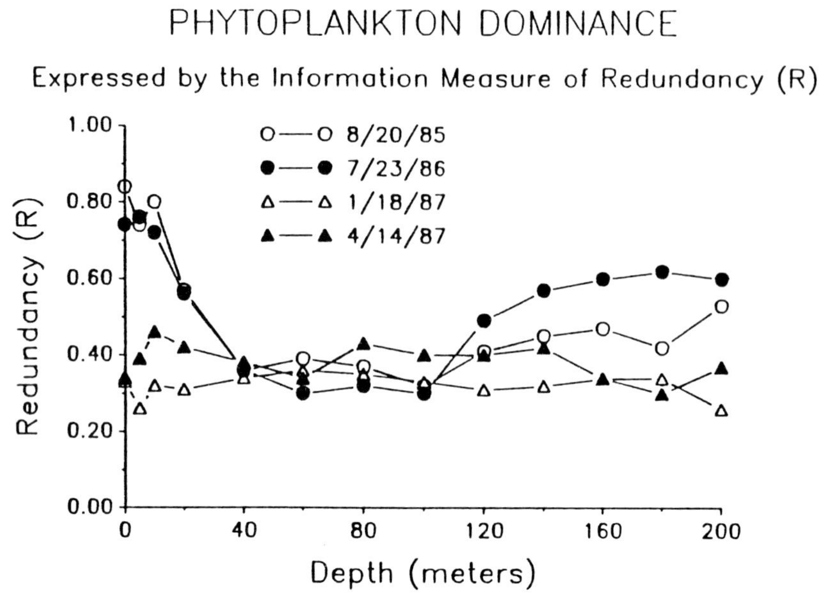

Figure 7. Dominance in phytoplankton assemblages at various depths in Crater Lake on August 20, 1985, July 23, 1986, January 18, 1987, and April 14, 1987. Dominance is expressed by the information measure of redundancy (*R*).

From a community perspective, the phytoplankton of Crater Lake can be described as a sparse but diverse assemblage which is spatially and temporally modified periodically by local variations in the abundance of the more dominant species. The most conspicuous modification takes place during the summer months in the epilimnion when *Nitzschia gracilis* can represent over 50% of the total phytoplankton biovolume. This pronounced increase in *N. gracilis* in the upper 20 m causes changes in such community properties as species diversity (H' <0.8) and dominance (R >0.6). Another interesting modification of the flora occurs in the spring when *Gymnodinium fuscum, Stephanodiscus hantzschii, Tribonema affine*, and *Synedra delicatissima* have relatively high cell densities throughout the water column to a depth of 180 m. Because of the prominence of more than one taxon, species diversity in the spring is higher than that found in the epilimnion during the summer. The co-dominance of *Dinobryon sertularia, Asterionella formosa*, and three dinoflagellates (*Gymnodinium fuscum, G. inversum*, and *Peridinium aciculiferum*) in the zone of maximum primary production (40 m to 100 m) represents another modification in the flora between June and September. However, community structure in the hypolimnetic region of the euphotic zone during the summer and early fall apparently is more variable from year to year than in the epilimnion.

Our data for 1986 and 1987 indicated that the total cell biovolume of phytoplankton in the water column of Crater Lake was at a minimum in January and reached periodic maxima between April and September. During the period from March to September 1986, the fluctuations were operating on a 2-month time resolution, a pattern that was not evident in the summer of 1985. The increase in the abundance of *Nitzschia gracilis* in the epilimnion during the summer months accounted for a more predictable periodicity of total cell biovolume in the upper 20 m of the water column. While the sampling frequency of one month apparently was adequate to reveal the summer maximum each year in the near-surface water, the temporal resolution of the total phytoplankton biomass in deeper water was not clearly defined.

In summary, the phytoplankton of Crater Lake is a complex assemblage of organisms that exhibits spatial and seasonal changes in distribution and abundance. Because of a shortage of plant nutrients (Larson 1987), production maxima are located at depths that represent trade-offs between light and nutrient limitation in relation to the physiological adaptations of individual taxa. The observational data set discussed in this paper provides baseline information for hypothesis generation and the design of experimental studies. In particular, studies of zooplankton-phytoplankton interactions and an investigation of the nutrient requirements of individual taxa would help explain distributional patterns observed in the field.

ACKNOWLEDGMENTS

This work is part of a series of Crater Lake limnological studies supported by the National Park Service under the direction of Gary L. Larson. We are particularly indebted to Dr. Larson for his coordination and supervision of the sampling program in the field. Also, we extend our sincere appreciation to Mark Buktenica, Elena Karnaugh, and Jerry McCrea for invaluable asistance in the field, and to Brad Smith for providing computer programs and help with the data analysis.

LITERATURE CITED

Brode, J. S. 1938. The denizens of Crater Lake. Northwest Sci. 12:50-57.

Hasler, A. D. 1938. Fish biology and limnology of Crater Lake, Oregon. Jour. Wildl. Manage. 2:94-103.

Kemmerer, G., J. F. Bovard, and W. R. Boorman. 1923. Northwestern lakes of the United States: Biological and chemical studies with reference to the possibilities in production of fish. Biol. Bur. Fish. 39:944.

Larson, D. W. 1984. The Crater Lake study: Detection of possible optical deterioration of a rare, unusually deep caldera lake in Oregon, U.S.A. Verh. Internat'l Verein. Limnol. 22:513-517.

Larson, D. W., C. N. Dahm, and N. S. Geiger. 1987. Vertical partitioning of the phytoplankton assemblage in ultraoligotrophic Crater Lake, Oregon, U.S.A. Freshwater Biol. 18:429-442.

Larson, G. L. 1987. A review of the Crater Lake limnological programs. Pages 58-69 *in* T. P. Boyle, ed., New Approaches to Monitoring Aquatic Ecosystems. ASTM STP 940. Amer. Soc. Testing & Materials, Philadelphia, Pa.

Lund, J. W. G., C. Kipling, and E. D. LeCren. 1958. The inverted microscope method of estimating algal numbers and the statistical basis of estimations by counting. Hydrobiologia 11:143-170.

McIntire, C. D, and W. S. Overton. 1971. Distributional patterns in assemblages of attached diatoms from Yaquina Estuary, Oregon. Ecology 52:758-777.

Taylor, W. D., J. L. Wee, and R. G. Wetzel. 1986. A modification of the Utermohl sedimentation technique for improved identification and cell enumeration of diatoms and silica-scaled Chrysophyceae. Trans. Amer. Micros. Soc. 105:68-72.

Utterback, C. L., L. D. Phifer, and R. J. Robinson. 1942. Some chemical, planktonic, and optical characteristics of Crater Lake. Ecology 23:97-103.

CRATER LAKE ECOSYSTEM

SAMPLING STRATEGY AND A PRELIMINARY DESCRIPTION OF THE PELAGIC ZOOPLANKTON COMMUNITY IN CRATER LAKE

Elena Karnaugh[1]
Department of Fisheries and Wildlife
Oregon State University
Corvallis, Oregon 97331

The pelagic zooplankton community of Crater Lake was evaluated for spatial and temporal distributional patterns and taxonomic structure between July 1985 and September 1987. Sampling methods were compared and a 0.5-m net towed vertically was selected as the method for monitoring the community. The community consisted of two cladoceran and nine rotifer species. No predatory species were present. Rotifers were numerically dominant, *Keratella cochlearis* being the most abundant species. *Bosmina longirostris* was numerically dominant to *Daphnia pulicaria*. The abundance of *Daphnia*, however, appeared to increase during the study. The majority of the zooplankton populations occurred in hypolimnion.

Very little is known about the pelagic zooplankton community in Crater Lake. Between 1896 and 1969 four studies of the community were undertaken (Table 1). Although methods and techniques varied, the community was dominated by crustaceans; few rotifers were collected. As part of the ten-year limnological study of the lake (Larson 1990), a study of the zooplankton was initiated in 1985. The objectives of this paper are to: (1) describe the spatial and temporal distribution patterns of the zooplankton community at the main monitoring station [13] from July 1985 through September 1987; (2) determine the representativeness of station 13 as a zooplankton monitoring site; and (3) evaluate the available sampling methods of pelagic zooplankton for incorporation into the lake monitoring program.

[1] Present address: Crater Lake National Park, Crater Lake, OR 97604.

METHODS

Zooplankton samples at Station 13 (deepest basin of the lake at 589 m) were taken using vertically towed nets to 200 m from 23 July 1985 to 5 September 1987. A comparison of sampling methods was undertaken in the summer of 1986 (Karnaugh 1988). The results from this test were used to establish sampling procedures for the zooplankton monitoring program. During the summer of 1986, Station 23 (second deepest basin at 479 m) was sampled to evaluate the representativeness of Station 13. Night tows were taken in 1986.

The test comparing sampling methods included: (1) a 0.75-m net with a 64-micron mesh and closing apparatus, (2) a 0.50-m net with 64-micron mesh; this net was not equipped with a closing apparatus until summer 1986, and (3) Van Dorn bottles with a 4-l capacity. Water samples obtained from the Van Dorn bottles were strained through either 35- or 64-micron mesh. To estimate water volume sampled, Tsurumi Seiki Kosakusho (TSK) flowmeters were attached on the inside and outside of the net mouths. Samples were preserved in a 4% sucrose-formaldehyde solution and enumerated by subsampling.

Community structure was analyzed using programs AID1 and AIDN (Overton *et al.* MS 1987). Simpson's Diversity Index (SDI, AID1, AIDN) was used to express community structure in terms of species richness and relative densities. A resemblance measure, SIMI (AIDN), was used to deter-

TABLE 1. SUMMARY OF PREVIOUS ZOOPLANKTON STUDIES AT CRATER LAKE, OREGON.

Investigator(s)	Year(s) of Study	Depth Sampled (m)	Species noted
Evermann 1897	1896	Surface, littoral	*Daphnia pulex pulicaria* Forbes , *Cyclops (Macrocyclops) albidus* Jurine, *Cyclops serrulatus (Eucyclops agilis)* Fisher, *Allorchestes dentata* Smith[1]
Kemmerer et al. 1924	1913	0-590	*Daphnia pulex*[2], *Bosmina longispina*[2], *Asplanchna, Notholca (Kellicottia) longispina, Anuraea oculeata (Keratella quadrata)*
Hoffman 1969	1967, 1968	0-125	*Daphnia pulicaria*[3], *Bosmina longirostris*[3]
Malick 1971	1969	0-100	*Daphnia pulicaria*[3], *Bosmina longirostris*[3]

[1] This species is misidentified; *Allorchestes* is a marine amphipod.
[2] Although reported as *Daphnia pulex* and *Bosmina longispina*, these species probably were *Daphnia pulicaria* and *Bosmina longirostris*.
[3] These species originally were reported as *Daphnia pulex* and *Bosmina longispina*.

mine similarities in taxonomic structure between paired samples. To compare absolute densities between Stations 13 and 23, the profile analysis was used (Johnson &Wischern 1982). To compare absolute densities of zooplankton obtained from different sampling methods, I used the unpaired *t*-test (Sokal & Rohlf 1981). Equality of population variances between samples were tested using the *F*-test (Sokal and Rohlf 1981).

Rotifer species collected during 1985 and 1986 were identified by Dr. Walter Koste of West Germany. Cladocera specimens from 1968, 1969, 1985, and 1986 samples were identified by Dr. Vladimír Korínek of the Katedra Parasitologie a Hydrobiologie, Czechoslovakia. Cladocera from the 1960s were verified as *Daphnia pulicaria* and *Bosmina longirostris*, rather than *D. pulex* and *B. longispina* as previously reported by Hoffman (1969) and Malick (1971). Dr. Edward S. Deevey, Jr., University of Florida, confirmed the bosminid identification.

RESULTS

The two criteria for determining whether differences existed between the zooplankton structures at Stations 13 and 23 were based on comparing absolute densities and community compositions (i.e., species richness and relative densities). Profile analysis revealed that for July and September there were no significant differences in absolute densities between Stations 13 and 23 (Table 2). The analysis also indicated that zooplankton numbers were not equally distributed among the three depth intervals. Because the profile analysis for August indicated a significant difference between the two stations, an unpaired *t*-test was applied to each of the three pairs of intervals. At the 0.05 level of significance, no difference occurred between the mean absolute densities at Station 13 and 23 for the 80 to 120 m and 120 to 200 m intervals. However, for the 20 to 80 m interval the mean densities between Stations 13 and 23 were significantly different. Taxonomic structure was nearly identical (SIMI >0.973) when samples from corresponding depth intervals for the two stations were paired for a specific date with one exception, the 20 to 80 m interval in August (SIMI = 0.733) (Table 3). Nonetheless, when all samples within a given station for a specific date were pooled to represent the water column as a whole, structure was nearly identical between the two stations for all three sampling dates (SIMI >0.991)(Table 3). Thus, in this comparison of zooplankton densities and taxonomic structure on three separate occasions in 1986, only one interval (20-80 m) on one date showed a significant difference in absolute densities and taxonomic structure; this difference was due to

TABLE 2. PROFILE ANALYSIS FOR STATIONS 13 AND 23 TO DETERMINE IF DIFFERENCES BETWEEN MEAN DENSITIES EXIST.

This statistical approach incorporates three tests. If for any test the hypothesis of no difference is rejected, results of the remaining test(s) cannot be used or interpreted. the 5% level of significance was used.

Test Question	Sampling Data		
	2 Jul	4 Aug	2 Sep
1. Are profiles parallel?	Yes	No	Yes
2. Are profiles the same?	Yes	—	Yes
3. Are profiles level?	No	—	No

an unusual decrease in numbers of *K. cochlearis* at Station 23.

As with the comparisons between Stations 13 and 23, the criteria for determining whether differences occurred among sampling methods were based on absolute densities and community composition. Mean values for each sampling method were 40,485/m^3 for the 0.75-m net, 38,843/m^3 for the 0.50-m net, 13,940/m^3 for the Van Dorn bottles with water samples strained through 64-micron mesh, and 22,268/m^3 for the Van Dorn bottles using 35-micron mesh. Population variances tested for equality except for the variances occurring between samples obtained for the 0.75-m net and the Van Dorn bottles with 35-micron mesh (p =.05). Therefore, tests for differences between means for that particular pair of sampling devices could not be done. At the 5% level of significance no differences occurred between the mean densities of the 0.75-m and 0.50-m nets, the 0.50-m net and the Van Dorn bottles with 35-micron mesh, and the Van Dorn bottles with 35- and 64-micron meshes; mean densities were significantly different between the 0.75-m net and the Van Dorn bottles with 64-micron mesh and the 0.50-m net and

TABLE 3. SIMILARITY INDICES (SIMI) BETWEEN STATIONS 13 AND 23.

A value of 1.0 indicates species richness and relative abundances are identical; a value of 0 denotes no similarity.

Depth Interval (m)	Sampling Date		
	2 Jul	4 Aug	2 Sep
20 to 80	0.994	0.733	0.994
80 to 120	0.999	0.999	0.999
120 to 200	0.973	0.990	0.931
20 to 200[1]	0.993	0.991	0.997

[1] Intervals pooled to represent water column as a whole

Van Dorn bottles with 64-micron mesh (Table 4). Taxonomic structures were similar between samples taken using the 0.75-m net and 0.50-m nets (SIMI >0.884), the 0.50-net and the Van Dorn bottles with both mesh sizes (SIMI>0.949), and between the two mesh sizes using the Van Dorn bottles (SIMI>0.985) (Table 4).

Because the 0.75-m net sampled the greatest volumes of water, rare species were well represented;

TABLE 4. DIFFERENCES IN MEAN DENSITIES AND SIMILARITY INDICES (SIMI) BETWEEN SAMPLING METHODS FOR SAMPLES TAKEN FROM 40 M TO 80 M AT STATION 13, CRATER LAKE, ON 29 JULY 1986

Methods Being Compared	Difference in Mean?	SIMI
0.75-m net and 0.50-m net	No	0.884
0.75-m net & Van Dorn, 64-micron net	Yes	0.754
0.75-m net & Van Dorn, 35-micron net	—	0.769
0.50-m net & Van Dorn, 64-micron net	Yes	0.949
0.50-m net & Van Dorn, 35-micron net	No	0.975
Van Dorn, 64-micron mesh & Van Dorn, 35-micron mesh	No	0.985

however, the size of the net made it awkward to operate and, as a result, this method was time consuming. Also, filtration efficiency was lowest using this method because of the amount of water sampled (data not shown). Although the Van Dorn bottles have the advantage of sampling discrete parcels of water and thereby avoid the clogging problems of towed nets, they were the most laborious and time consuming of the three types of gear. Also, because they sampled the smallest volumes of water, rare species were not captured. The 0.50-m net showed no density differences when compared to either the 0.75-m net or the Van Dorn bottles with 35-micron mesh. Community structure of samples obtained with the 0.50-m net was highly similar to those of the other three methods. In particular, high similarity values were obtained between the 0.50-m net samples and those obtained using the Van Dorn bottles, which represent 100% filtration efficiency. Filtration efficiency of the 0.50-m net was improved over that of the 0.75-m net because a lesser volume of water was filtered.

Two species of cladocerans and nine species of rotifers were collected from 1985 to 1987 (Table 5).

Nine of the zooplankton species were collected during all sampling periods. *Collotheca* was collected only in the summer of 1985; *Conochilus* was collected during the latter part of the summer of 1986. No other zooplankton species were sampled in the pelagia during night tows, and no invertebrate predators were sampled in day or night tows.

The highest absolute densities of zooplankton occurred during March 1986 and the summers of 1985 and 1986; lowest densities occurred during May 1986, January and April 1987, and the summer of 1987 (Karnaugh 1988). Zooplankton densities varied from a low of 19,219/m^3 in September 1987 to a high of 116,781/m^3 in March 1986. The mean of zooplankton densities for summer sampling periods varied from 28,798/m^3 in 1987 to 72,908/m^3 in 1986; the coefficient of variation, which is the ratio of standard deviation to mean and provides a relative measure of dispersion, was high in 1987 (CV = 0.34), but low in 1985 and 1986 (CV = 0.14 and 0.11, respectively).

TABLE 5. SPECIES LIST OF PELAGIC ZOOPLANKTON, CRATER LAKE, OREGON, 1985-1987

Phylum or Order	Species
Cladocera	*Daphnia pulicaria* (Forbes 1983), (emend. Hrbacek, 1959)
	Bosmina longirostris (*sensu lato*)
Rotifera	*Keratella cochlearis* (Gosse 1851) *morphe macracantha* (Lauterborn 1900)
	Keratella quadrata var. *dispersa* (Carlin 1943)
	Kellicottia longispina longispina (Kellicott 1897)
	Polyarthra dolichoptera dolichoptera (Idelson 1925)
	Philodina cf. *acuticornis* (Murray 1902)
	Filinia terminalis (Plate 1886)
	Synchaeta oblongata (Ehrenberg 1831)
	Conochilus unicornis (Rousselet 1892)
	Collotheca pelagica pelagica (Rousselet 1893)

Rotifers numerically were dominant, ranging from 91.4 to 99.1% of the total sample (Table 6). Of the two cladoceran species, *Bosmina* was numerically dominant, ranging from 93.2 to 100% of the cladoceran numbers sampled (Table 6). *K. cochlearis* was numerically dominant from July 1985 through April 1987, ranging from 51.1 to 98.3% of the total sample (Table 7). During the summer of 1987, however, *Philodina* numerically was dominant and represented 38.2% of the total season densities, while *K. cochlearis* represented 24.8%.

During the summers of 1985 and 1986, 99+% of the zooplankton numbers occurred in the 20 to 200 m portion of the water column (Karnaugh 1988). Vertical zonation of the species was apparent during the summer seasons (Fig. 1). In general, the majority of individual species abundances occurred in the 80 to 120 m interval; species showing highest numbers in this interval were: *K. cochlearis*, *K. quadrata*, *Filinia*, *Kellicottia*, *Synchaeta*, and *Collotheca*. Of these species, *K. quadrata* and *Collotheca* also were abundant in the 120 to 160 m interval and *Synchaeta* in the 40 to 80 m interval. *Bosmina* and *Daphnia* had highest densities in the 40 to 80 m interval, and *Conochilus* (a solitary form) was collected only in this interval. *Polyarthra* was the only species with high densities in the 0 to 20 m interval in addition to high numbers in the 40 to 80 m interval, and *Philodina* was the only species with high densities in the 160 to 200 m interval.

TABLE 6. PERCENT OF ZOOPLANKTON TYPES ESTIMATED IN SAMPLES COLLECTED FROM SURFACE TO 200 M AT STATION 13, CRATER LAKE, FROM 1985 TO 1987

Sampling Period	Zooplankton		Cladocerans	
	Rotifers	Cladocerans	*Bosmina*	*Daphnia*
Summer 1985	91.8	8.2	100.0	0
Winter 1986	99.1	0.9	100.0	0
Summer 1986	97.9	2.1	99.3	0.7
Winter 1987	97.1	2.9	96.6	3.4
Summer 1987	91.4	8.6	93.2	6.8

DISCUSSION

Tests of different sampling methods indicated that the 0.5-m net was the least time consuming, easiest to apply and operate, and compatible results were obtained between the 0.75-m and 0.5-m nets in terms of absolute zooplankton densities and community structure. For these reasons, the 0.5-m net

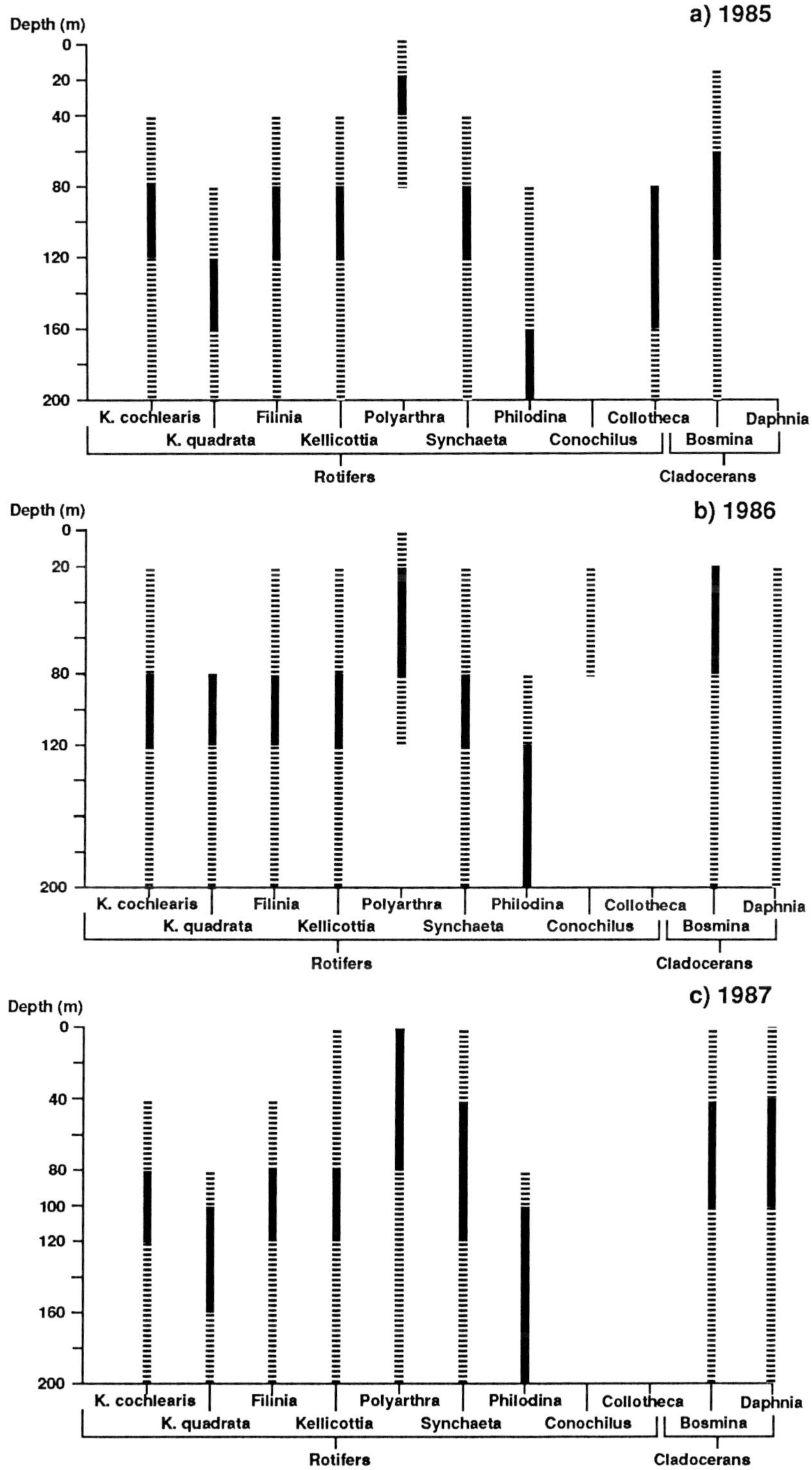

Figure 1. A general representation of species-specific depth occurrences during summer (1985-87) at Station 13, Crater Lake. Solid lines indicate the depth at which the majority of individual abundances occurred. Broken lines indicate depths where lesser individual numbers occurred throughout the summer or depths where individual numbers increased during a portion of the summer.

TABLE 7. RELATIVE DENSITIES OF SPECIES COLLECTED FROM SURFACE TO 200 M AT STATION 13, CRATER LAKE FROM SUMMER 1985 TO SUMMER 1987.
Values presented are percent of sample for a specific sampling period

Species	Sampling Period Summer 1985	Winter/Spring January 1986	Winter/Spring May 1986	Summer 1986	Winter/Spring January 1987	Winter/Spring April 1987	Summer 1987
Bosmina	8.2	0.1	1.8	2.1	3.5	2.2	8.0
Daphnia	<0.05	<0.05	<0.05	<0.05	0.1	0.1	0.6
K. cochlearis	59.3	98.3	90.8	70.2	67.6	51.1	24.8
K. quadrata	3.8	0.4	1.4	1.0	0.1	0.3	0.5
Filinia	5.5	<0.05	1.6	3.8	4.3	5.3	2.4
Kellicottia	9.3	0.4	0.9	1.7	1.3	2.5	1.9
Polyarthra	10.5	0.3	1.9	13.6	1.6	1.8	8.8
Synchaeta	2.4	0.5	0.9	1.7	6.5	10.8	14.8
Philodina	0.7	<0.05	0.7	5.9	10.4	25.8	38.2
Conochilus	—	—	—	<0.05	4.6	0.1	—
Colletheca	0.3	—	—	—	—	—	—

was adopted in the ongoing monitoring program in 1987.

From July 1985 to September 1987 the pelagic zooplankton taxonomic structure of Crater Lake was relatively simple (consisting of two cladocerans and nine rotifers) and uncomplicated because of no invertebrate predation. None of the rotifers was predaceous, and no other invertebrate predators occurred in the pelagia (e.g., cyclopoids, mysids, Chaoborus larvae). Only one vertebrate predator, the zooplanktivorous kokanee (*Oncorhynchus nerka kennerly*), occurred in the open waters (Mark Buktenica, personal communication).

During summer months, the majority of zooplankton occurred in the hypolimnion (i.e., below 20 m). Only *Polyarthra* appeared to effectively inhabit the upper 20 m of the water column. Within the hypolimnion, zooplankton species displayed depth-specific preferences.

From July 1985 to April 1987, *K. cochlearis* was numerically dominant, and this dominance was closely related to high total zooplankton densities and low species diversity. However, in the summer of 1987 *K. cochlearis* densities decreased; in conjunction with this decline, total zooplankton densities decreased and diversity increased.

A gradual increase of *Daphnia* densities occurred from 1985 to 1987. Some evidence exists that this species may be cyclic in Crater Lake, being abundant in some years and rare in others. Evermann (1897), Kemmerer *et al.* (1924), Brode (1938), and Hasler (1938) reported high daphnid densities; however, Hasler and Farner (1942) reported that in 1940 *Daphnia* was absent. Hoffman (1969) reported that in 1967 *Bosmina* and *Daphnia* accounted for 98% and 2%, respectively, of total season densities, and in 1968 *Bosmina* and *Daphnia* accounted for 59% and 41%, respectively, of total season densities. Malick (1971) reported that *Bosmina* and *Daphnia* accounted for 2% and 98%, respectively, of total season densities.

For Crater Lake, changes in pelagic zooplankton community structure (i.e., rotifer- to daphnid-dominated structures) most likely relate to trophic factors such as fish predation and competition. A continual increase, through time, in zooplankton absolute densities could suggest possible eutrophication (Karnaugh 1988). However, these factors—structure and abundances—should always be interpreted within a whole-system context and on a long-term basis. In this light, the degree to which pelagic zooplankton—and biotic interactions in general—may influence water clarity can be evaluated along with other factors that also may contribute to changes or fluctuations in lake transparency.

LITERATURE CITED

Brode, J. S. 1938. The denizens of Crater Lake. Northwest Sci. 12:50-57.

Evermann, B. W. 1897. United States Fish Commission Investigations at Crater Lake. Mazama 1:230-238.

Hasler, A. D. 1938. Fish biology and limnology of Crater Lake, Oregon. Jour. Wildl. Manage. 2:94-103.

Hasler, A. D., and D. S. Farner. 1942. Fisheries investigations in Crater Lake, Oregon, 1937-1940. Jour. Wildl. Manage. 6:319-327.

Hoffman, F. O. 1969. The horizontal distribution and vertical migrations of the limnetic zooplankton in Crater Lake. M.S. Thesis. Oregon State Univ., Corvallis, Ore. 60 pp.

Johnson, R. A. and D. W. Wischern. 1982. Applied multivariate statistical analysis. Prentice-Hall, Inc., New Jersey. 594 pp.

Karnaugh, E. N. 1988. Structure, abundance, and distribution of pelagic zooplankton in a deep, oligotrophic caldera lake. M.S. Thesis. Oregon State Univ., Corvallis, Ore. 167 pp.

Kemmerer, G., J. F. Bovard, and W. R. Boorman. 1924. Northwestern lakes of the United States: Biological and chemical studies with reference to possibilities in production of fish. Bull. U. S. Bur. Fish. 39:51-140.

Larson, G. L. 1990. Status of the ten-year limnological study of Crater Lake, Crater Lake National Park. Pages 7-18 *in* E. T. Drake *et al.*, Crater Lake: An Ecosystem Study. Pacific Div., Amer. Assn. Advance. Sci., San Francisco, Calif.

Malick, J. G. 1971. Population dynamics of selected zooplankton in three oligotrophic Oregon lakes. M.S. Thesis. Oregon State Univ., Corvallis, Ore.. 112 pp.

Overton, W. S., B. G. Smith and C. D. McIntire. 1987. Aid programs (analysis of information and diversity). Oregon State Univ., Corvallis, Ore. 37 numb. pages (manuscript).

Sokal, R. R., and F. J. Rohlf. 1981. Biometry, 2 ed. W. H. Freeman & Co., San Francisco, Calif. 859 pp.

ECOLOGY OF KOKANEE SALMON AND RAINBOW TROUT IN CRATER LAKE

Mark Buktenica[1]
Department of Fisheries and Wildlife
Oregon State University, Corvallis, Oregon 97331
and
Gary L. Larson
Cooperative Park Studies Unit, College of Forestry
Oregon State University, Corvallis, Oregon 97331

Crater Lake, originally barren of fish, was stocked on an irregular basis from 1888 through 1941 with several species of salmonids. Two species occur in the lake today—kokanee salmon (*Oncorhynchus nerka*) and rainbow trout (*Salmo gairdneri*). This study was initiated in the summer of 1986 to evaluate the ecology of adult fish in terms of length, weight, age, growth, morphology, food habits, and distribution in Crater Lake relative to the lake's limnological characteristics. Age determinations from scale analysis, supported by modal progressions in length-frequency histograms, indicated that the kokanee salmon population was dominated, in number, by the 1984 year class. The rainbow trout population was composed of multiple age groups.

Food sources were partitioned in that kokanee salmon generally fed on small-bodied taxa (mean dry weight 1.3 mg) from the midwater column and from the lake bottom; rainbow trout fed on large-bodied taxa (mean dry weight 9.8 mg) from the lake surface and the lake bottom. Hydroacoustic studies during the first week of September 1987 showed that the fish underwent diel migrations within and between the nearshore (0 m to 100 m contour) and offshore (100 m to 589 m contour) zones of the lake. Based on capture records, it appeared that kokanee were primarily offshore and in deep water during the day, and then they moved shoreward into shallower water at night. Rainbow trout appeared to remain nearshore in shallower water during the day than at night. The maximum depth for an acoustic target was 98.5 m. The maximum depth of capture for kokanee in Crater Lake was 86.25 m.

The ecology of these two species is a dynamic response to Crater Lake's limnologic variables. Trophic interactions suggest that this is not a simple benthic-rainbow trout and limnetic-kokanee salmon system. A complex series of interactions are catalyzed by these species operating within and between limnetic and benthic communities.

Crater Lake covers the floor of the Mt. Mazama caldera that formed about 6,800 years ago (Bacon 1983). It is a closed lake system in that the surface inlets originate inside of the caldera, and there is no surface outlet. Though originally barren of fish, several species of salmonids were stocked on an irregular basis from 1888 through 1941. The species included rainbow trout (*Salmo gairdneri*), brown trout (*S. trutta*), cutthroat trout (*S. clarki*), steelhead trout (*S. gairdneri*), and coho or "silverside" salmon (*Oncorhynchus kisutch*). However, Wallis and Bond (1950) identified six kokanee salmon (*O. nerka*) that had been collected in 1939 and 1947 and presumed to be coho salmon. Since there is no stocking record of kokanee in Crater Lake, it is unclear if early fisheries investigations collected one or both of the Pacific salmon species. No coho salmon have been identified from Crater

[1]Present address: Crater Lake National Park, Crater Lake, Oregon 97604

Lake since 1950. Today only two fish species are known to inhabit the lake, kokanee salmon and rainbow trout.

Fisheries investigations of Crater Lake date back to 1896 (Buktenica 1989). Most were conducted during the stocking era. Many of these studies were restricted by small sample sizes and were of short duration (e.g., one day) because the investigations were dependent on samples from fishermen's creels. Although these studies provide a historic data base for future studies, it is clear that very little is known about fish in the Crater Lake ecosystem.

Optical and phytoplankton studies conducted from 1978 to 1981 raised concern that the process of eutrophication had been accelerated in Crater Lake (Larson 1984). This concern led to a comprehensive ten-year limnological investigation of Crater Lake of which this fish study was a part. The principal goal of the study was to develop a better understanding of the ecological roles of adult fish in Crater Lake relative to length, weight, age, growth, morphology, food habits and distribution (Buktenica 1989). This paper summarizes these findings.

STUDY AREA

Crater Lake is an ultraoligotrophic caldera lake located at 1,882 m in elevation on the crest of the Cascade Mountains in southwestern Oregon. The lake is widely known for its exceptional water clarity. Crater Lake is the eighth deepest lake in the world and is one of the clearest freshwater lakes (Hutchinson 1957).

Crater Lake, with a distinctive regime of environmental conditions, provides an unusual habitat for its biota. The caldera walls are composed of rock cliffs and precipitously steep talus slopes rising 250 to 600 m above the lake surface and continuing at slightly reduced slopes below the water line where they eventually flatten out into three main basins at 450 m, 550 m, and 589 m depth. The littoral zone is comprised of a very narrow band around the 48 km^2 lake that widens slightly around Wizard Island where nearly one-third (by surface area) of the littoral region exists.

A shallow epilimnion typically forms to a depth of 5 -20 m from late July to September; a well-defined thermocline may not develop until September as occurred in 1986 (Larson 1987). The water column is well oxygenated, displaying a slight decrease in dissolved oxygen (D.O.) in the surface strata with increasing temperature, and a slight decrease in D.O. at 550 m. Total alkalinity is generally uniform with depth ranging from 25 to 27 mg/1 $CaCO_3$. Conductivity ranges from 112 to 120 micromhos/cm and increases slightly with depth.

The photic zone extends to great depths as evidenced, in part, by Secchi disc transparency readings approaching 40 meters (37.2 m, July 1985) and 1% incident surface light intensities extending 80-100 m in depth (Larson 1986). It is further evidenced by the unusually deep depth distributions of the lake's flora and fauna. During summer months the chlorophyll maxima typically occur between 100 m and 140 m (Larson 1986). Zooplankton abundance was greatest between 40 m and 120 m in depth and was dominated in abundance by rotifers in 1986 and 1987 (Karnaugh 1988). The freshwater moss, *Drepanocladus*, has been found growing in thick mats to 120 m (Hasler 1938).

MATERIALS AND METHODS

Capture Methods

Fish were collected from early June through September at weekly intervals in 1986 and 1987. In addition, fish were collected once in April 1987, October 1987, and January 1988. Horizontal gill nets were set overnight on the limited shelf-like areas around Crater Lake down to 20 m to capture fish near the shore (Fig. 1). Both floating and sinking multifilament nets measuring 38 m x 3 m were used. Mesh sizes ranged from 19 mm (3/4 in.) to 51 mm (2 in.) square mesh, in five 7.6 m panels.

A sportsmen's downrigger (fine cable on a handwinch) was modified to troll for fish in the offshore areas (from the 100 m to the 589 m contour). An artificial lure or a lure and flasher (for attraction) was attached to the cable at five-meter intervals to 100 m. A similar lure was trolled behind the boat to fish the near-surface depths. Angling with rod and reel was employed along the shoreline with artificial lures.

Field Measurements

Fork length and total length were measured to the nearest millimeter on an 0.6 m measuring board. Whole fish weights were determined on a Homs top-loading temperature compensated spring dial scale (1000 gm in 2 gm increments) to the nearest gram. Scales for aging were taken, by scraping with

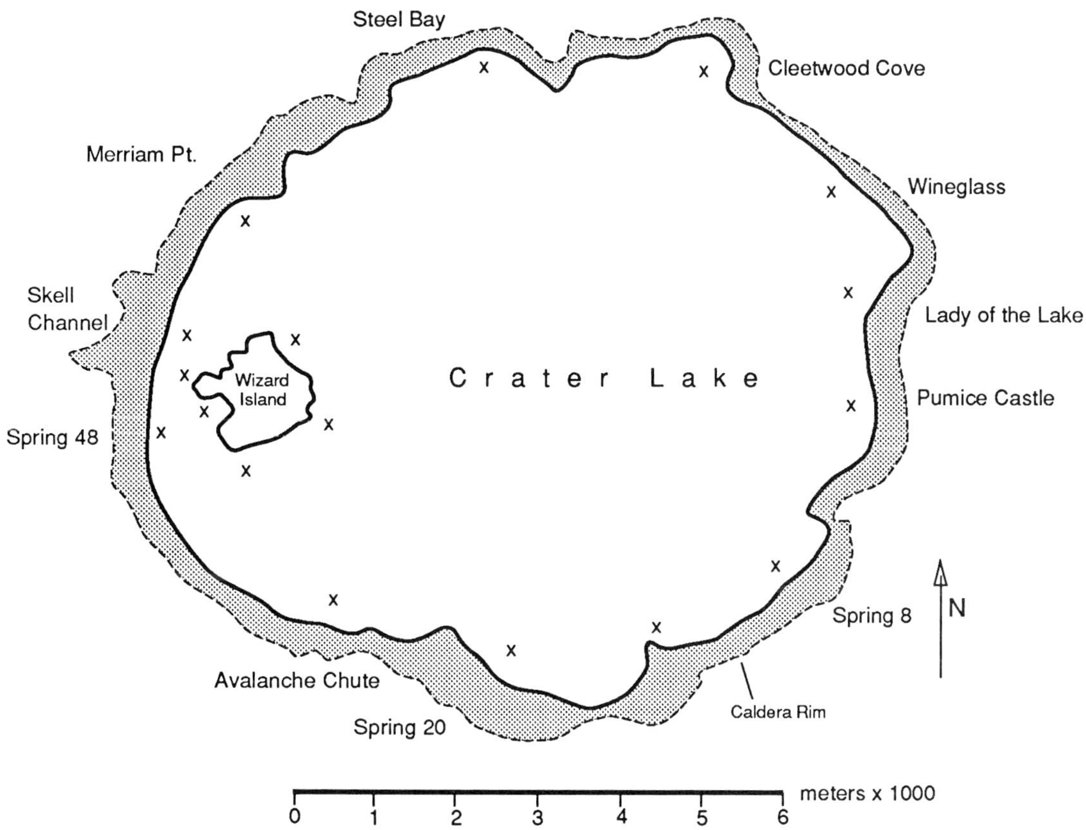

Figure 1. Location of horizontal gill net sets (X) on Crater Lake, Oregon.

the blunt edge of a scalpel blade, just below the posterior margin of the dorsal fin and above the lateral line (Jearld 1983). After the body cavity was slit ventrally, the fish were sexed (ripening ova were removed, preserved and archived) and the specimens were preserved in 10% formalin.

Meristics and Morphometrics

All fish collected were stratified by length class, and a subsample of ten fish was chosen randomly out of each length class for rainbow trout, and kokanee salmon. Mouth width was measured as the greatest ventral distance across the mouth opening. Gill raker measurements and counts were taken on the first left gill arch, under magnification. Gill raker counts included all projections lying in a near-linear series along the arch (Nelson 1968). Gill raker length was measured, using pointed dial calipers (.05 mm dial graduations), as the distance from the tip to the ventral margin of the base of the longest anterior gill raker (Kliewer 1970). Gill raker spacing was measured as the distance from center to center, from the origin of the base of the longest anterior gill raker to the same location on the gill raker ventral to it. Analysis of variance was used to analyze the meristic data statistically.

Age

Six scales from each fish were cleaned and mounted between two microscope slides. The mounted scales were viewed through a microfiche projector for age determination. Distances between the scale focus, annuli, and the scale margin were taken on a subsample of scales utilizing a BioSonics Optical Pattern Recognition System (OPRS). The OPRS consists of a microscope, a video camera and monitor, a real-time video frame grabber, a digitizing tablet, and a microcomputer and software. Individual circuli distances and optical images were stored on floppy disk and archived. Length-frequency analysis and modal-progression analysis (e.g., following the relative abundance of a dominant year class from year to year) were used as age validation techniques (Jearld 1983).

The internationally accepted convention for aging fish is to designate January 1 as the birth date of fish in the Northern Hemisphere, whether or not the annular ring or slow growth zone is complete by this date (Jearld 1983). No scales were analyzed from fish captured between October and April at Crater Lake due to the inaccessibility of the lake in the winter; therefore, the terminal edge of the annular ring was used to designate age class instead of scale development on January 1. Fish in their first year of life, before completion of the first annular ring, were designated as members of the age-0 group. Fish captured after the completion of the first annular ring but prior to completion of the second annular ring were designated as members of the age-I+ group. The same convention was used to designate members of subsequent age groups (e.g., II+, III+, IV+, . . . VII+).

Food Habits

Whole fish were originally fixed in 10% formalin and later transferred to ethanol. After the stomachs were removed, the contents of each stomach were flushed from the esophagus to the pyloric sphincter and stored in 70% ethanol in individual containers for later enumeration. Food analysis included frequency of occurrence, percent composition by number, weight, vertical distribution in the water column (by weight), aquatic or terrestrial origin (by weight) and mean weight of the food items.

Distribution

Fish distributions were evaluated by hydroacoustic and fish capture methods. A Biosonics Model 101 420 KHz Scientific sounder with a 15°9transducer was used in the acoustic survey. This survey was limited to a one-week sampling period in early September 1987. A stratified random sampling design was used. The strata were nearshore (0-100 m contours) and offshore (100-589 m contours) zones of the lake and six four-hour periods of time, day and night (Buktenica 1989).

RESULTS AND DISCUSSION

Capture

Horizontal gill nets were the most productive method for capturing kokanee salmon and rainbow trout; however, if equal effort were afforded the two methods, angling may prove more effective for capturing rainbow trout in Crater Lake (Table 1). The downrigger and angling methods were valuable in that they provided samples from the offshore zone and at times of the day when the gill nets were not effective. Sampling methods targeted adult fish. No age-0+ fish of either species were captured. Age I+ rainbow trout were caught on hook and line and by gill nets. No age-I+ kokanee salmon were captured.

TABLE 1. FISH CAPTURES FOR CRATER LAKE KOKANEE SALMON AND RAINBOW TROUT BY SAMPLING METHOD AND BY YEAR

		Sampling Method			
Species	Year	Horiz. gillnet	Angling	Down rigger	Total
Kokanee	1986	55	22	27	104
salmon	1987	171	4	4	179
Rainbow	1986	23	37	0	60
trout	1987	50	21	0	71
Totals		299	84	31	414

Weight-Length

The lengths of kokanee salmon and rainbow trout ranged from 181-302 mm and 174-453 mm, respectively. Regression lines were fit to log-transformed length and weight data to test the hypothesis that kokanee salmon and rainbow trout have the same length-weight relationship. The hypothesis was rejected with a p value <.001 (analysis of covariance—significant at the .001 level), and this result indicated that rainbow trout were heavier than kokanee for a given body length.

For many years, the widely held belief that the fish in Crater Lake are starving because Crater Lake is a nearly sterile body of water has persisted. Previous and current fish investigations contradict this belief. Hubbard (1933, unpublished rep.) reported:

> There is no truth to the often heard statement that Crater Lake trout are starved to death. The fish have plenty to eat in the lake and eat it . . . The rainbows are a nice, plump oval. The fish are well-filled out, and the flavor is unexcelled.

Hasler (1938) noted, "Growth of trout and salmon in the lake is exceptionally rapid." Hasler and Farner

(1942) noted that growth in Crater Lake fishes was nearly at an optimum. Fish captured in 1986 and 1987, even those fish captured in April, consistently possessed fat reserves. These fish experienced comparable growth rates as determined by length at age and back-calculated length at age (Buktenica 1989), relative to other populations in oligotrophic lakes (Ricker 1938; Carlander 1969; Lindsay and Lewis 1978), but slightly slower growth rates compared to those found by previous investigations on Crater Lake (Hasler 1938; Hasler and Farner 1942). It also appears that there were fewer large rainbow trout in 1986 and 1987 than in previous years (Hasler 1938).

Meristics and Morphometrics

There was a clear difference in gill raker number, gill raker length, and gill raker space between kokanee salmon and rainbow trout; the ranges of these characteristics for the two species did not overlap, and the standard deviations were small (Table 2). Kokanee had significantly narrower mouths than did rainbow trout, though the difference was not as great as for the other characteristics. Gill raker length, gill raker space, and mouth width measurements are presented as a percentage of fork length to account for the linear increase in each characteristic with an increase in fork length (Buktenica 1989). Gill raker counts were within the expected ranges for each species (Mottley 1936; Vernon 1957; Lindsey 1958; Bidgood and Berst 1967; Nelson 1968; Kurenkov 1977).

Age

Age frequency histograms suggest one dominant year class for kokanee (Fig. 2). For rainbow trout, however, no clear patterns are readily apparent for age frequency (Fig. 2), although the II+ and III+ fish did appear to be the most abundant age class in 1986 and 1987, respectively. Length frequency distributions support these trends for both species (Buktenica 1989). These results indicate a far more complex population structure for rainbow trout than that for kokanee salmon.

Although it is not known why year class dominance occurred amongst kokanee salmon, this is not the first time this has been observed in Crater Lake (Kibby 1966, unpublished rep.). Dominant year classes in kokanee populations also have been reported in other lakes, e.g., Odell Lake (Lewis 1971)

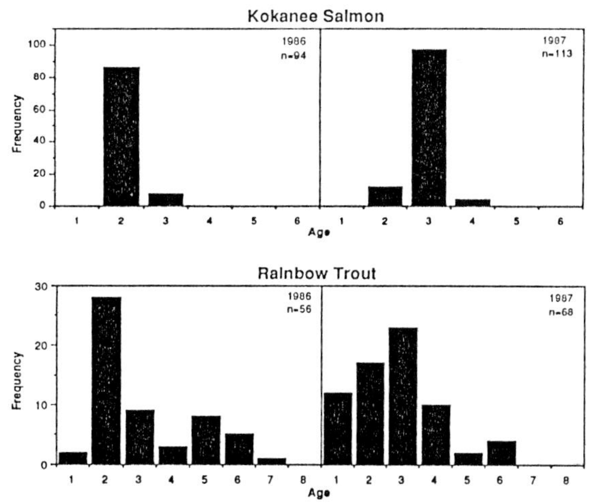

Figure 2. Age frequency histograms for kokanee salmon and rainbow trout captured in 1986 and 1987, from Crater Lake, Oregon. Note the different scales on the Y axes.

and Flathead Lake (Hanzel 1984), though the extent of the dominance was not as great as in Crater Lake.

Food Habits

Kokanee salmon primarily fed on four food groups: Chironomidae, Trichoptera, Amphipoda, and Cladocera. Chironomid larvae and pupae together accounted for 51% of their diet by weight. Trichoptera and Amphipoda accounted for 13% and 9% of their diet by weight, respectively. Cladocerans were almost exclusively represented by *Daphnia pulicaria*. *Bosmina longirostris* occurred in one stomach sample along with the *Daphnia*.

Due to partial digestion and fragmentation, *Daphnia* were only countable in 63 out of 99 stomachs. Therefore, *Daphnia* were under-represented in percent composition by weight (15%). Kokanee stomach samples were strongly characterized by having a few food types per stomach. One or two food types typically dominated a sample; many stomachs were stratified with layers of chironomid larvae and chironomid pupae. These well-defined strata may suggest alternating feeding periods, perhaps in different locations.

Rainbow trout fed heavily on Trichoptera, Hymenoptera, Chironomidae pupae, terrestrial Coleoptera, Diptera, aquatic Coleoptera, Ephemeroptera (≥30% occurence), Gastropoda and

TABLE 2. MERISTIC AND MORPHOMETRIC SUMMARY STATISTICS FOR KOKANEE SALMON AND RAINBOW TROUT

	GILL RAKER NUMBER		GILL RAKER LENTH*		GILL RAKER SPACE*		MOUTH WIDTH*		Sample
	Range	Mean+/-STD	Range	Mean+/-STD	Range	Mean+/-STD	Range	Mean+/-STD	Size
Kokanee Salmon	31-36	32.97+/-1.25	2.27-3.54	2.89+/-0.30	0.38-0.59	0.46+/-0.05	4.71-7.24	6.05+/-0.61	59
Rainbow Trout	18-21	19.56+/-1.01	1.51-2.26	1.82+/-0.19	0.60-1.08	0.81+/-0.12	5.34-8.21	6.85+/-0.74	32

* Values for Gill Raker Length, Gill Raker Space, and Mouth Width are Presented as Percent Fork Length.

terrestrial Hemiptera (≥25% occurence). Trichoptera (25%) and Hymenoptera (14%) were the only orders that represented more than 10% composition by weight; the remaining 61% composition by weight was accounted for by 17 additional food types. Rainbow trout were more likely to have a large variety of prey species in a single stomach sample than were kokanee, though many stomachs contained primarily one food type (e.g., Amphipoda or Gastropoda).

Kokanee as a group fed almost solely on aquatic food types. Fifty-four percent of their diet was assumed to have been taken from the midwater column, while 41% was taken from the benthos. The mean dry weight of the individual prey in the kokanee stomach samples was 1.3 milligrams (Buktenica 1989).

Rainbow trout also fed heavily on aquatic food items, but more food items from terrestrial origin were eaten. Thirty-seven percent of their diet was assumed to have been taken from the lake surface, 11% from the midwater column and 52% from the benthos. The mean dry weight of individual prey in the rainbow trout stomach samples was 9.8 milligrams (Buktenica 1989).

Kokanee are considered to be zooplanktivores and are found to rely heavily on zooplankton in their diets (Cordone *et al.* 1971; Rieman and Bowler 1980). When zooplankton abundance is low, however, they have been found to feed on chironomid larvae and pupae and, to a lesser degree, other benthic organisms (Northcote and Lorz 1966). Rainbow trout typically feed on benthic and terrestrial invertebrates, although piscivory is common, and planktivory may occur (Scott and Crossman 1973).

Results from previous food habits studies at Crater Lake varied considerably; however, some patterns may be noted. *Daphni*a have fluctuated greatly in importance in the diet of "silversides" and kokanee. Hubbard (1933, unpublished rep.) stated that *Daphnia* comprised 14% of the fish food. Brode (1935) found *Daphnia* in 74% of the stomachs (mostly "silversides") and 62% of the volume in 1934; for the years 1934-1936, 51% of the stomachs (mostly "silversides") contained *Daphnia*, and they comprised about 34% by volume (Brode 1938). Brode (1937, unpublished rep.) discussed the food habits of 224 fish (214 were silversides) in Crater Lake and hypothesized that there were four food habitat groups: Group 1, a plankton feeding group that fed primarily on *Daphnia*; Group 2, a shore feeding group that fed primarily on benthic macroinvertebrates; Group 3, that fed on plankton (*Daphnia*) and shore organisms, in this case almost solely amphipods; and Group 4 that fed on terrestrial insect adults taken from "wind streaks" on the lake surface. In 1940, Hasler and Farner (1942) found only one silverside stomach to contain *Daphnia*. They noted that the absence of *Daphnia* in 100 m to surface plankton tows was in contrast to their abundance in 1937 (Hasler 1938). Patten and Thompson (1957, unpublished rep.) did not record any *Daphnia* in stomach samples of kokanee (although this may have been a result of sample preparation). They concluded that the important food types (by frequency and by volume) for kokanee were Amphipoda, Diptera (Chironomidae), and Trichoptera, and those for rainbow trout were Trichoptera, Hymenoptera, Gastropoda, Coleoptera, and Diptera (Chironomidae). Patten and Thompson also stated that "the kokanee usually preferred the smaller forms of the insect orders" and that "the stomach contents indicate that rainbow feed as actively at the surface as below . . . while most of the kokanee foods were taken below the surface."

Although the information from prevous studies is sometimes sketchy, the general food habits for trout and salmon were similar to those recorded in 1986 and 1987. Kokanee fed heavily on taxa found in the midwater column and off the lake bottom. Rainbow trout fed off the lake bottom and on insects from the lake surface. Kokanee fed on a few food types, while rainbow trout fed on a wide variety of food types. Where their diets overlapped, kokanee tended to take smaller-bodied taxa. For example, over 95% of the chironomid larvae eaten by kokanee were of the genus *Heterptrissocladius*, while only four rainbow stomachs were found to contain prey of this genus. It should be noted that salamanders and small fish were found in the stomachs of trout in earlier studies, but none were observed in the 1986 and 1987 samples.

The confusion over the identification of the silversides in studies prior to 1950 prohibits direct comparison of food habits for kokanee in the present study. That the importance of *Daphnia* in the diet of the salmon apparently varied with its abundance in the lake is consistent with what was found in 1986 and 1987. An increase of *Daphnia* in the kokanee diet in 1987 (Buktenica 1989) corresponded with an increase in *Daphnia* abundance in the lake in 1987 (Karnaugh 1988). Many of the same food groups important to the diet of the salmon in past studies also were important in 1986 and 1987, though the relative importance fluctuated among the studies. This also was the trend for rainbow trout in Crater Lake.

Distribution

Fish exhibited a pattern of diel vertical migration in the offshore zone (Buktenica 1989). Median fish depth determined acoustically ranged from 75 m during the day to 19 m at night, a 56 m daily vertical migration. The maximum depth of detection was 98.5 m. An opposite vertical migration is indicated by the nearshore data. Median nearshore fish depth ranged from 2 m in the day to 17 m at night.

Fish capture techniques were used in conjunction with hydroacoustic data to assess fish distribution. Downrigger captures support the deeper occurrence of offshore fish during the day. Thirty-one fish were captured; all were kokanee. The median depth of capture was 65 m. The maximum depth of capture was 86.25 m. Angling occurred along the shoreline during the day and in the evening. Most of the fish angled were rainbow trout. Twenty-one out of the twenty-five kokanee angled were captured after 1800 hours; twenty were captured after 1900 hours. Overnight gill net sets were successful in capturing kokanee salmon and rainbow trout.

The capture data, when combined with the acoustic data, suggests that the kokanee are mainly deep and offshore during the day, and that they migrate to shallower water and possibly shoreward at night. The rainbow trout are nearshore during the day and night. The acoustic data indicates an upward migration during the day.

Distributions of fish are variable among and within lakes and represent responses to dynamic interactions among physical, chemical, and biological components of a lake ecosystem. Vertical and horizontal migrations of fish are common and are believed to be responses to the interaction among some combination of these components. It is difficult, however, to evaluate the relative importance of these variables to the distribution patterns of fish because the interactions are not static and are very complex. In fact, distributions of kokanee salmon and rainbow trout in lakes are quite variable. Northcote *et al.* (1964) observed vertical diel migration of kokanee to reverse in succeeding years. In Lake Tahoe, relative abundance of kokanee and rainbow trout that were captured in offshore and nearshore gill net catches fluctuated seasonally (Cordone *et al.* 1971).

One of the most interesting aspects of the distribution of Crater Lake fish is the unusually deep occurrence and the great magnitude of diel migrations. The maximum depths of occurrences for kokanee reported in the literature are considerably shallower than those in Crater Lake, e.g. 36.5 m in Lake Tahoe, CA (Cordone *et al.* 1971) and 30.0 m in Odell Lake, OR (Lewis 1972).

The deep-water occurrences of offshore fish may be related to the deep penetration of light in Crater Lake and perhaps the deep-water occurrences of chlorophyll and zooplankton. There does not appear to be a relationship between fish distribution and temperature, dissolved oxygen or pH, because the thermocline was shallow, and dissolved oxygen and pH were fairly uniform throughout the water column. It appears that the distributions of Crater Lake fish were most closely associated with feeding habits. Kokanee salmon are believed to feed primarily on cladocerans and chironomids in the deep-water offshore zone during the day and to migrate to shallower water and shoreward to feed between dusk and dawn. Rainbow trout appeared to feed primarily in the nearshore area of the lake, rising to shallower depths during the day and feeding heavily on insects from the lake surface.

CONCEPTUAL FRAMEWORK

Based on the results, we conclude that the ecology of the two fish species is different in Crater Lake. Kokanee salmon and rainbow trout differed most notably in morphology, age structure, food habits, and distribution in the lake. Differences in feeding habits and distributions are likely a response, in part, to morphological and behavioral differences between kokanee salmon and rainbow trout. Direct and indirect interactions between the two species may also account for some proportion of the observed differences. Kokanee salmon and rainbow trout likely play different ecological roles within the lake community primarily as an expression of their trophic relations.

Recently, it has been recognized that the structure and organization of lake communities is a dynamic response to controls and interactions from both the top down and the bottom up (Carpenter *et al.* 1985; Benndorf *et al.* 1986). McQueen *et al.* (1986) found the bottom up controls to be strongest at the bottom of the food web (nutrients to phytoplankton to zooplankton) and to weaken as one moves up the food web. Conversely, top down effects are strongest between top consumers and weaken as one moves down the food web. The predictability of responses of community subunits weakens as one moves away from the controls.

The limnetic invertebrate community in Crater Lake is relatively simple and devoid of invertebrate predators. Recently, the limnetic zooplankton community has been dominated in number by rotifers, although *Daphnia* appear to be increasing in number. Previous zooplankton investigations between 1896 and 1969 found *Daphnia* abundance to range from rare to very abundant and dominance to shift among rotifers, *Bosmina* and *Daphnia* (see Karnaugh 1988). The effect of kokanee on the zooplankton community structure in Crater Lake is unknown; however, the current community structure is consistent with that of a zooplankton community under predatory control; preferential feeding by zooplanktivorous fish has been shown to cause a shift in zooplankton community structure from large-bodied cladocerans to small-bodied cladocerans and rotifers (Brooks and Dodson 1965).

Little is known about the benthic community in Crater Lake. Periphyton and macrophytes are relatively sparse along the shoreline, and moss exists to depths of 120 m. Most of the available knowledge about benthic macroinvertebrates has come almost solely from food habit analysis of fish. Since rainbow trout and kokanee salmon interact with the benthic community, they may have altered the benthic community structure and organization through a variety of poorly understood pathways. These changes may be expressed as local extinctions (Reimers 1979), changes of within-lake distributions (Macan 1966a, 1966b), reduced mean weights of prey species, or as reduced biomass of benthic prey populations (Post and Cucin 1984).

The food web in Crater Lake is not simply a benthic-rainbow trout and limnetic-kokanee salmon system. It is much more complicated. In fact, kokanee salmon greatly increase the diversity of the interactions among the limnetic, benthic, and terrestrial components of the food web, especially prey of small body size. Rainbow trout also are important to the food web, but the focus is on prey of large body size from the benthic and terrestrial components.

While this conceptual discussion may be heuristically useful, it must be emphasized that lake systems have a high level of complexity. Interactions among life history types and habitat types are not static, but are dynamic and ever-evolving. Exhibited life history patterns are in part an expression of the developmental environment of the population and may change with evolving environments as well as in sympatry with other species or groups of species (William Liss, Oregon State Univ., Dep. Fish. & Wildl., pers. comm.).

The concepts presented above are of interest in fisheries sciences because the stocking of fish and invertebrates is a well-practiced management technique to supplement local faunas in order to improve fishing (Li and Moyle 1981) and more recently in order to improve water quality in environmentally degraded systems (Benndorf *et al.* 1984). The ecological implications of fish introduction are poorly understood (Goetze *et al.* 1988) and are of growing concern as naturally fishless areas continue to be stocked and are diminishing in number.

An interesting opportunity exists at Crater Lake to follow the apparently cyclic abundances of kokanee salmon and members of the limnetic zooplankton community. Investigation of abundance, distribution, and diel migrations of fish and zooplankton should be expanded, in conjunction with continued food habits analysis, throughout the summer sampling season and during other times of the year if possible. Attempts to sample juvenile fish by previous investigators were unfruitful; nonetheless, the collection of juveniles and a description of reproduction would greatly expand the ecologic knowledge on Crater Lake fish.

CONCLUSIONS

Crater Lake fish attain comparable size and growth rates to fish of the same species in other Northwest oligotrophic lakes.

The kokanee salmon population in Crater Lake was dominated in number by the 1984 year class, while the Rainbow trout population structure was more complex.

Food resources were partitioned in that kokanee salmon generally fed on small-bodied taxa (mean dry weight 1.3 mg) from the mid-water column and from the lake bottom, and rainbow trout fed on large-bodied taxa (mean dry weight 9.8 mg) from the lake surface and the lake bottom.

Vertical diel migrations of fish were recorded in the nearshore and offshore zones of the lake. Kokanee salmon appeared to be found primarily offshore in deep water during the day and to migrate shoreward and to shallower water at night. Rainbow trout appeared to remain nearshore, shallower during the day than at night.

The ecology of these two species is a dynamic response to Crater Lake's limnologic variables. Trophic interactions suggest that this is not a simple benthic-rainbow trout and limnetic-kokanee salmon system. A complex series of interactions are catalyzed by these species operating within and between limnetic and benthic communties.

ACKNOWLEDGMENTS

We thank the staff of Crater Lake National Park for their support of the project and assistance in the field. We also thank Steve Brady, Mike Hurley, and Will Cameron for their help in the field. Thanks also are given to Richard Thorne, David Marino, Steve Dilly, Courtney Loomis, and Abigail Buktenica for their help with various aspects of the project.

LITERATURE CITED

Bacon, C. R. 1983. Eruptive history of Mount Mazama and Crater Lake Caldera, Cascade Range, U.S.A. Jour. Volcanol. Geotherm. Res. 18:57-115.

Benndorf, J., H. Kneschke, K. Kossatz, and E. Penz. 1984. Manipulation of the pelagic food web by stocking with predacious fishes. Internat'l Rev. Gesamten. Hydrobiol. 69:407-438.

Bidgood, B. F. and A. H. Berst. 1967. Phenotypic characteristics of rainbow trout in the Great Lakes. Jour. Fish. Res. Bd. Can. 24(4):887-892.

Brode, J. S. 1935. Food habits of Crater Lake fish. Crater Lake Nature Notes 8:11-13.

Brode, J. S. 1937. Food habits of Crater Lake fish. Nat'l Park Serv., Crater Lake, Oregon. 7 pp. (Unpublished preliminary rep.)

Brode, J. S. 1938. The denizens of Crater Lake. Northwest Sci. 12(3):50-57.

Brooks, J. L., and S. I. Dodson. 1965. Predation, body size, and composition of plankton. Science 150:28-35.

Buktenica, M. 1989. Ecology of kokanee salmon and rainbow trout in Crater Lake, a deep ultra-oligatrophic caldera lake (Oregon). M.S. Thesis. Dep. Fisheries & Wildl., Oregon State Univ., Corvallis, Ore. 89 pp.

Carlander, K. D. 1969. Handbook of freshwater fishery biology. Volume 1. Life history data on freshwater fishes of the United States and Canada, exclusive of the Perciformes. Iowa State Univ. Press, Ames, Iowa. 752 pp.

Carpenter, S. R., J. F. Kitchell, and J. R. Hodgson. 1985. Cascading trophic interactions and lake productivity. BioScience 35(10):634-639.

Cordone, A. J., S. J. Nicola, P. H. Baker, and T. C. Frantz. 1971. The kokanee salmon in Lake Tahoe. Calif. Fish & Game 57(1):28-43.

Goetze, B., W. J. Liss, and G. L. Larson. 1988. Ecological implications of fish introductions into temperate lakes: A review. Nat'l Park Serv., Cooperative Park Studies Unit, College of Forestry, Oregon State Univ. 61 pp. (unpublished ms.)

Hanzel, D. A. 1984. Measure annual trends in recruitment and migration of kokanee populations and identify major factors affecting trends. Montana Dep. Fish, Wildl. & Parks, Fish. Div., Job Completion Rep., Project No. F-33-R-18, Job No. I-b.

Hasler, A. D. 1938. Fish biology and limnology of Crater Lake, Oregon. Jour. Wildl. Manage. 2(3): 94-103.

Hasler, A. D., and D. S. Farner. 1942. Fisheries investigations in Crater Lake, Oregon, 1937-1940. Jour. Wildl. Manage. 6:319-327.

Hubbard, A. C. 1933. Fact and fancy about Crater Lake fish. Nat'l Park Serv., Crater Lake, Ore. 50 pp. (unpublished rep.)

Hutchinson, G. E. 1957. A treatise on limnology. Vol. 1. John Wiley & Sons, Inc., New York. 1015 pp.

Jearld, A., Jr. 1983. Age determination. Chapter 16 *in* L. Nielson and D. Johnson, eds., Fisheries Techniques. Amer. Fish. Soc., Bethesda, Maryland.

Karnaugh, E. N. 1988. Structure, abundance, and distribution of pelagic zooplankton in a deep, oligotrophic caldera lake. M.S. Thesis. Oregon State Univ., Corvallis, Ore. 171 pp.

Kibby, H. 1966. A study in fish ecology of Crater Lake, Oregon. Nat'l Park Service, Crater Lake, Ore. 1120 pp. (unpublished rep.)

Kliewer, E. V. 1970. Gill raker variation and diet in lake whitefish *Coregonus clupeaformis* in northern Manitoba. Pages 147-165 *in* C. C. Lindsey and C. S. Woods (eds.), Biology of Coregonid Fishes. Univ. Manitoba Press, Winnipeg, Manitoba, Can. 560 pp.

Kurkenkov, S. I. 1977. Two reproductively isolated groups of kokanee salmon *Oncorhynchus nerka kennerly*, from Lake Kronotskiy. Jour. Ichthyol (Engl. transl. Vopr. Ikhtiol.) 17(4):526-534.

Larson, D. W. 1984. The Crater Lake Study: Detection of possible optical deterioration of a rare unusually deep caldera lake in Oregon, U.S.A. Verh. Internat'l Verein. Limnol. 22:513-517.

Larson, G. L. 1986. Crater Lake Limnological Program. 1985 Ann. Rep.. Nat'l Park Serv., Cooperative Park Studies Unit, Oregon State Univ., Corvallis, Ore.

Larson, G. L. 1987. Crater Lake Limnological Program. 1986 Ann. Rep., Nat'l Park Serv., Cooperative Park Studies Unit, Oregon State Univ., Corvallis, Ore.

Lewis, S. L. 1971. Life history and ecology of kokanee in Odell Lake. Oregon State Game Com., Res. Div., Job Progress Rep. F-71-R-6. 48 pp.

Lewis, S. L. 1972. Life history and ecology of kokanee salmon in Odell Lake. Oregon Wildlife Com., Res. Div., Job Progress Rep. JF-71-R-8. 71 pp.

Li, H. W. and P. B. Moyle. 1981. Ecological analysis of species introductions into aquatic systems. Trans. Amer. Fish. Soc. 110:772-782.

Lindsey, C. C. 1958. Modification of meristic characters by light duration in kokanee *Oncorhynchus nerka*. Copeia 1958(2):134-136.

Lindsay, R. B. and S. L. Lewis. 1978. Lake and reservoir investigations (kokanee ecology). Oregon Dep. Fish & Wildl., Fish Div., Job Final Rep. F-71-R. 39 pp.

Macan, T. T. 1966a. The influence of predation on the fauna of a moorland fish pond. Arch. Hydrobiol. 61:432-452.

Macan, T. T. 1966b. Predaton by *Salmo trutta* in a moorland fish pond. Internat'l Verein. Theor. Angew. Limnol. Verh. 16:1081-1087.

McQueen, D. J., J. R. Post, and E. L. Mills. 1986. Trophic relationships in freshwater pelagic ecosystems. Can. Jour. Fish. & Aquatic Sci. 43:1571-1581.

Mottley, C. McC. 1934. A biometrical study of the kamloops trout of Kootenay Lake, *Salmo kamloops* Jordan. Jour. Biol. Bd. Can. 2(4):359-377.

Nelson, J. S. 1968b. Variation in gill raker number in North American kokanee, *Oncorhynchus nerka*. Jour. Fish. Res. Bd. Can. 25:(2):415-420.

Northcote, T. G., H. W. Lorz, and J. C. MacLeod. 1964. Studies on diel verticle movement of fishes in a British Columbia lake. Proc. Internat'l Assn. Theor. Appl. Limnol. 15:940-946.

Northcote, T. G., H. W. Lorz. 1966. Seasonal diel changes in food of adult kokanee (*Oncorhynchus nerka*) in Nicola Lake, British Columbia. Jour. Fish. Res. Bd. Can. 23:1259-1263.

Patten, B. G., and R. B. Thompson. 1957. Food studies of a small sample of rainbow trout (*Salmo gairdneri*) and kokanee salmon (*Oncorhynchus nerka*) from Crater Lake, Oregon. National Park Service, Crater Lake, Ore. 14 pp. (unpublished manuscript)

Post, J. R., and D. Cucin. 1984. Changes in the benthic community of a small Precambrian lake following the introduction of yellow perch, *Perca flavescens*. Can. Jour. Fish. & Aquatic Sci. 41: 1496-1501.

Reimers, N. 1979. A history of a stunted brook trout population in an alpine lake: A life span of 24 years. Calif. Fish & Game 65(4):106-215.

Ricker, W. E. 1938. "Residual" and kokanee salmon in Cultus Lake. Jour. Fish. Res. Bd. Can. 4(3): 192-218.

Rieman, B. E., and B. Bowler. 1980. Kokanee trophic ecology and limnology in Pend Oreille Lake. Idaho Dep. Fish & Game, Fish. Bull. 1. 27 pp.

Scott, W. B., and E. J. Crossman. 1973. Freshwater fishes of Canada. Fish. Res. Bd. Can., Bull. 184.

Vernon, E. H. 1957. Morphometric comparison of three races of kokanee within a large British Columbia lake. Jour. Fish. Res. Bd. Can. 14(4):573-598.

Wallis, O. L., and C. E. Bond. 1950. Establishment of kokanee in Crater Lake, Oregon. Jour. Wildl. Manage. 14:190-193.

LIMNOLOGICAL RESPONSE OF CRATER LAKE TO POSSIBLE LONG-TERM SEWAGE INFLUX

Douglas W. Larson[1], Clifford N. Dahm[2], N. Stan Geiger[3]
[1]Hydraulics and Hydrology Branch, U. S. Army Corps of Engineers, 319 SW Pine Street, Portland, Oregon 97204
[2]Department of Biology, University of New Mexico, Albuquerque, New Mexico 87131
[3]Scientific Resources Incorporated, 11830 SW Kerr Parkway, Lake Oswego, Oregon 97035

Average Secchi disk transparency readings in Crater Lake, Oregon, diminished from 38 m and 37 m in 1937 and 1969, respectively, to 29 m in 1978. This trend has persisted: a Secchi reading of 21.8 m in August 1984 is the lowest on record. This reduction in lakewater visibility is attributed to biostimulation of phytoplankton populations in the lake's 0-40 m stratum. Investigators reported finding no phytoplankton at these depths in 1940, whereas recent surveys (1978-1984) have routinely found densities greater than 700,000 cells liter^{-1}. The biostimulation possibly stems from caldera springs contaminated by inadequate sewage and wastewater treatment facilities in Crater Lake National Park. One spring in particular, which emerges along the lake's caldera wall directly below park tourist facilities, discharges at the rate of 2 liters sec^{-1} and has nitrate concentrations of between 150 and 320 μg liter^{-1}. These levels are 5-10 times higher than those measured for ground waters emerging below undeveloped localities in the park.

Crater Lake appears to be nitrogen-deficient; summertime concentrations of inorganic nitrogen are usually less than 1 μg liter^{-1} throughout the 0-200 m stratum. Concentrations of dissolved inorganic phosphorus are generally 15-20 times greater. Consequently, the influx of nitrogen from sewage-contaminated ground waters may be substantial over the long run and indeed significant insofar as the availability of nitrogen to phytoplankton populations is concerned.

Crater Lake, Oregon, deepest lake in the United States and seventh deepest in the world at 589 m, has long been celebrated for its intensely blue color and extraordinary water transparency (Kemmerer *et al.* 1924; Atwood 1937; Smith *et al.* 1973). Utterback *et al.* (1942) measured downwelling light to a depth of 120 m, and Hasler (1938) recorded Secchi transparency depths of 36, 39, and 40 m in August 1937. The record Secchi depth for lakes, 44 m, was obtained with a 100 cm-diameter disk at Crater Lake in July 1969 (Larson 1972).

These rare optical properties were expected to last for "ages" because of the lake's "relative inaccessibility, unique morphometry, and protection by the National Park Service" (Smith *et al.* 1973). But limnological data obtained at Crater Lake in the late 1970s indicated that the lake's optical quality was possibly diminishing (Larson 1983, 1984a, 1984b; Larson and Forbes 1980). The suggestion that optical deterioration had occurred was based primarily on Secchi disk readings which, when compared with Secchi data recorded in 1937 (Hasler 1938) and in 1968-69 (Larson 1972), showed that the lake had undergone a 25-30% reduction in Secchi disk transparency. Additionally, repeated deployments of an underwater irradiance meter provided further evidence suggesting optical deterioration. The results of these measurements, when compared with 1940 and 1968-69 irradiance data, demonstrated an increase in the rate of light attenuation with depth as well as a shift in the transmissivity at various wavebands (Dahm *et al.* 1990).

Subsequent limnological surveys of Crater Lake, conducted by the National Park Service, have continued to show diminished Secchi disk transparency

(Larson *et al.* 1985, 1986, 1987, 1988). Secchi readings, however, have remained fairly stable since 1978. A reading of 21.8 m, the lowest on record, was obtained in August 1984.

Reduced Secchi transparency is often caused by increased phytoplankton biomass (Wetzel 1975), which is regarded as a precursor of accelerated eutrophication in oligotrophic lakes (Hasler 1969). Various investigators (Pettit 1936; Utterback *et al.* 1942; Smith *et al.* 1973) attributed the unusual optical features of Crater Lake to the paucity of suspended particulate matter in its waters, which minimized the opportunity for light to be scattered. Indeed, Hutchinson (1957) described the lake as "almost optically pure," in which the maximum transmittance of light (i.e., around wavelength 450 nm) was due almost solely to molecular scattering. The lake's deep-blue color, for instance, results from the molecular back-scattering of downwelling light, predominantly the short wavelengths in the visible spectrum (Pettit 1936; Hutchinson 1957).

It is reasonable to believe, therefore, that reduced Secchi transparency in Crater Lake has resulted from a concomitant increase of suspended particulate matter in the lake's 0-40 m stratum. Furthermore, phytoplankton and phytoplanktonic detritus may constitute a major portion of this suspended material. In fact, an estimated 60% of the typical summertime particulate load in Crater Lake is organic (J. Dymond, Oregon State Univ., pers. comm.).

In a series of reports to the National Park Service (Larson 1978, 1980, 1981, 1983, 1984a), it was hypothesized that: (1) the waters of Crater Lake had become less transparent because of increased phytoplankton abundance; (2) more phytoplankton were now present in the lake's water column, particularly in the 0-40 m stratum, because of an influx of algal nutrients capable of supporting greater phytoplankton biomass; and (3) nutrients derived from sewage are entering Crater Lake via ground waters emanating from septic tank-drainfield facilities located on the south side of the caldera rim. Our research was aimed at determining whether this set of hypotheses provides a correct assessment of limnological and anthropogenic conditions which might be affecting the Secchi transparency of Crater Lake. This research also contributed original data describing important features of the Crater Lake environment, including the lake's phytoplankton community and the nutrient chemistry of ground water inflows and lakewater. Prior to 1978, when this research began, relatively little was known about these and other limnological attributes of Crater Lake.

METHODS

Limnological surveys of Crater Lake were conducted each summer between 1978 and 1984. This work proceeded at three lake stations identified as 13, 16, and 23 (Fig. 1). The lake was inaccessible during winter. The following research tasks were undertaken:

(1) Light attenuation data were obtained with a Kahlsico underwater irradiameter (model 268WA390, selenium barrier-layer photocell, spectral range in sunlight of 430-670 nm). Lakewater transparency was estimated with 20 cm- and 100 cm-diameter Secchi disks. Secchi observations were usually made by two or more persons under the most favorable conditions possible, i.e., smooth lake surface, clear sky, and midday deployment. Photometer data were obtained under similar conditions.

(2) Water samples for biological and chemical determinations were collected from discrete depths in the water column with 4-liter, messenger-activated, PVC Van Dorn bottles. Samples for phytoplankton and chlorophyll-*a* determinations were collected between lake surface and 200 m. Water-chemistry sampling was extended to 300 m and occasionally to near lake-bottom.

(3)Samples for nutrient chemistry were filtered (Whatman GF/F glass fiber filters, precombusted for 4 hr at 450°C) and transferred to acid-washed, 1-liter plastic bottles. Sample bottles were immediately placed in an ice-chest. Analyses were usually completed within 24 hr after the samples were collected.

(4) Chemical analyses were done in accordance with Hager *et al.* (1972), as follows: (a) *nitrate-nitrogen*: colorimetric, cadmium reduction, diazotization, automated; (b) *ammonium-nitrogen*: colorimetric, phenate, automated; (c) *orthophosphate*: colorimetric, ascorbic acid, phosphomolybdate; and (d) *silica*: colorimetric, ascorbic acid reduction to molybdate blue, automated.

(5) Water samples for phytoplankton determinations were dispensed into 1-liter plastic bottles and

fixed with 3% formalin. Subsamples (10 ml) were gently filtered through 0.45 µm, MF-Millipore membrane filters. Filters were made transparent by placing on immersion oil and warming to draw oil into filter by water evaporation (Lind 1974). Mounted filters were observed using phase and brightfield microscopy at 1000X. Discrete algal particles (cells or colonies) having distinct chromatophores were identified to species and counted along filter transects. For supplemental identifications of fragile species, observations were made with an inverted microscope (Wild M40) on samples which had been fixed in the field with Lugol's solution (Lund *et al.* 1958).

(6) Water samples for chlorophyll-*a* determinations were dispensed into 1-liter plastic bottles and treated with saturated magnesium carbonate. Samples were Millipore-filtered (0.45 µm, HA-type) within 4-6 hr after sample collection. Extraction and measurement of chlorophyll-*a* was done in accordance with Strickland and Parsons (1968). Percent absorbance by pigment extracts was measured on a Bausch and Lomb Spectronic 70 spectrophotometer.

Ground water inflows (springs) were sampled throughout the summers of 1983 and 1984. These springs are located mostly in the south-southeast corner of the caldera between Skell Head and the Rim Village-lodge area (Figs. 1 and 2). Roughly 40 springs were found, but not all were sampled. Springwater samples were collected in acid-washed, 1-liter plastic bottles. Samples were analyzed for nitrate-nitrogen, ammonium-nitrogen, orthophosphate, and silica, using the methods described above. Instantaneous flow measurements were obtained for some springs by simply using a calibrated bucket and stopwatch, i.e., timed volume technique (U. S. Geol. Surv. 1977).

RESULTS

Optical measurements

Secchi disk readings, obtained since 1978, are summarized in Table 1. Secchi data collected before 1978 are included for comparison. These readings, in addition to the photometer data reported in the previous paper by Dahm *et al.* (1990), are evidence suggesting that the optical properties of Crater Lake have changed over the past 20 years. While stating this, however, we assume that optical measurements obtained by earlier investigators (e.g., Hasler 1938;

TABLE 1. SECCHI DATA FOR CRATER LAKE, OREGON

Year of Record	20 cm-diameter disk			100 cm-diameter disk		
	n	Average (m)	Range (m)	n	Average (m)	Range (m)
1937 (Hasler 1938)	3	38.3	36-40			
1968 (Larson 1972)	2	33.5[1]	31-36[2]			
1969 (Larson 1972)	3	36.6[2]	32-39	1	44[3]	
1978	7	29.3	28.0-30.0			
1979	4	27.8	22.9-31.0	3	32.6	28.5-35.0
1980	9	33.7	27.9-36.5	6	36.7	33.8-39.6
1981	5	29.6	28.7-31.1	5	32.2	30.4-33.3
1982	38	26.8	21.9-30.7	1	33[3]	
1983	25	28.6	22.0-32.0			
1984	18	29.7	21.8-32.5	18	35.2	32.0-39.2
1985[4]	26	29.3	25.3-37.5	3	30.1	not available
1986[4]	86	29.9	25.9-33.5			
1987[4]	90	32.7	27.1-37.0			

[1] This average does not include a relatively low value (18 m) recorded on 27 August during a heavy rainstorm.
[2] Below average readings of 31 m on 14 June 1968 and 32 m on 31 August 1969 were due to high, hazy overcast, which reduced incident solar radiation.
[3] Single value.
[4] Data collected by the National Park Service (Larson 1986, 1987, 1988).

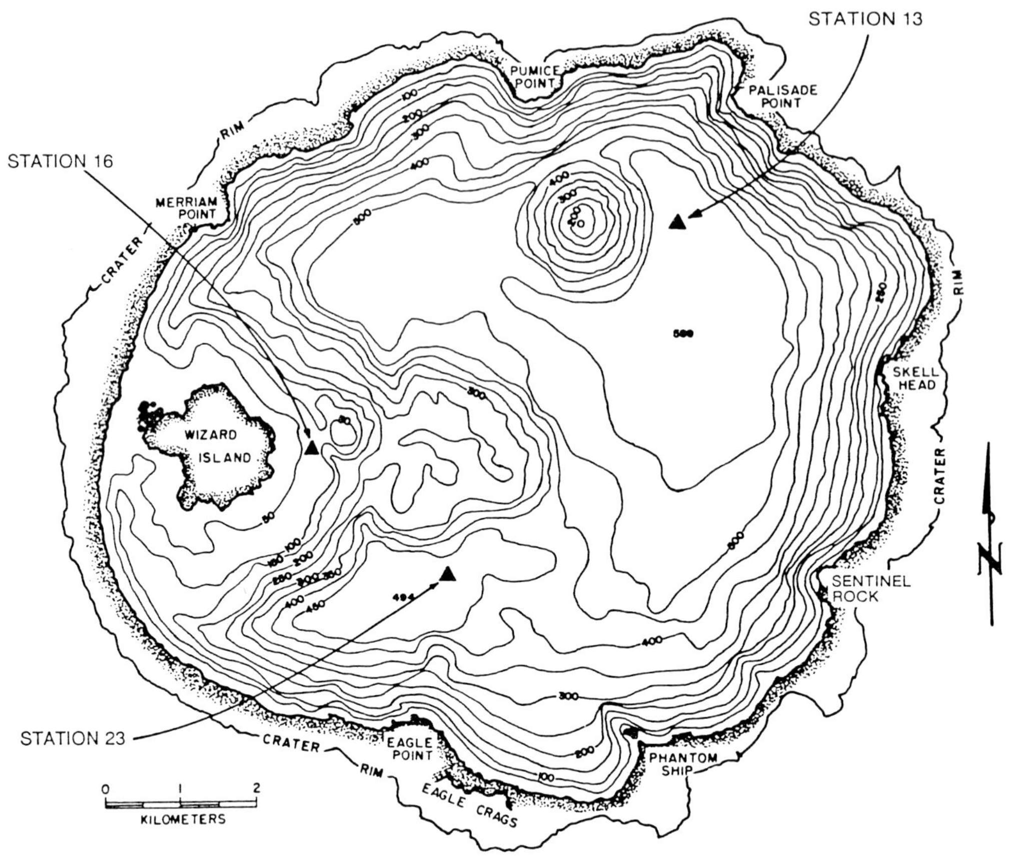

Figure 1. Bathymetric map of Crater Lake, Oregon, showing primary limnological sampling stations.

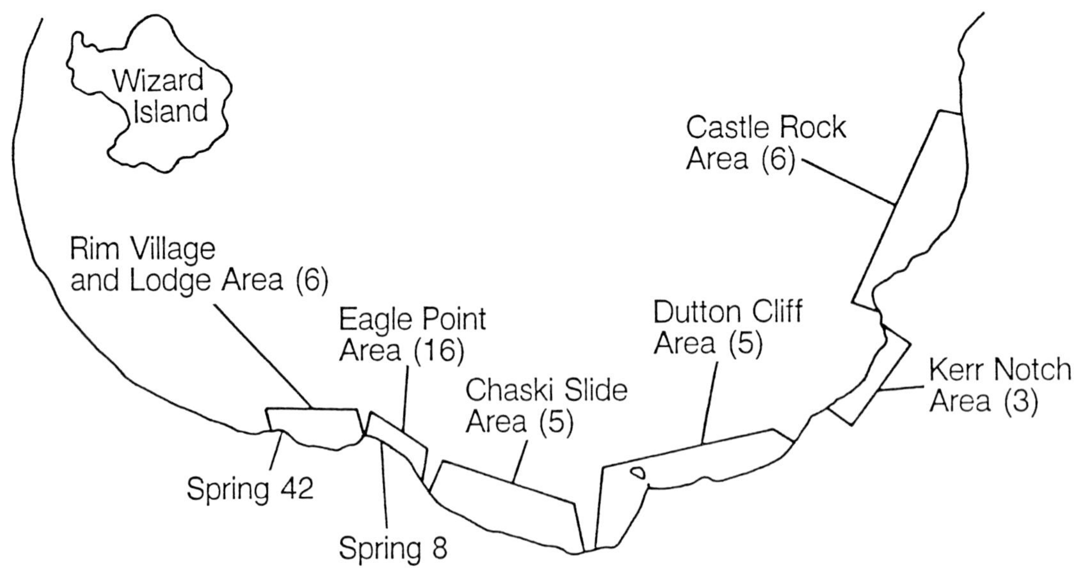

Figure 2. General locations of caldera springs, Crater Lake National Park.

Utterback *et al.* 1942; Larson 1972) accurately characterized Crater Lake's transparency and spectral transmissivity, and therefore are sound reference points with which to compare recent optical data.

Lake chemistry

Lakewater concentrations of inorganic nitrogen were extremely small, particularly in the euphotic zone (0-200 m) where nitrogen was essentially depleted by phytoplankton nutritional demands (Table 2). Concentrations of orthophosphate in the 0-200 m stratum were several times larger than inorganic nitrogen levels (Table 2), suggesting further that nitrogen, not phosphorus, is the limiting algal nutrient in Crater Lake. Nitrogen-deficiency is also implied by the predominance of nitrogen-fixing algae in the lake's sublittoral epilithic periphyton community (Loeb and Reuter 1981). Silica, on the other hand, is one of the most abundant elements in the lake (Phillips and Van Denburgh 1968). Concentrations of silica ranged from 15.7 to 16.0 mg liter^{-1} throughout the water column, attesting to the lake's silica-rich condition.

Phytoplankton assemblage

The phytoplankton community in Crater Lake, consisting of at least 140 species, was described in the previous paper by Geiger and Larson (1990) and elsewhere (Larson and Geiger 1980; Geiger and Larson 1981; Larson *et al.* 1987). Roughly one-third of these species have been observed in water samples collected from the 0-40 m stratum (Geiger and Larson, unpublished data). The predominant species in this layer is *Nitzschia gracilis*, a pennate diatom which exhibits prolific growth in August and September. *N. gracilis* is barely present in early summer, but soon becomes numerically preeminent among species comprising the lake's entire phytoplankton community (Larson *et al.* 1987). *N. gracilis* densities tend to peak in mid-August, reaching numbers that typically exceed 5 x 10^5 cells liter^{-1} (Figs. 3 and 4; Table 3). The observed maximum density in 1984 was nearly 1 x 10^6 cells liter^{-1} (Fig. 4), which accounted for more than 95% of the total phytoplankton biomass in the 0-40 m stratum. Still, these are comparatively small phytoplankton densities, not nearly large enough to constitute nuisance algal

TABLE 2. INORGANIC NUTRIENT CONCENTRATIONS IN CRATER LAKE DURING SUMMERS 1983 AND 1984[1]

	Nitrate-N (μg liter^{-1})				Ammonium-N (μg liter^{-1})				Orthophosphate (PO_4-P)(μg liter^{-1})			
Depth (m)	15Jul83	10Aug83	15Aug84	23Aug84	15Jul83	10Aug83	15Aug84	23Aug84	15Jul83	10Aug83	15Aug84	23Aug84
Surface	<1[2]	<1	<1	<1	<1	<1	<1	<1	13	17	11	12
20	<1	<1		<1	<1	<1		<1	13	14		15
60	1	<1		<1	<1	<1		<1				11
100	<1	<1	<1	<1	<1	<1	<1	1	12	14	12	14
120	<1	<1		<1	<1	<1		1				13
160	2	<1		<1	2			<1				8
200	4	2	<1	<1	1	2	<1	<1	11	12	13	9
250	8	6		1	<1	<1		<1	13	15		9
300		9	6	4		1	<1	<1		17	16	15
350			8	8			<1	<1			16	10
400			10	9			<1	<1			17	10
450			10	9			<1	<1			18	20
500			11	11			<1	<1			19	13
550			11				<1	<1			19	
580				12			<1	<1				12

[1] Selected profiles.
[2] Minimum detection limits for NO_3-N and NH_4-N are 1.0 μg liter^{-1}.

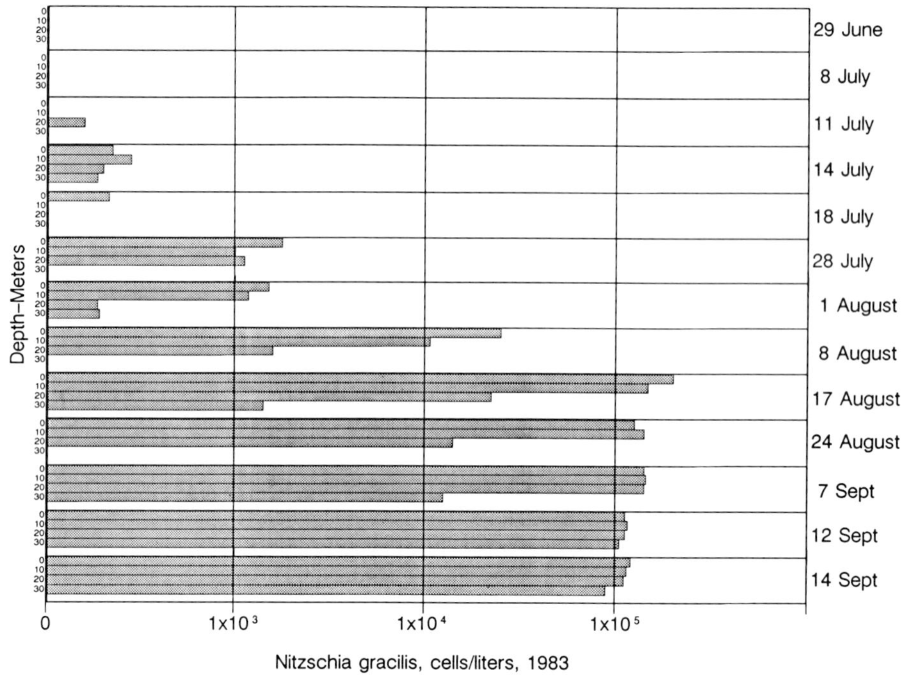

Figure 3. Pattern of *N. gracilis* abundance in the 0-30 m stratum of Crater Lake during summer 1983. Data from Station 13.

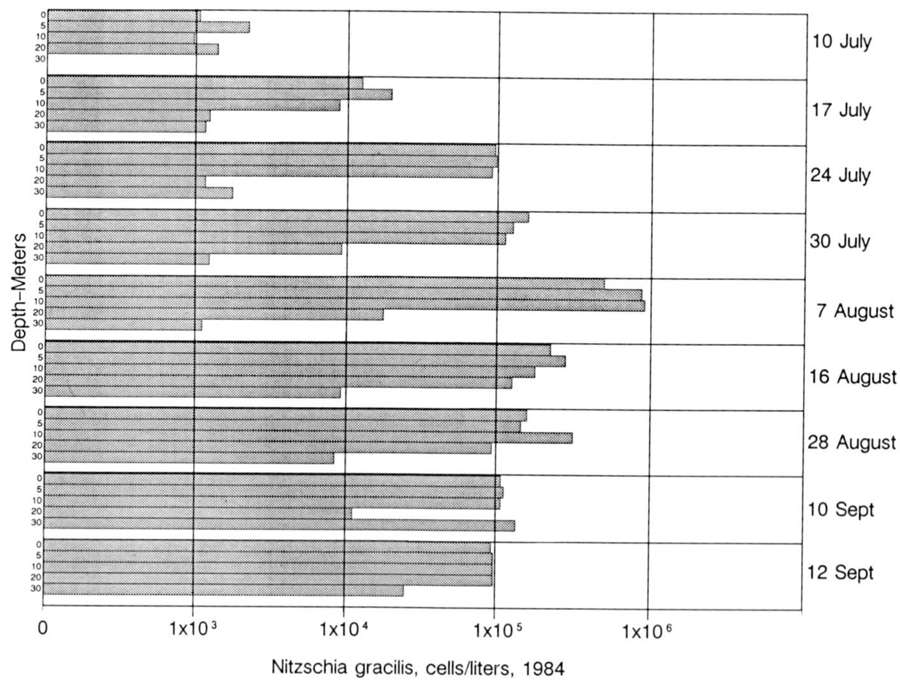

Figure 4. Pattern of *N. gracilis* abundance in the 0-30 m stratum of Crater Lake during summer 1984. Data from Station 13.

TABLE 3. TYPICAL PATTERN OF SUMMERTIME BUILDUP OF *NITZSCHIA GRACILIS* IN CRATER LAKE
Data for 1980

Date	Depth (m)	Density (cells liter^{-1})	% of Total Density	No. Species Present
10 Jul	surface	7,039	15	9
	20	7,049	15	10
	40	6,058	8	11
24 Jul	surface	211,722	88	9
13 Aug	surface	764,999	96	2
	20	630,825	93	5
	40	42,605	61	10
26 Aug	surface	59,253	83	9
	20	65,000	86	21
10 Oct	surface	27,677	98	3
	20	149,993	92	7
	40	40,932	70	9

blooms like those described by Paerl (1988) and Stockner (1988) for hypereutrophic lakes.

Phytoplankton chlorophyll

The vertical distribution of chlorophyll-*a* in Crater Lake was described first in 1969 (Larson 1972), and later for this study by Larson *et al.* (1987). Generally, chlorophyll is found throughout the euphotic zone (0-200 m), with maximum concentrations occurring in the 120-140 m stratum. Minimum concentrations of chlorophyll occur in shallower waters (0-40 m) despite the presence of relatively large numbers of *N. gracilis* cells. These cells, greatly magnified in SEM photomicrographs, appear to have relatively small chloroplasts per unit biovolume, i.e., low chlorophyll content (Geiger and Larson 1990). This characteristic is probably an adaptation to the prevailing high light intensities in shallow waters during summer. Studies have shown that phytoplankton cells subjected to intense ambient light often contain much less chlorophyll than cells growing under reduced light conditions (Kirk 1983). Thus, chlorophyll data alone are apparently not a reliable indicator of the magnitude of phytoplankton biomass present in the 0-40 m stratum.

The data presented in Table 4 indicate that concentrations of chlorophyll-*a* in Crater Lake are, for the most part, greater now than they were 20 years ago. This suggests that the lake has recently become more productive biologically. It can be argued, however, that the 1969 chlorophyll data—consisting of only three vertical profiles—is hardly a comprehensive, reliable point of reference with which to compare post-1978 chlorophyll levels. Therefore, the suggestion that Crater Lake has shifted to a higher trophic state, based solely on the data in Table 4, is offered here with considerable caution.

Substantially more chlorophyll was present in the water column at Station 23 than in waters sampled

TABLE 4. SUMMARY OF CHLOROPHYLL-*a* DATA OBTAINED AT STATION 13, CRATER LAKE
Only July and August data are presented.

Date	mg Chl-*a* m^{-2}		mg Chl-*a* m^{-3}	
	0-200m[1]	0-50m[2]	0-200m	0-50m
16 Jul 69[3]	59.95	4.64	0.300	0.084
5 Aug 69	35.00	9.84	0.175	0.179
31 Aug 69	8.39	0.012	0.042	0.0002
1 Aug 79	77.69	10.35	0.388	0.207
28 Aug 79	69.59	9.04	0.349	0.164
13 Aug 80	111.16	18.13	0.556	0.363
26 Aug 80	132.47	25.03	0.662	0.501
13 Jul 81	89.90	11.08	0.450	0.222
27 Jul 81	103.12	18.33	0.516	0.367
25 Aug 81	119.61	21.20	0.598	0.424
15 Jul 82	99.70	9.20	0.499	0.184
21 Jul 82	70.58	15.88	0.353	0.318
29 Jul 82	59.73	4.68	0.299	0.094
5 Aug 82	56.40	12.20	0.284	0.244
23 Aug 82	67.85	8.15	0.339	0.163
11 Jul 83	35.85	6.65	0.179	0.133
18 Jul 83	60.01	12.10	0.300	0.242
26 Jul 83	48.97	6.54	0.245	0.131
1 Aug 83	26.36	3.86	0.132	0.077
8 Aug 83	27.17	—	0.194	—
24 Aug 83	52.09	6.15	0.261	0.123
31 Jul 84	68.37	6.53	0.342	0.131
7 Aug 84	30.69	2.42	0.154	0.048
22 Aug 84	41.41	4.01	0.207	0.080
23 Jul 85[4]	28.9	—	0.15	—
20 Aug 85	54.5	—	0.27	—
23 Jul 86	103.5	—	0.52	—
20 Aug 86	122.3	—	0.61	—
23 Jul 87	96.3	—	0.48	—
20 Aug 87	111.5	—	0.56	—

[1] 0-200 m stratum.
[2] 0-50 m stratum.
[3] Data for 1969, reported by Larson (1972).
[4] Data for 1985, 1986, and 1987 collected by the National Park Service (Larson 1986, 1987, 1988).

at Station 13; this was particularly evident during 1983 (Table 5). Station 23 is located in the lake's south basin, approximately 2 km offshore of the Rim Village-lodge area (Fig. 1). The disproportionately larger chlorophyll concentrations at Station 23 indicates, perhaps, that phytoplankton productivity is greater in the south basin than it is on the opposite side of the lake at Station 13. Another possibility is that chlorophyll and other forms of particulate matter are circulated into the south basin and concentrated there by wind-induced water movements, e.g., seiches and strong surface currents. Surface-current patterns in Crater Lake are complex, with current velocities reaching 10 cm $sec.^{-1}$ (Kibby *et al.* 1968).

TABLE 5. CHLOROPHYLL-*a* CONCENTRATIONS COMPARED BETWEEN STATIONS 13 AND 23, 0-200 M STRATUM, CRATER LAKE

Date	Station 13		Station 23	
	mg m^{-2}	mg m^{-3}	mg m^{-2}	mg m^{-3}
11 Aug 81	89.31	0.447	208.99	1.045
21 Jul 82	70.58	0.353	91.68	0.458
1 Sep 82	80.30	0.402	74.93	0.375
11 Jul 83	35.85	0.179	59.45	0.297
18 Jul 83	60.01	0.300	67.00	0.335
26 Jul 83	48.97	0.245	92.45	0.462
1 Aug 83	26.36	0.132	68.49	0.343
8 Aug 83	27.17	0.194	74.50	0.373
24 Aug 83	52.09	0.261	60.50	0.303
2 Sep 83	75.81	0.379	101.82	0.509
7 Sep 83	40.06	0.200	100.70	0.504
14 Sep 83	72.55	0.363	152.10	0.761

Ground Water chemistry

Phillips and Van Denburgh (1968) speculated that ground water inflows are an important source of dissolved solids influencing, to some extent, the chemistry of Crater Lake. We speculated further that some ground water inflows are enriched with nitrogen, possibly due to sewage contamination, which stimulates the growth of *N. gracilis* in nitrogen-deficient lakewaters. In fact, concentrations of inorganic nitrogen are appreciably higher in ground waters emerging along the caldera wall directly below the Rim Village-lodge area than in other caldera springs tested (Table 6). One spring in particular, identified as Spring 42 (Fig. 2), had average nitrate concentrations of 238 and 265 μg $liter^{-1}$ in 1983 and 1984, respectively. These values were 5-10 times higher than average nitrate values for ground waters located, for example, in the Eagle Point and Chaski Slide areas (Table 6).

Nitrate concentrations in Spring 42 waters increased sharply throughout the summer (Figs. 5 and 6), reflecting perhaps a seasonal reduction in the rate of ground water dilution by snowmelt from a steadily diminishing snowpack on the caldera rim (Redmond 1988). Silica, chloride, and sulfate also became more concentrated in waters representing Spring 42 (Table 7). Over these summertime periods, however, concentrations of orthophosphate remained fairly constant (Table 7). As a matter of fact, orthophosphate concentrations in Spring 42 waters were similar, more or less, to concentrations measured in most other caldera springs (Table 6).

DISCUSSION AND CONCLUSIONS

Origin of excess nitrate in Spring 42

The National Park Service has continued to observe unusually high nitrate concentrations in Spring 42 waters every summer since 1984 (Larson 1986, 1987, 1988). These concentrations averaged roughly 250 μg $liter^{-1}$ all summer and possibly during fall and winter as well (Larson 1988). Spring 42 discharges at a constant rate of 2 liters $sec.^{-1}$, at least during summer. Assuming that the average summertime concentration of nitrate in Spring 42 waters is 250 μg $liter^{-1}$, the nitrate-loading rate for this spring alone is 3.9 kg $summer^{-1}$. Based on this loading estimate, Spring 42 has contributed around 80 kg of nitrate to Crater Lake surface waters during summer months only over the past 20 years. For an entire year, the amount of nitrate entering Crater Lake via Spring 42 may be as much as 12-15 kg, which, when extrapolated for a 20-year period, totals 240-300 kg of nitrate.

Several other springs emanate from the Rim Village-lodge area, including springs that may enter below the lake's surface. These springs also contain relatively high concentrations of inorganic nitrogen (Larson 1984a) and, thus, may each contribute as much nitrogen to the lake as Spring 42. An important objective in this study, one that unfortunately became increasingly controversial and divisive, was to determine why these particular springs contain significantly more nitrate than others in the caldera.

TABLE 6. CHEMICAL DATA SUMMARIZED FOR GROUND WATER INFLOWS AT THREE LOCATIONS IN THE CRATER LAKE CALDERA, 10 AUGUST-14 SEPTEMBER 1983
Concentrations in μg liter^{-1}

Location	NO_3-N			NH_4-N			PO_4-P		
	n	Average	Range	n	Average	Range	n	Average	Range
Rim Village area (3 springs)	12	163.7	63-299	12	16.9	0-68	12	53.8	28-126
Eagle Point area (14 springs)	28	25.3	2-59	28	4.3	0-16	28	50.7	22-103
Chaski Slide area (4 springs)	6	48.8	21-73	6	3.8	2-6	6	53.0	43-60
Spring 42	5	238	150-299	5	2.8	0-6	5	45.6	40-48

We suggested, starting in 1978 (Larson 1978), that Spring 42 is one of several springs enriched with nitrogen originating from inadequate sewage-disposal facilities in Crater Lake National Park. There were compelling reasons for proposing this scenario. First, we were aware of earlier problems with sewage-escapement in the Park, notably an incident in 1975 when raw sewage grossly contaminated a spring which supplied the Park with most of its drinking water. This contamination caused more than 1000 cases of diarrhea and other waterborne diseases among Park visitors and staff (Rosenburg *et al.* 1977; Craun 1981). Additionally, there were other reports of occasional minor problems with sewage-disposal, but these apparently were never publicized (Stohr-Gillmore 1983). Second, we discovered from maps and other information furnished by the National Park Service that Spring 42 may be part of a ground water aquifer that underlies the Park's main sewage-disposal system. This system, merely a septic tank that discharges into a drainfield (National Park Service plans and specifications on file at Crater Lake National Park), collects sewage and other wastewater produced in the lodge and at various Rim Village sites (Fig. 7). The drainfield, or percolation gallery area, is perched 200-250 m above the surface of Crater Lake, and is situated less

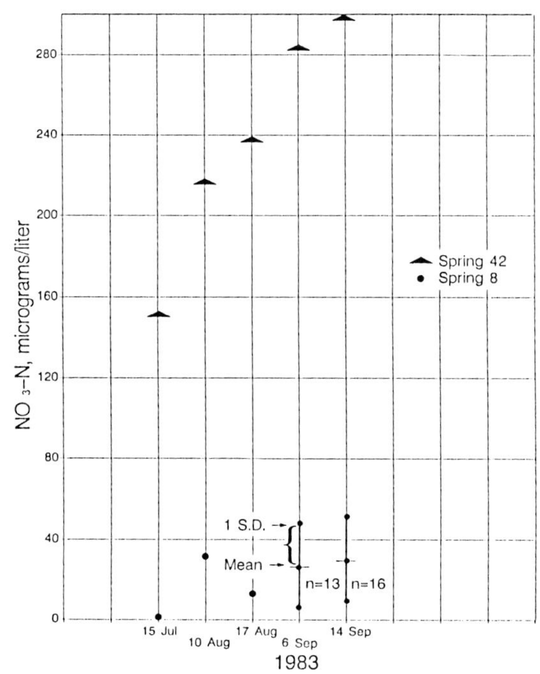

Figure 5. Summertime concentrations of nitrate-nitrogen compared between Spring 42 and Spring 8 (control), Crater Lake National Park, 1983.

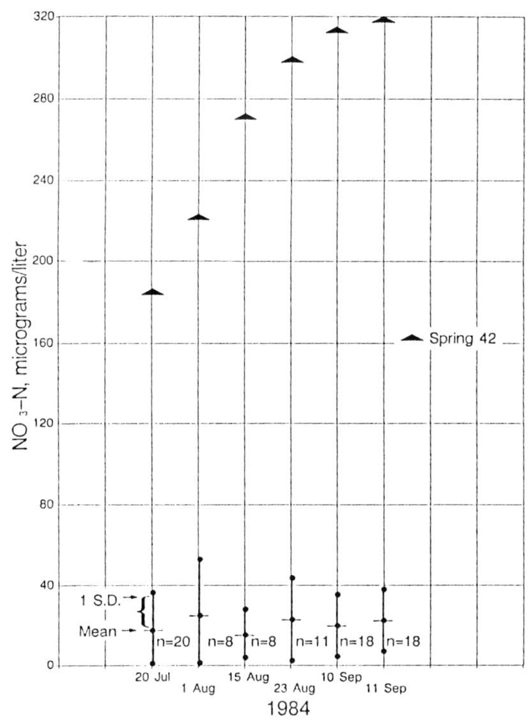

Figure 6. Summertime concentrations of nitrate-nitrogen compared between Spring 42 and control springs, Crater Lake National Park, 1984.

TABLE 7. SUMMERTIME PATTERNS OF VARIOUS IONIC CONCENTRATIONS IN SPRING 42 WATERS, CRATER LAKE

Date	NO_3-N	PO_4-P	Si[1]	SO_4	Cl
15 Jul 83	150	48			
10 Aug 83	217	47			
17 Aug 83	239	40			
6 Sep 83	285	47			
14 Sep 83	299	46			
12 Jul 84	148	46	30.1		
20 Jul 84	158	44	30.2		
1 Aug 84	234	46	34.3		
15 Aug 84	258	44	43.8		
23 Aug 84	298	47	43.2		
10 Sep 84	319	46	40.8		
11 Sep 84	313	44	41.1		
16 Jul 85[2]				390	
30 Jul 85				409	635
22 Aug 85				502	846
29 Aug 85				560	947

[1] Concentrations in μg liter^{-1} except Si expressed in mg liter^{-1}.
[2] Unpublished data for 1985 provided by D. W. Larson.

than 500 m from where Spring 42 emerges along the caldera wall (Fig. 7). Pumice soils predominate throughout the drainfield and surrounding area; Phillips and Van Denburgh (1968) described these soils as "so highly permeable that in places all precipitation infiltrates where it falls." Each summer, this drainfield receives millions of gallons of domestic wastewater generated by 500,000 or more Park visitors. The system, however, was designed and built in the mid-1940s to accommodate far fewer people, probably around 200,000 users which then was the average number of Park visitors per year (Stohr-Gillmore 1983). The system was improved shortly after the 1975 sewage-pollution incident (J. Jarvis, National Park Service, pers. comm.), an action that was long overdue since septic tank-drainfield systems are designed to last only 10-15 years (Canter *et al.* 1987). Still, the system continues to be the Park's principal means of sewage-disposal despite an upsurge in the number of visitors over the past 20 years (Fig. 8). This burgeoning demand on the system has probably reduced its capacity to adequately contain and biologically degrade the mounting quantity of septic wastewater. Consequently, during summer, drainfield soils are overburdened with largely untreated wastewater which, because of its considerable volume, infiltrates

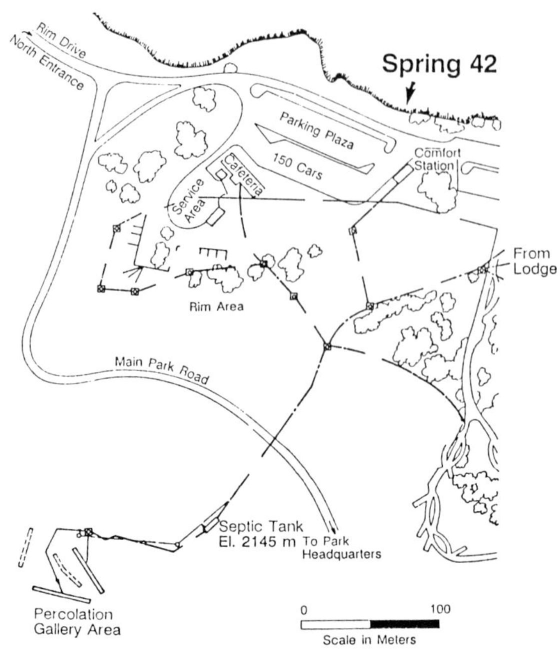

Figure 7. Proximity of Spring 42 to Rim Village septic tank-drainfield system, Crater Lake National Park.

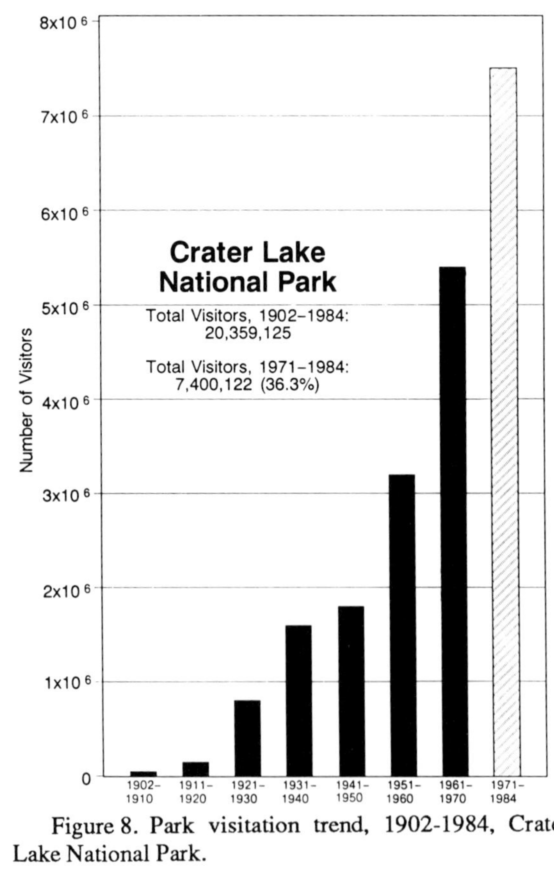

Figure 8. Park visitation trend, 1902-1984, Crater Lake National Park.

readily into surrounding soils and underlying ground waters.

The actual volume of wastewater discharged to this particular drainfield each summer is not known. In general, the per capita volume of sewage wastewater produced in western alpine recreational areas is about 115 liters day^{-1} (McGauhey *et al.* 1963). Assuming that the Park's visitation rate between mid-June and mid-September now averages 6,000 people day^{-1}, then the estimated average daily flow of septic wastewater into the drainfield is 6.9 X 10^5 liters. At this rate, which is a conservative estimate, roughly 62 X 10^6 liters of wastewater enters the drainfield over the entire summer. As expected, this wastewater is highly enriched with nitrogen and phosphorus: typical septic tank effluent contains 40 mg $liter^{-1}$ of total nitrogen and 15 mg $liter^{-1}$ of total phosphorus (Canter and Knox 1985; Canter *et al.* 1987). Therefore, based on these concentrations, roughly 2500 kg of total N and 930 kg of total P are loaded into the drainfield each summer.

Ammonia and urea are the most abundant forms of nitrogen in domestic wastewater, comprising 80-90% of the total N present (Goering 1972). Moreover, urea hydrolyzes readily to ammonia, resulting in an even greater ammonia fraction (Goering 1972). Presumably, then, most of the nitrogen present in the Park's septic-tank effluent is ammonia. In an aerobic soil environment ammonia is oxidized to nitrate by nitrifying bacteria (Canter *et al.* 1987). Because aerobic conditions prevail in soils with high permeability (Canter *et al.* 1987), the drainfield soil column is probably oxygenated to a great extent. Oxygenated soils allow nitrification to proceed, which converts effluent ammonia to an equal amount of nitrate. Assuming that most of the effluent ammonia undergoes nitrification, perhaps as much as 2,000 kg of nitrate is produced in this manner each summer. Thus, nitrate becomes the predominant form of nitrogen in septic wastewater percolating through drainfield soils.

Nitrate is extremely mobile in soils that are highly permeable and oxygenated (Canter *et al.* 1987). Consequently, nitrate can migrate long distances and, ultimately, leach into ground waters. Phosphorus, on the other hand, is readily adsorbed by soil particles and, thus, is largely retained in the soil column (Canter *et al.* 1987). This explains, perhaps, why nitrate-enriched waters from Spring 42 are not similarly enriched with orthophosphate (Table 7). Nevertheless, the limiting nutrient in Crater Lake is apparently nitrogen, an abundance of which likely occurs in ground waters underlying the Park's septic drainfield and feeding springs that discharge through the caldera wall below Rim Village, including Spring 42. The combined nitrogen-loading capacity of these ground water inflows, discharging continuously for decades perhaps, could be substantial and therefore an important supplement to the lake's extremely tight nitrogen budget. This process of nutrient loading has been studied at Lake Tahoe, California: Loeb and Goldman (1979) found that certain ground waters entering the lake are highly enriched with nitrate and soluble P, which they believe may have contributed significantly to the lake's eutrophication.

Finally, it should be noted that the National Park Service has itself acknowledged that sewage wastewater is probably entering Crater Lake via ground waters emanating from the Rim Village-lodge area. The agency states the following in a 1987 environmental assessment of proposed Park developments (U.S. Department of the Interior 1987):

> Data collected in 1983 indicates that spring water entering the lake from below the Rim Village area contains nutrients (primarily nitrates), which could affect clarity of the water. A panel of scientists who reviewed the elevated nitrate concentrations in the springs concluded that there was no apparent natural cause and, therefore, the elevated concentrations are likely caused by human activities in the area. The most obvious source is the leachfield in the Rim Village, and it was recommended that its use be discontinued.

Despite this conclusion, however, hard evidence proving that these springs are sewage-contaminated is still lacking. A tracer study to possibly obtain such evidence was recommended in November 1983 by Dr. J. F. Quinlan, National Park Service Research Geologist (Larson 1984a), but the work was never attempted. This lack of evidence has fueled a lengthy debate over whether the Park's sewage is contaminating Crater Lake. Regrettably, the dispute has delayed called-for action to promptly halt the lake's pollution, if that in fact is occurring.

Lake response to suspected anthropogenic nitrogen-loading

Most of the nitrogen taken up by phytoplankton in Crater Lake is recycled in the euphotic zone (Dymond and Collier 1986), suggesting that ongoing nitrogen loading from allochthonous sources, e.g., ground water inflows, direct precipitation, and surface runoff, has resulted in a progressive net-gain of nitrogen in the euphotic layer. Possibly, *N. gracilis* has responded to this enhanced nitrogen-availability with steadily increasing population densities. The environmental impetus for prolific growths of *N. gracilis* during summer, however, may have been decades-long nitrogen loading by sewage-contaminated springs emerging below the Rim Village-lodge area.

N. gracilis typically reaches its maximum abundance during the first half of August. Population densities tend to remain fairly large and stable for 3-4 weeks following this peak (Figs. 3 and 4). This prolific, sustained growth, causing an increase in the concentration of suspended particulate matter in the lake's 0-40 m stratum, appears to coincide with the seasonal reduction in Secchi disk transparency (Fig. 9). We conclude, therefore, that Secchi depth is inversely related to the increased presence of *N. gracilis*.

Unfortunately, no record exists prior to 1978 concerning the status of *N. gracilis* or, for that matter, any phytoplankton species in Crater Lake. Our basis for suggesting that *N. gracilis* has become more abundant in the lake's 0-40 m stratum over the past 20-30 years is twofold. First, rates of phytoplankton primary production in the 0-30 m stratum during the early 1980s (1980-1983) averaged 3.6 mg C m^{-2} hr^{-1}, which was 50% greater than the average productivity occurring in this layer 12-15 years earlier, i.e., 1967-1969 (Dahm *et al.* 1990). Second, during the summer of 1940, Utterback *et al.* (1942) reported finding "practically no phytoplankton" in the 0-20 m stratum, and only "a few thousand cells per liter" between depths of 20 and 30 m. The Utterback research team included L.D. Phifer, an experienced marine algologist with the Oceanographic Laboratory at the University of Washington. Dr. Phifer collected phytoplankton by vertically towing a "20X" plankton net from depths of 100 and 200 m, and by centrifuging water samples obtained throughout the water column with Nansen-type bottles. Captured phytoplankton were identified to

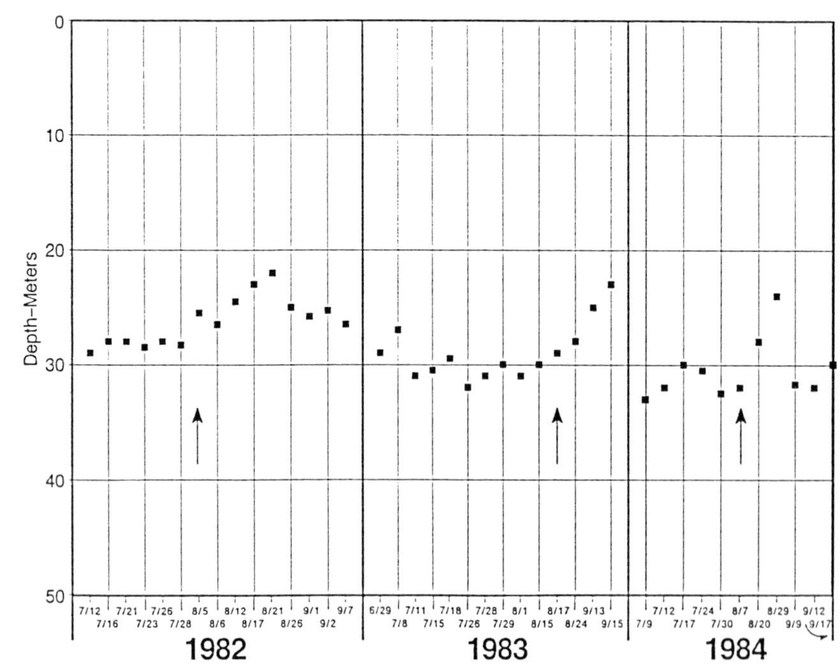

Figure 9. Onset of peak growth by *N. gracilis* in Crater Lake (vertical arrows) corresponds to diminishing mid-summer Secchi disk transparency (black squares), 1982-1984.

genus; only six genera were reported (Utterback *et al.* 1942). However, this work was done in mid-July (on 18 July) when *N. gracilis* densities are still relatively small (Figs. 3 and 4; Table 3). These collections, had they been made a month later when *N. gracilis* numbers usually peak, would have provided decisive information about the status of this particular species during a period when Secchi readings were routinely 39-40 m (Hasler 1938), and when visitor use of Park facilities and resources was considerably less than present-day usage (ninety percent of the roughly 22 million people who have ever visited Crater Lake did so after 1940; Fig. 8). Here then is a prime example of how the absence of a long-term, systematic limnological monitoring program at Crater Lake has forced scientific investigators and the National Park Service to speculate about whether or not the lake has changed as the result of anthropogenic encroachment.

Lastly, there is general agreement that the Park's septic tank-drainfield system near Rim Village is probably the source of excess nitrate in Spring 42 and in neighboring springs. There is no consensus, however, regarding the relative importance of sewage-derived nitrate in the lake's overall nitrogen budget. Other sources of nitrogen such as precipitation may be far more important. At this point though, no one to our knowledge has made a definitive study accounting for all the various sources of nitrogen supplying Crater Lake. A reviewer of this paper has estimated that the lake receives as much as 9,000 kg of nitrate from precipitation each year. Scientists who actually work on Crater Lake believe that the amount of nitrate received through precipitation is much less, probably around 5,000 kg yr^{-1}. But even this value is uncertain because of limited information about the lake's nitrogen budget and about the nitrogen-loading capacity of annual precipitation.

There is also disagreement regarding the possible effect of sewage-derived nitrogen input on the lake's biological productivity. It has been argued, for example, that nitrogen-enriched ground water inflows are greatly diluted by the lake's relatively large volume of water (16 km^3) and, therefore, are trophically inconsequential. But on the basis of the data presented here, it is more likely that long-term nitrogen loading by springs emerging below the Rim Village-lodge area has contributed significantly to increased phytoplankton productivity in nitrogen-deficient lakewaters. Consequently, because of greater phytoplankton biomass, lake optical properties have diminished. Additionally, studies have shown that oligotrophic lakes are extremely sensitive and responsive to even minimal nutrient additions (Goldman 1981). Phytoplankton populations in lakes that are nutrient-impoverished appear to be very efficient in the uptake, utilization, and recycling of scarce nutrients (Goldman and Wetzel 1963). Conceivably, phytoplankton in Crater Lake are predominantly of this type, i.e., well-adapted for low nutrient conditions. If so, seemingly negligible rates of nutrient loading could actually stimulate greater phytoplankton activity, leading perhaps to a higher trophic status for the lake. Consider Lake Tahoe, for example. There, in a period of less than 30 years, nutrient loading once thought to be relatively insignificant has caused this lake to shift from an ultra-oligotrophic condition to one that is more eutrophic (Goldman 1981). This accelerated change in trophic status, occurring despite the lake's considerable volume (156 km^3) and alpine setting (elevation: 1898 m), eventually caused a sharp reduction in Secchi disk transparency (Goldman *et al.* 1982). Concluded Dr. Charles Goldman, principal limnological investigator at Lake Tahoe since 1958: "Twenty years ago few were willing to believe that a mountain lake like Tahoe could change so quickly" (Goldman 1981). Ten years from now one can hope that Dr. Goldman's requiem for Lake Tahoe is not repeated for Crater Lake.

ACKNOWLEDGMENTS

We thank members of the National Park Service who supported and encouraged our research at Crater Lake, including J. Salinas, G. Larson, J. Larson, H. Tanski, R. Cranson, J. Jarvis, M. Stohr-Gillmore, M. Forbes, J. Rouse, and R. Benton. We are especially grateful to Dr. Ellen Drake, Senior Editor of this Symposium Proceedings, for her considerable patience and understanding during our preparation of this paper several months after the deadline for completion.

LITERATURE CITED

Atwood, W. W. 1937. Crater Lake and Yosemite through the ages. Nat'l Geog. 71 (3):326-342.

Canter, L. W., and R. C. Knox. 1985. Septic tank system effects on ground water quality. Lewis, Chelsea, Mich.

Canter, L. W., R. C. Knox, and D. M. Fairchild. 1987. Ground water quality protection. Lewis, Chelsea, Mich. 562 pp.

Craun, G. F. 1981. Outbreaks of waterborne diseases in the United States:1971-1978. Jour. Amer. Water Works, Manage. Operations 73(July):360-369.

Dahm, C. N., D. W. Larson, N. S. Geiger, and L. K. Herrera. 1990. Secchi disk, photometry, and phytoplankton data from Crater Lake: Long-term trends and relationships. Pages 143-151 in E. T. Drake *et al.*, ed., Crater Lake: An Ecosystem Study. Pacific Division, Amer. Assn. Advance. Sci., San Francisco, Calif.

Dymond, J., and R.W. Collier. 1986. Geochemistry and limnology of Crater Lake. Page 73 *in* G. L. Larson, ed., Crater Lake Limnological Studies. 1985 Ann. Rep., Nat'l Park Serv., Coop. Park Studies Unit, College of Forestry, Oregon State Univ., Corvallis, Ore. 73 pp. (with appendices).

Geiger, N. S., and D. W. Larson. 1981. Crater Lake: Its planktonic algae. Mazama 63:54-59.

Geiger, N. S., and D. W. Larson. 1990. Phytoplankton species distribution in Crater Lake, Oregon, 1978-1980. Pages 153-165 *in* E. T. Drake *et al.*, eds., Crater Lake: An Ecosystem Study. Pacific Division, Amer. Assn. Advance. Sci., San Francisco, Calif.

Goering, J. J. 1972. The role of nitrogen in eutrophic processes. Pages 43-68 *in* R. Mitchell, ed., Water Pollution Microbiology. John Wiley-Interscience, New York. 416 pp.

Goldman, C. R. 1981. Lake Tahoe: Two decades of change in a nitrogen deficient oligotrophic lake. Verh. Internat'l Verein. Limnol. 21:45-70.

Goldman, C. R., and R. G. Wetzel. 1963. A study of the primary productivity of Clear Lake, Lake County, California. Ecology 44(2):283-294.

Goldman, C. R., R. L. Leonard, R. P. Axler, J. E. Reuter, and S. L. Loeb. 1982. Interagency Tahoe Monitoring Program, Second Annual Rept., Water Year 1981. Tahoe Research Group, Inst. Ecol., Univ. California, Davis. 193 pp.

Hager, S. W., E. L. Atlas, L. I. Gordon, A. W. Mantyla, and P. K. Park. 1972. A comparison at sea of manual and autoanalyzer analyses of phosphate, nitrate, and silicate. Limnol. Oceanogr. 17:931-937.

Hasler, A. D. 1938. Fish biology and limnology of Crater Lake, Oregon. Jour. Wildl. Manage. 2:94-103.

Hasler, A. D. 1969. Cultural eutrophication is reversible. Bioscience 19:425-431.

Hutchinson, G. E. 1957. A treatise on limnology. Vol. 1. Geography, physics, and chemistry. John Wiley, New York. 1015 pp.

Kemmerer, G., J. F. Bovard, and W. R. Boorman. 1924. Northwestern lakes of the United States: Biological and chemical studies with reference to possibilities in production of fish. Bull. U. S. Bur. Fish. 39:51-140.

Kibby, H. V., J. R. Donaldson, and C. E. Bond. 1968. Temperature and current observations in Crater Lake, Oregon. Limnol. Oceanogr. 13(2): 363-366.

Kirk, J. T. O. 1983. Light and photosynthesis in aquatic ecosystems. Cambridge Univ. Press, Cambridge. 401 pp.

Larson, D. W. 1972. Temperature, transparency, and phytoplankton productivity in Crater Lake, Oregon. Limnol. Oceanogr. 17 (3):410-417.

Larson, D. W. 1978. Limnology of Crater Lake, with emphasis on diel vertical distributions of algae. Investigator's Ann. Rep., Natural Sciences Research, Nat'l Park Serv., Crater Lake National Park, Ore. (October 1978).

Larson, D. W. 1980. Limnology of Crater Lake, with emphasis on the distribution and abundance of phytoplankton. Investigator's Ann. Rep., Natural Sciences Research, Nat'l Park Serv., Crater Lake National Park, Ore. (December 1980).

Larson, D. W. 1981. Limnology of Crater Lake, with emphasis on the distribution and abundance of phytoplankton. Investigator's Ann. Rep., Natural Sciences Research, Nat'l Park Serv., Crater Lake National Park, Ore. (December 1981).

Larson, D. W. 1983. Crater Lake limnological studies, 1982. First annual report on the limnology and water quality monitoring program at Crater Lake National Park, Oregon. Nat'l Park Serv., Pacific Northwest Region, Seattle, Wash. 36 pp. (with appendices).

Larson, D. W. 1984a. Crater Lake limnological studies, 1983. Second annual report on the limnology and water quality monitoring program at Crater Lake National Park, Oregon. Nat'l Park Serv., Pacific Northwest Region, Seattle, Wash. 33 pp. (with appendices).

Larson, D. W. 1984b. The Crater Lake study: Detection of possible optical deterioration of a rare, unusually deep caldera lake in Oregon, U.S.A. Verh. Internat'l Verein. Limnol. 22:513-517.

Larson, D. W. and N. S. Geiger. 1980. Species composition and vertical distribution of pelagic zone phytoplankton in Crater Lake, Oregon: 1940-1979. Pages 96-104 *in* Proc. 2nd Conf. Sci. Res. in Nat'l Parks. Vol. 2, Aquatic Biology. Nat'l Park Serv., Washington, DC. 192 pp.

Larson, D. W. and M.E. Forbes. 1980. Optical properties of Crater Lake, Oregon: Variation in Secchi disk transparency, 1937-1979. Pages 615-617 *in* Proc. 2nd Conf. Sci. Res. in Nat'l Parks. Vol.5, Physical Sciences. Nat'l Park Serv., Washington, DC. 690 pp.

Larson, D. W., C. N. Dahm, and N. S. Geiger. 1987. Vertical partitioning of the phytoplankton assemblage in ultraoligotrophic Crater Lake, Oregon, U.S.A. Freshwater Biol. 18: 429-442.

Larson, G. L. 1985. Crater Lake limnological studies, 1984 Ann. Rep., Nat'l Park Serv., Coop. Park Studies Unit, School of Forestry, Oregon State Univ., Corvallis, Ore. 55 pp. (with appendices).

Larson, G. L. 1986. Crater Lake limnological studies, 1985 Ann. Rep., Nat'l Park Serv., Coop. Park Studies Unit, School of Forestry, Oregon State Univ., Corvallis, Ore. 45 pp. (with appendices).

Larson, G. L. 1987. Crater Lake limnological studies, 1986 Ann. Rep., Nat'l Park Serv., Coop. Park Studies Unit, School of Forestry, Oregon State Univ., Corvallis, Ore. 81 pp.

Larson, G. L. 1988. Crater Lake limnological studies, 1987 Ann. Rep., Nat'l Park Serv., Coop. Park Studies Unit, School of Forestry, Oregon State Univ., Corvallis, Ore. 174 pp.

Lind, O. 1974. Handbook of common methods in limnology. Mosby & Co., St. Louis, Mo.

Loeb, S. L., and C. R. Goldman. 1979. Water and nutrient transport via ground water from Ward Valley into Lake Tahoe. Limnol. Oceanogr. 24(6):1146-1154.

Loeb, S. L., and J. E. Reuter. 1981. The epilithic periphyton community: A five-lake comparative study of community productivity, nitrogen metabolism, and depth-distribution of standing crop. Verh. Internat'l Verein. Limnol. 21:346-352.

Lund, J. W. G., G. M. Kipling, and E. D. LeCren. 1958. The inverted microscope method of estimating algal numbers and the statistical basis of estimates made by counting. Hydrobiologia 11: 143-170.

McGauhey, P. H., R. Eliassen, G. Rohlich, H. F. Ludwig, and E. A. Pearson. 1963. Comprehensive study on protection of water resources of Lake Tahoe Basin through controlled waste disposal. Report prepared for the Board of Directors, Lake Tahoe Area Council, Lake Tahoe, Calif. 157 pp.

Paerl, H. W. 1988. Nuisance phytoplankton biomass in coastal, estuarine, and inland waters. Limnol. Oceanogr. 33 (4), part 2:823-847.

Pettit, E. 1936. On the color of Crater Lake water. Proc. Nat'l Acad. Sci. 22 (2):139-146.

Phillips, K. N., and A. S. Van Denburgh. 1968. Hydrology of Crater, East, and Davis lakes, Oregon. U. S. Geol. Surv. Water-Supply Pap. 1859-E. 60 pp.

Redmond, K. 1988. Climate, climate variability, and Crater Lake. Pages 69-81 *in* G. L. Larson, ed., Crater Lake Limnological Studies. 1987 Ann. Rep., Nat'l Park Serv., Coop. Park Studies Unit, School of Forestry, Oregon State Univ., Corvallis, Ore. 174 pp.

Rosenburg, M. L. *et al.* 1977. Epidemic diarrhea at Crater Lake from enterotoxigenic *Escherichia coli*. Ann. Intern. Med. 86:714-730.

Smith, R. C., J. E. Tyler, and C. R. Goldman. 1973. Optical properties and color of Lake Tahoe and Crater Lake. Limnol. Oceanogr. 18:189-199.

Stockner, J. G., and K. S. Shortreed. 1988. Response of *Anabaena* and *Synechococcus* to manipulation of nitrogen:phosphorus ratios in a lake fertilization experiment. Limnol. Oceanogr. 33(6, part 1): 1348-1361.

Stohr-Gillmore, M. W. 1983. Environmental management appraisal of Crater Lake, Oregon. M.S. Thesis. Univ. Oregon, Eugene, Ore. 101 pp.

Strickland, J. D. H., and T. R. Parsons. 1968. A practical handbook of seawater analysis. Bull. Fish. Res. Bd. Can. 167.

U. S. Department of the Interior. 1987. Crater Lake National Park, Mazama Campground/Rim Village Corridor, Oregon. Supplement to the 1984

Environmental Assessment, Development Concept Plan, Nat'l Park Serv., Pacific Northwest Region, Seattle, Wash. 80 pp.

U. S. Geological Survey. 1977. National handbook of recommended methods for water-data acquisition, Chapt. 2, ground water flow measurements, p. 2-16, U. S. Geol. Surv., Reston, Va.

Utterback, C. L., L. D. Phifer, and R. J. Robinson. 1942. Some planktonic and optical characteristics of Crater Lake. Ecology 23:97-103.

Wetzel, R.G. 1975. Limnology. Saunders, Philadelphia, Pa. 743 pp.

SUMMARY OF CRATER LAKE STUDIES AND COMPARISON WITH THE EARLY STAGES OF EUTROPHICATION OF LAKE TAHOE

Charles R. Goldman
Division of Environmental Studies
University of California
Davis, CA 95616

This symposium brings together an array of studies focusing on limnological and geological aspects of this important caldera lake. Public law established a ten-year study which began in 1982 as a result of observed loss in water clarity from early Secchi measurements. These indicated a drastic decline in transparency from 38 m in 1937 to 21 m in 1978. Extensive geological investigations by USGS scientists and by Oregon State University oceanographers have been directed towards the possibility of geothermal inputs by remote, geochemical, and sediment studies of the basin. The ratio of Cl to Li and He isotope ratios provides additional support for the input of thermal springs. Variable hydrothermal activity during the last 6900 years may have modified nutrient loading. Blooms of a single diatom species, *Nitzschia gracilis*, among the five highly diverse phytoplankton communities contributes most to the transparency loss and higher primary productivity rates than were encountered in the late 1960s. Two cladocerans and nine rotifers make up the zooplankton with possible "top-down" effects of kokanee predation influencing cyclic phenomena of *Daphnia*. Food supply is partitioned between rainbow trout and kokanee with almost 100 m of water column utilized. Long-term weather data have been used to evaluate the system's water balance. Seasonal heating and cooling and what appears to be thermal input from the lake floor produce low stability in deep water. Like Lake Tahoe, nitrogen limitation of algal growth is indicated. Spring water contaminated by sewage seepage has been suggested as a possible cause of the loss of transparency. The intensive data set from Lake Tahoe provides comparison of nutrient response, primary productivity, and transparency change useful in predicting future conditions and management strategies for Crater Lake. Recommendations for additional studies are also included.

I will first present a short review of my observations of some of the papers presented; then I hope to provide some insight into the Crater Lake investigation by a comparison between the Crater Lake work with what is known about Lake Tahoe which I have studied for thirty years.

I will begin by noting that the symposium organizers, Ellen Drake and Gary Larson, did a remarkable job of assembling the participants who were able to amass a considerable breadth of limnological, geological, and meteorological information presented in the papers in this session. When I accepted the job of summarizing this symposium I realized that I was dealing, in part, with the controversy as to whether or not geothermal inflow into Crater Lake exists. This kind of scientific argument is healthy, helps to sharpen wits, and improves the opportunities for both research and funding for future study. Unfortunately, it also has the potential of polarizing the contestants in the ensuring debate. As more data begin to accumulate, it is important to keep an open mind and allow the actual research results to answer the questions. This requires the most objective interpretation of the data. The ever-present danger of "missing the forest for the trees" is particularly hazardous because working in our own laboratories, we may become over-confident that we are headed in exactly the right direction. As we examine the increasing store of data on Crater Lake, we will do well to recall the words of the famous English statistician, G. E. P. Box (1954), who noted that "no amount of artistry in statistical design can compensate for the omission of the most important factor." As illustrated by some of the data on the depth of spring mixing in Lake Tahoe that is presented later, it is sometimes the unexplainable "outlier data point" which, after more careful evaluation, eventually proves to be the most informative. If we are too focused or myopic in our view we are more likely to miss the few, but often extraordinary, opportunities for gaining greater understanding of how the physical-chemical environment influences the function of aquatic ecosystems.

Sediment Studies

Lakes are, after all, "reservoirs of history" in the sense that their sediments contain a historical record of the events which have taken place in the lake itself, as well as in the associated watersheds and airsheds. Lake sediments may contain forest fire or volcanic ash which can serve as time markers. In addition to organic and inorganic materials which form varves of annual deposition in lake sediments, the remains of fish, diatoms, pollen, and macrofossils from the drainage basin provide clues to the history of climatic changes and activities on the watershed (Dymond and Collier 1990). Jack Dymond's sophisticated sedimentation collection devices have been used both at Crater Lake and at Lake Tahoe. It has been a useful concept to view both lakes in this respect as a kind of model ocean. I have been impressed with the similarities that a large lake like Tahoe has with a small ocean such as the Mediterranean.

Hydrothermal Investigations

Small thermal gradients which exceed the adiabatic gradient have been observed in Crater Lake by several investigators. These observations suggest the possibility of geothermal heating. Anomalous levels of some chemical constituents have also been found in Crater Lake. The Collier/Dymond submersible exploration in the summer of 1988 and is again scheduled for the summer of 1989 enable researchers to make direct observations, conduct detailed sampling, and carry out temperature measurements within the region of anomalous water compositions on the bottom of Crater Lake. The discovery of bacterial mats during one of the manned dives may have significance as to the existence or non-existence of hydrothermal or springwater sources.

There may well be some confusion regarding what can be considered a hot spring or a cold spring. One could argue that a cold spring flowing into a colder lake becomes a hot spring. Since lakes tend to vary considerably in temperature, particularly during years when they freeze, a cold spring in summer could logically be classed as a "hot spring" in winter due to the temperature differential. Such designations, of course, must be related to the relative temperatures of the receiving water and the inflowing water. At Castle Lake, near the northern border of California, springs which drain a very porous slope discharge from the lake bottom, providing cold water spawning habitat for brook trout. These "cold springs" of summer become the "warm springs" of winter causing holes in the lake ice to develop above their outflow. More exploration and analysis are needed to resolve the questions of whether there is hydrothermal input or whether relatively warm cold spring or some combination is causing the temperature anomalies observed.

Long-term Climatic Effects

The greenhouse effect resulting from increased carbon dioxide concentrations may already be upon us. Any climatic change has enormous potential for altering the trophic status of lakes throughout the United States as well as the Great Lakes area and Canada. Even a one-degree rise in average annual temperature has a far-reaching effect on global climate. It is important to plan now for collecting the essential baseline data to make it possible to identify, evaluate, and anticipate climatic change. The research site at Crater Lake could be particularly valuable in this respect.

The importance of long-term data collection cannot be over-emphasized. Redmond's (1990) extensive evaluation of the climatic records dating back to 1931, including the World War II data gap, showing the five-meter water level fluctuations in Crater Lake, has considerable potential for further analysis and interpretation. The precipitation/evaporation/seepage questions still remains rather open and can continue to be the subject of further, exciting research. Crater Lake does act as a macro-rain gauge which, if expanded beyond its lake status, might even serve as a "micro-ocean." Some direct measurements using evaporation-pans on the lake would be very interesting in view of the cool water and the high-evaporation rates reported.

Lake Mixing

Lake mixing is another key element in the dynamics of Crater Lake. Crater Lake, despite its depth, freezes during the coldest winters. With occasional freezing of the surface it may be possible to investigate a number of physical processes which result from restricted mixing. Many years ago I predicted that if the pollution of Tahoe continued, the lake might freeze, due simply to the loss of transparency. As the turbidity of a system increases, the depth of light penetration is reduced, resulting in greater surface warming and an increased evaporation rate. A net loss of heat results. A turbid system therefore should be a colder system. All other things being equal, an eutrophication of Crater Lake should promote a more frequent freezing of the surface and possibly increased mobilization of Mn and Fe from bottom sediments if anoxic conditions develop at the sediment-water interface.

Phytoplankton

The existence of a diatom bloom (*Nitzschia* sp.) is particularly interesting, especially since the bloom is reported to persist in the absence of detectable levels of nitrate. The exhaustion of nitrate is also commonly found in the euphotic zone of Tahoe. The phytoplankton simply take up all the available dissolved nitrogen transforming it to a particulate organic form. The entire phytoplankton assemblage is supported by nutrient recycling, through zooplankton and fish excretion. The careful characterization of the phytoplankton assemblage by some of the participants in this symposium and reported in this volume is exactly what is needed on a continuing basis for the monitoring of Crater Lake.

Zooplankton

Two cladocerans and nine rotifers exist as prey for a variety of planktonic fish that, according to the views of Carpenter and Kitchell (1984, 1987), might provide a means of top-down control of the system. Professor Shapiro at Minnesota has termed this situation "bio-manipulation." Some Tahoe basin developers have even made the claim that Lake Tahoe is becoming more eutrophic solely because the introduced opossum shrimp, *Mysis relicta*, has decimated populations of the cladocerans *Daphnia* sp. and *Bosmina* sp. This suggestion was put forward despite the increasing trend in primary productivity before the appearance of *Mysis* in the lake. In similarly low fertility Crater Lake, manipulation of the food chain by heavy fish stocking or through some natural increase in the planktivore population, as has occurred in 1987, would only be expected to slightly reduce the transparency of the lake.

Nitrogen excretion of zooplankton in oligotrophic systems are less important in affecting phytoplankton densities. This situation is in contrast to lakes of higher productivity such as Lake Michigan, Clear Lake, California, and Castle Lake, California, whose phytoplankton populations are more influenced by changes in the abundance of zooplankton from year to year. With no fish stocking at Crater Lake since 1941, Crater Lake will tend to be similar to Tahoe, showing more "bottom up" than the "top down" control demonstrated in more productive systems (Elser *et al.*, in press).

Effect of Sewage

As Douglas Larson and others (1990) pointed out, the transparency of Crater Lake may also have been affected by sewage and waste water management practices used by its legal guardian, the National Park Service. It is quite natural that the National Park Service would be disturbed by this possibility. When one considers the relatively slow evolution of sewage treatment technology during the last few decades, it is not surprising that some effluent may have entered the lake. In any event, this issue is being addressed and should be resolved as quickly as possible. Exportation of sewage effluent should be undertaken if, in fact, any sewage-derived nitrogen or phosphorus is getting into the lake. Lake Tahoe phytoplankton have been shown to respond to as little as one part per billion nitrogen. Although data collected at Crater Lake during the spring indicate that only a small amount of nitrate may be involved, one should at least calculate what effect the addition of a few kilograms of nitrogen a day may have on the transparency of this spectacular lake over the period of years of possible pollution.

CRATER LAKE AND LAKE TAHOE COMPARISON

Physical Comparisons

While Crater Lake is situated in a volcanic caldera, Tahoe was formed in a graben fault basin during the uplift of the Sierra Nevada. The sediments in Tahoe appear almost exactly as shown in the pictures of the Crater Lake sediments presented by Dymond. This is not surprising because they are both ultra-oligotrophic systems at nearly the same elevation, both with small watersheds. Lake Tahoe, at a maximum depth of 505 meters, is not quite as deep as Crater Lake's 589 meters, but it is about ten times as large as Crater Lake in surface area (Table 1). The lake basin contains 156 km^3 of water and is shaped like a giant horse trough, dropping off rapidly around its shore to 450 meters of nearly flat bottom with a few lake mounts along some of the faults found along the floor.

FACTORS AFFECTING LAKE TRANSPARENCY

Crater Lake, like Tahoe, is renowned for its high transparency. Historically, Crater Lake has been slightly more transparent than Tahoe. A study was conducted in 1970 which compared the optical properties of the two lakes with a submersible spectroradiometer (Smith *et al.* 1973). This device measures absolute values of spectral irradiance. I expect similar equipment will be available to repeat these measurements and make it possible to evaluate what changes may have occurred over a period of about twenty years.

Tahoe, unfortunately, is now losing its transparency and there is also some evidence of transparency loss at Crater Lake. In the winter, when Tahoe waters are more transparent, the transparency decline is currently half a meter per year. This means that in another 30 to 40 years, Tahoe will lose its unique cobalt blue color and become a more ordinary lake. Despite the steady decline in water clarity (Goldman 1988), from Secchi annual average depths of around 30 meters in 1958 to about 20 meters today, Secchi transparencies of 40 meters still occasionally occur in the winter during an upwelling event. Also, during the drought years of 1976 and 1977 when algal growth rate declined there is a temporary improvement in transparency. At the other extreme, we measured a transparency of only 9 meters at the mid-lake station during the El Niño southern oscillation event of 1983.

Primary Production

The fluctuations in transparency from year to year, although great, have a very close relationship to primary productivity (Goldman 1988). Year-round carbon-14 measurements of primary productivity

TABLE 1. COMPARATIVE FEATURES OF CRATER LAKE, OREGON AND LAKE TAHOE, CALIFORNIA-NEVADA

	Lake Tahoe	Crater Lake
Formation	Graben fault; volcanic dam ~ 2 mybp	Mount Mazama caldera ~ 6600 ybp (Fryxell 1965) ~6900 ybp (Collier & Dymond 1988)
Elevation (in m)	1898	1882
Maximum depth (in m)	505	589
Mean depth (in m)	313	325
Relative depth (Z_r)	1.8′	6.6′
Surface area (km^2)	499	53.2
Volume (km^3)	156	16
Length (km)	34.7	9.7
Width (km)	19.3	8.2
Watershed area (km^2)	800	14.6
Lake area to watershed area (flate map) ratio	0.625	3.6
Surface outflow	Truckee River	None

are made using carbon-14 on a weekly or bi-weekly basis to a depth of 105 meters. This depth is approximately the extent of the euphotic zone for phytoplankton in Tahoe. This zone is exhibiting a progressive shortening as eutrophication proceeds. If annual Secchi depths between 1968 and 1984 are plotted against primary production, we find that as productivity increases, transparency diminishes at a rate of about 0.4 meters per year. It is important to note that the euphotic zone extends to four or five times the Secchi depth in the transparent waters of Tahoe. Light extinction measurements from the euphotic zone have also been regressed on this same productivity data set.

The importance of long-term data collection is illustrated in Table 2. A typical grant period of three years would not have shown significant change in Lake Tahoe productivity. Due to considerable interannual variation, seven years of data collecting were required to establish a trend at the .001 level. In 1979, the maximum depth of primary production was 32.1 meters. This depth was decreased to 26.6 m in 1982 and 24.0 m in 1986. During the 1983 El Niño year it further decreased to 23.5 meters as the lake lost transparency.

The value of archived samples spanning over a long time period also cannot be over-emphasized. For example, the dominant phytoplankton species have shifted markedly over the years, going first toward smaller forms (Goldman 1981) which tend to reduce transparency, and most recently toward large colonial green alga, which do not decrease transparency as much as the same biomass of smaller species. Further, *Fragilaria crotonensis*, a ubiquitous diatom, was dominant in Lake Tahoe 20 years ago and then disappeared. I recommend that a collection of archived samples and photomicrographs of the species present be maintained so that such comparisons and identifications can be made in Crater Lake.

Human Population

The Tahoe basin, as well as the lake, has undergone considerable change since development began to escalate in the late 1950s. The lake has drawn developers' interest due to its proximity to large centers of population and, of course, the magnetism of the casinos at the south end as well as extensive housing and ski development at both the north and south ends of the lake. The population explosion,

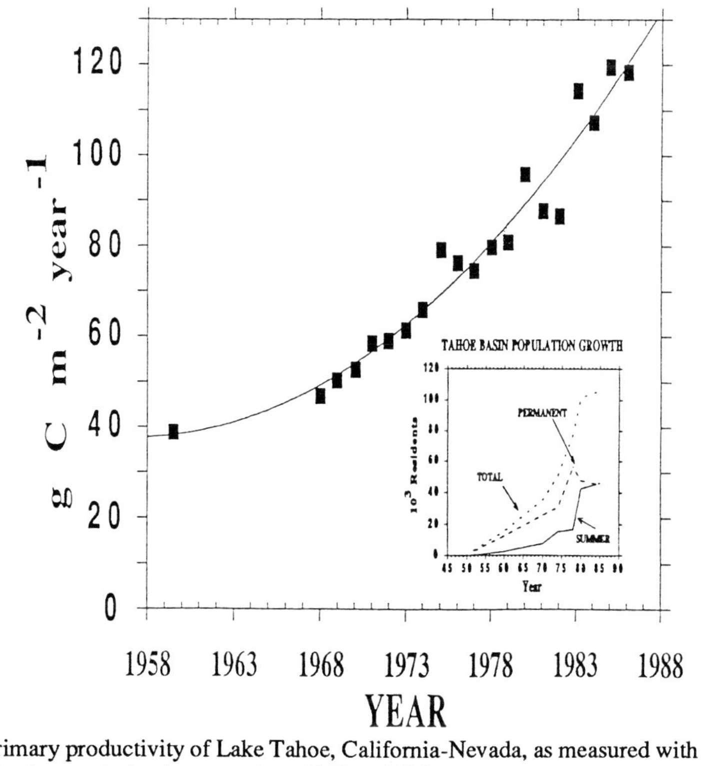

Figure 1. Annual primary productivity of Lake Tahoe, California-Nevada, as measured with ^{14}Carbon between 1959 and 1987. Inset: Tahoe basin population increase from 1952 to 1985.

which began just after World War II and has continued to date, has had a major impact on Lake Tahoe. If one overlays the curve which depicts the phytoplankton productivity increase with the figure showing the rising population density, the similarity is striking and reflects the inescapable human involvement in watershed disturbance and the decline of lake water quality (Fig. 1). In fact, eutrophication of the world's lakes has become more the rule than the exception during the last quarter-century. The existence of nutrient-sensitive, nitrogen-starved phytoplankton populations is another shared trait of Tahoe and Crater Lake. Both systems are severely nitrogen-limited. Tahoe, with its 700-year retention time, is gradually becoming less nitrogen-limited and progressively more phosphorus-sensitive as nitrogen accumulates in the system. This has measurably increased the N:P ratio over the last two decades.

Developers in the Tahoe basin have become more sophisticated in their understanding of how lakes function and have pointed out that perhaps all lakes in the western United States are becoming more eutrophic. They argue that increasing eutrophication of Tahoe might not be due to development of the basin watersheds. A continuous, 30-year data set collected at Castle Lake, a small subalpine lake in northern California, however, has served as an experimental control for work done at Tahoe. These data show no trend of increasing productivity.

Sources of Nutrient Input

Unlike Crater Lake watershed caldera, the Tahoe basin has been subject to prolonged and severe disturbance. Extensive erosion of its steep and fragile slopes has taken place. During the development of Incline Village at the north end of the lake, creeks draining the area were laden with sediment and nutrients. Stream discharge, reflecting precipitation in the basin, correlates well with the change in productivity in analyses of the inter-annual variation in primary productivity. Although approximately the same amount of precipitation fell during 1982 and 1983, productivity was high in 1983 and much lower in 1982 (Goldman and de Amezaga 1984). This 1982 outlier in the data turned out to be extremely important since it directed us to a comparison of the depth of mixing. We are able to determine the depth of mixing by following the descent of the summer nitrocline during late winter (Paerl *et al.* 1975). During 1982, mixing was shallow as opposed to 1983 when the lake was completely mixed by a late winter storm. Thus, although annual runoff is important in the steady accumulation of nutrients in the system, it is the internal loading from deep mixing that accounts for most of the inter-annual variability in productivity at Lake Tahoe (54%) and undoubtedly in other deep lakes like Crater Lake (Goldman and Jassby, in press).

TABLE 2. STATISTICAL TRENDS IN LAKE TAHOE

(A) ANNUAL ALGAL GROWTH RATE

Years	*# of Years*	*F-ratio*	*DF*	*Significance*
1959.5-				
-1969	3	17.3	1, 1	NS
-1970	4	62.2	2, 1	NS
-1971	5	89.9	2, 2	$p<.025$*
-1972	6	59.6	2, 3	$p<.005$***
-1973	7	63.2	2, 4	$p<.001$****
.				
-1986	20	173.6	2,17	$p<.001$****

(B) AVERAGE ANNUAL SECCHI DEPTH

Years	*# of Years*	*F-ratio*	*DF*	*Significance*
1968-				
-1970	3	0.17	1, 1	NS
-1971	4	1.09	1, 2	NS
-1972	5	4.94	1, 3	NS
-1973	6	14.52	1, 4	$p<.025$*
-1974	7	12.31	1, 5	$p<.025$*
-1975	8	18.21	1, 6	$p<.01$**
-1976	9	11.96	1, 7	$p<.025$*
-1977	10	7.88	1, 8	$p<.025$*
-1978	11	11.40	1, 9	$p<.01$**
-1979	12	11.69	1,10	$p<.01$**
-1980	13	18.69	1,11	$p<.005$***
-1981	14	12.62	1,12	$p<.005$***
-1982	15	19.21	1,13	$p<.001$****
-1983	16	27.96	1,14	$p<.001$****
-1984	17	38.27	1,15	$p<.001$****
-1985	18	42.11	1,16	$p<.001$****
-1986	19	45.82	1,17	$p<.001$****

Incomplete vertical mixing of lakes is a function of the lake's depth in relation to its surface area exposed to wind. This "relative depth" (Z_r) is defined as the maximum depth expressed as a percentage of the mean surface diameter and is an indication of mixing tendency. As shown in Table 1, Z_r for

Crater Lake is greater than 6 while that for Tahoe is about 2. Crater Lake, therefore, may be less likely to mix regularly than Lake Tahoe.

Other sources of nutrient input to Lake Tahoe include the legacy from old septic tanks and their leach fields. Spring melting of the accumulated snowpack brings the groundwater to the soil surface at Tahoe. The nitrogen and phosphorous from the leach fields may be carried to the surface, resulting in a "terrestrial" bloom of filamentous algae on the saturated soil surface. During the investigation of groundwater in the Tahoe basin, high nitrate levels were found in some of the wells. These high concentrations were probably the result of nutrients, derived from old septic tank leach fields, slowly seeping into the groundwater (Loeb and Goldman 1979). More recent studies show an improvement in the well-water quality, suggesting that the contaminated groundwater has now reached the lake. Without active septic tank use in the basin, groundwater is also becoming cleaner (Loeb 1980).

Air pollution has been a matter of concern in the Tahoe basin for many years. About half of the nitrogen loading at Tahoe is derived from air pollution, most of which is generated within the basin by the enormous concentration of vehicular traffic around the lake. Pollutants are derived not only from auto exhaust but also from road dust suspended in the atmosphere by traffic over both paved and unpaved roads. Application of cinders, salt, and sand during winter increases this problem. Air pollution at South Tahoe, near the casinos, is at times like that found in the Los Angeles basin. In fact, the lead concentrations in the casino parking lots were actually approaching southern California levels. What was once considered by local inhabitants and visitors to be "forest haze" is now known to be largely air pollution trapped in the basin by an inversion layer.

As part of an interagency monitoring program on the lake, the Tahoe Research Group deployed spar buoy samplers and began to make areal comparisons of atmospheric deposition. Comprehensive data have been collected which characterize the wet and dry fallout on the lake. Stations located on the north-south transect show maximum air pollution near the south shore, an expected improvement in the mid-lake regions, and then a decline in air quality toward the north shore of the lake where population density increases.

Evidence of Eutrophication

The first obvious evidence of eutrophication at Tahoe was the appearance of periphyton, giving Tahoe a distinctly green margin, especially during the spring months. Areas of heavy development were found to be associated with increased periphyton growth offshore (Loeb *et al.* 1984). In Crater Lake, it may also be possible to characterize nutrient inputs on the basis of the periphyton distribution. A high-sensitivity, *in situ* measurement of carbon-14 uptake by periphyton is probably the most effective way of detecting any stimulation of growth by nutrient sources. Although natural rock substrates are preferred (Loeb 1980), artificial substrates can also be utilized (Aloi 1986). In Lake Tahoe, the productivity of periphyton is lower along the west side of the lake where development has been lightest. Studies, based on 7 synoptic cruises with 35 stations sampled during each, indicate that population density together with stream inflow influence the phytoplankton productivity around the lake. The only area that is relatively unaffected is the middle of the lake (Goldman, 1974).

RECOMMENDATIONS FOR FUTURE CRATER LAKE STUDIES

To understand lake systems which exhibit significant inter-annual variation, I find it essential to conduct long-term data collection. To do so, more Federal and State agencies must become involved. Adequate sampling for a minimum of a decade is necessary, and protocols need to be established to eliminate the possibility of uncalibrated changes in sampling procedures and analytical methods.

Complete primary productivity profiles need to be measured at Crater Lake on a frequent basis. I would recommend weekly measurements during the summer months. These profiles will provide the best contemporary biological integration of physical and chemical factors at work. Some careful depth profiles of nitrate should be included during the late winter months or following iceout to determine the depth of spring mixing. This is an essential element for resolving the questions of thermal and chemical gradients near the lake bottom. I would predict from the high Z_r value for Crater Lake that it does not mix completely every year. This may well be the most important measurement to understand the physical dynamics of the lake (Goldman *et al.* 1989).

Taxonomic data will also become increasingly important in studies of biological processes in Crater Lake. In particular, autoradiography might be used to determine which phytoplankton species are primarily responsible for converting inorganic carbon to organic carbon in the system. Plankton samples should be archived so that disagreements can be resolved between contemporary and future taxonomists regarding species composition. In both Crater Lake and Tahoe nature has provided a marvelous opportunity to follow the classes of kokanee salmon (Buktenica and Larson 1990) and assess their predation on the zooplankton (Karnaugh 1990) as well as the zooplankton's possible impact on phytoplankton biomass and species composition (Geiger and Larson1990; Debacon and McIntire 1990).

With the recent emphasis on climatic change, it is very important to measure atmospheric inputs and establish and maintain a well-equipped weather station on site. For example, there should be dry and wet collectors operating twenty-four hours a day on Wizard Island, an ideal sampling site for Crater Lake. It is often surprising how much the atmospheric fallout on the lake surface differs from the fallout on the surrounding watershed. A very important link in the nutrient budget, the dry fallout component, is missing for Crater Lake. At Tahoe, now that we have buoy-mounted dry fallout collectors stationed in the middle of the lake, we are finding that about half of the nitrogen loading is actually derived from fallout from the atmosphere. Nitrogen levels in the dry and wet fallout are, of course, closely related to road traffic, dust, exhaust, and air pollution in its various forms, not to mention the naturally occurring ammonia and nitrate in the rainfall. This additional air data will be necessary if we hope to assemble a reliable nutrient budget for Crater Lake.

With regard to possible sewage leaks, some lysimetry might be done to track the flow of nitrate through the soil and fingerprinting sources may be achieved by isotopic ratios. If sewage leaks are found, it should be relatively easy to estimate the amount of nitrogen accumulated over the years and translate the result into phytoplankton biomass on the basis of estimated carbon and nitrogen content of cells. Researchers can then estimate the potential decrease in transparency due to the increased chlorophyll concentrations resulting from this nitrogen input. Previous Crater Lake bioassays have demonstrated the positive response of phytoplankton to EDTA enrichment. This chelator almost always stimulates phytoplankton, presumably by mobilizing, through solubilization, unavailable iron and trace elements. More studies of this nature should be pursued at Crater Lake in combination with nitrogen and phosphorus bioassay experiments.

Crater Lake is developing an impressive data base, the value of which is going to increase exponentially as research protocols are designed and as the frequency and regularity of sampling are increased. Like Tahoe, Crater Lake is a natural resource of unique beauty and extraordinary value. With careful study by the diversity of scientists represented at this symposium, it should be possible to develop and implement an effective management strategy that can protect Crater Lake for the benefit of this and future generations.

LITERATURE CITED

Aloi, J. E. 1986. The ecology and primary productivity of the eulittoral epilithon community: Lake Tahoe, California-Nevada. Ph.D. Thesis. Univ. California, Davis, Calif.

Box, G. E. P. 1954. The exploration and exploitation of response surfaces: Some general considerations and examples. Biometrics 16-60.

Buktenica, M., and G. L. Larson. 1990. Ecology of kokanee salmon and rainbow trout in Crater Lake. Pages 185-195 *in* E. T. Drake *et al.*, eds., Crater Lake: An Ecosystem Study. Pacific Div., Amer. Assn. Advance. Sci., San Francisco, Calif.

Carpenter, S. R,. and J. F. Kitchell. 1984. Plankton community structure and limnetic primary production. Amer. Nat. 124:159-172.

Carpenter, S. R., and J. F. Kitchell. 1987. The temporal scale of limnetic primary production. Amer. Nat. 129:417-433.

Debacon, M. K., and C. D. McIntire. 1990. Spatial and temporal patterns in the phytoplankton of Crater Lake (1985-1987). Pages 167-175 *in* E. T. Drake *et al.*, eds., Crater Lake: An Ecosystem Study. Pacific Div., Amer. Assn. Advance. Sci., San Francisco, Calif.

Dymond, J., and R. Collier. 1990. The chemistry of Crater Lake sediments: Definitions of sources and implications for hydrothermal activity. Pages 41-60 *in* E. T. Drake *el al.*, eds., Crater Lake: An

Ecosystem Study. Pacific Div., Amer. Assn. Advance. Sci., San Francisco, Calif.

Elser, J. J., H. J. Carney, and C. R. Goldman. (in press). Nutrient supply and demand in pelagic ecosystems: A comparison of three large lakes. *In* Proc. Internat'l Mountain Watershed Symp. Lake Tahoe, CA-NV.

Geiger, N. S., and D. W. Larson. 1990. Phytoplankton species distribution in Crater Lake, Oregon, 1978-1980. Pages 153-165 *in* E. T. Drake *et al.*, eds., Crater Lake: An Ecosystem Study. Pacific Div., Amer. Assn. Advance. Sci., San Francisco, Calif.

Goldman, C. R. 1974. Eutrophication of Lake Tahoe emphasizing water quality. EPA-660/3-74-034. U. S. Govt. Printing Office, Washington, DC. 408 pp.

Goldman, C. R. 1988. Primary productivity, nutrients, and transparency during the early onset of eutrophication in ultra-oligotrophic Lake Tahoe, California-Nevada. Limnol. Oceanogr. 33(6, part 1):1321-1333.

Goldman, C. R. and E. de Amezaga. 1984. Primary productivity and precipitation at Castle Lake and Lake Tahoe during twenty-four years, 1959-1982. Verh. Internat'l Verein. Limnol. 22:591-599.

Goldman, C. R. and A. Jassby. (in press). Spring mixing depth and annual productivity fluctuations in Lake Tahoe, a deep subalpine lake. *In* M. Tilzer and C. Serruya, eds., Functional and Structural Properties of Large Lakes. Sci.-Technol.

Goldman, C. R., A. Jassby, and T. Powell. 1989. Interannual fluctuations in primary production: Meteorological forcing at two subalpine lakes. Limnol. Oceanogr. 34(2):310-323.

Herdendorf, C. E. 1982. Large Lakes of the World. Jour. Great Lakes Res. 8(3):379-412.

Karnaugh, E. 1990. Sampling strategy and a preliminary description of the pelagic zooplankton community in Crater Lake. Pages 177-183 *in* E. T. Drake *et al.*, eds., Crater Lake: An Ecosystem Study. Pacific Div., Amer. Assn. Advance. Sci., San Francisco, Calif.

Larson, D. W., C. N. Dahm, and N. S. Geiger. 1990. Limnological response of Crater Lake tto possible long-term sewage influx. Pages 197-212 *in* E. T. Drake *et al.*, eds., Crater Lake: An Ecosystem Study. Pacific Div., Amer. Assn. Advance. Sci., San Francisco, Calif.

Loeb, S. L. 1980. The Production of the Epilithic Periphyton Community in Lake Tahoe, California-Nevada. Ph.D. Thesis. Univ. California, Davis, Calif.

Loeb, S. L. and C. R. Goldman. 1979. Water and nutrient transport via ground water from Ward Valley into Lake Tahoe. Limnol. Oceanogr. 24(6):1146-1154.

Loeb, S. L., P. Eloranta, and J. E. Reuter. 1984. Littoral phytoplankton productivity and biomass as indicators of different nutrient loading of Lake Tahoe. Verh. Internat'l Verein. Limnol. 22:605-611.

Paerl, H. W., R. C. Richards, R. L. Leonard, and C. R. Goldman. 1975. Seasonal nitrate cycling as evidence for complete vertical mixing in Lake Tahoe, California-Nevada. Limnol. Oceanogr. 20(1):1-8.

Redmond, K. T. 1990. Crater Lake climate and lake level variability. Pages 127-141 *in* E. T. Drake *et al.*, Crater Lake: An Ecosystem Study. Pacific Div., Amer. Assn. Advance. Sci., San Francisco, Calif.

Smith, R. C., J. E. Tyler, and C. R. Goldman. 1973. Optical properties and color of Lake Tahoe and Crater Lake. Limnol. Oceanogr. 18(2):176-188.

Strub, P. T., T. M. Powell, and C. R. Goldman. 1985. Climatic forcing: Effects of El Niño on a small, temperate lake. Science 227:55-57.